T0184582

SERIES EXPANSION METHODS FOR STRONGLY INTERACTING LATTICE MODELS

Perturbation series expansion methods are sophisticated numerical tools used to provide quantitative calculations in many areas of theoretical physics. This book gives a comprehensive guide to the use of series expansion methods for investigating phase transitions and critical phenomena, and lattice models of quantum magnetism, strongly correlated electron systems and elementary particles.

Early chapters cover the classical treatment of critical phenomena through high-temperature expansions, and introduce graph theoretical and combinatorial algorithms. The book then discusses high-order, linked cluster perturbation expansions for quantum lattice models, finite temperature expansions, and lattice gauge models. Numerous detailed examples and case studies are also included, and an accompanying resources website, www.cambridge.org/9780521842426, contains programs for implementing these powerful numerical techniques.

A valuable resource for graduate students and postdoctoral researchers working in condensed matter and particle physics, this book will also be useful as a reference for specialized graduate courses on series expansion methods.

JAAN OITMAA was born in Tallinn, Estonia in 1943. After the war his family migrated to Australia, where he has spent most of his life. He received his undergraduate and graduate education at the University of New South Wales, in Sydney, obtaining his Ph.D. in 1968. His early postdoctoral work was in lattice dynamics at the University of California, Irvine. During his second postdoctoral position, at the University of Alberta, he became interested in the field of critical phenomena, and learnt the techniques of series expansions from Donald Betts' group. On returning to Australia he held a Queen Elizabeth II Research Fellowship at Monash University and then rejoined his *alma mater* as a lecturer in Physics, in 1972. He was promoted to full Professor in 1991, and upon his retirement in 2003 was accorded the title of Professor Emeritus. Throughout his long career he has been an enthusiastic teacher at all levels, supervised many Ph.D. students, published over 170 research papers in top international journals, and served as Head of School for 6 years (1993–1998) and President of the Australian Institute of Physics for 2 years (1997–1998). He is a Fellow of both the Australian Institute of Physics and the American Physical Society.

CHRIS HAMER received a B.Sc. and M.Sc. from the University of Melbourne, and a Ph.D. from the California Institute of Technology in 1972. He began his research career in elementary particle physics, and studied series expansions in lattice gauge theory. His later interests moved towards statistical mechanics and condensed matter physics, including the theory of finite-size scaling, and linked cluster methods for series expansions. He held research positions at the Brookhaven National Laboratory, the Universities of Cambridge, Liverpool and Melbourne, and was a Senior Research Fellow at the Australian National University for 8 years from 1979, before taking up a position as Senior Lecturer at the University of New South Wales (UNSW) in 1987. He is now a Visiting Associate Professor at UNSW. He has authored about 150 research publications. He is a Fellow of the Australian Institute of Physics, and was editor of the AIP journal, *The Physicist*, for 5 years from 1998–2002.

WEIHONG ZHENG was born in Guangdong, China. He obtained a B.Sc. in 1984 and then a Ph.D. in 1989 at Zhongshan University in Guangdong, studying Hamiltonian lattice gauge theory. He was subsequently appointed as a lecturer at Zhongshan University. He moved to the University of New South Wales in 1990, where he began a long and productive collaboration with Jaan Oitmaa and Chris Hamer, working on linked cluster series expansion methods. He is presently a Senior Research Associate at UNSW, and has built up a worldwide reputation as an acknowledged expert in the techniques of perturbative series expansions. His early work was in the field of particle physics, but more recently he has concentrated on the theory of condensed matter systems. He has written around 100 research publications.

SERIES EXPANSION METHODS FOR STRONGLY INTERACTING LATTICE MODELS

JAAN OITMAA, CHRIS HAMER AND
WEIHONG ZHENG

School of Physics, The University of New South Wales, Sydney, NSW 2052, Australia.

CAMBRIDGE
UNIVERSITY PRESS

CAMBRIDGE UNIVERSITY PRESS
Cambridge, New York, Melbourne, Madrid, Cape Town, Singapore,
São Paulo, Delhi, Dubai, Tokyo, Mexico City

Cambridge University Press
The Edinburgh Building, Cambridge CB2 8RU, UK

Published in the United States of America by Cambridge University Press, New York

www.cambridge.org
Information on this title: www.cambridge.org/9780521143592

© J. Oitmaa, C. Hamer and W. Zheng 2006

This publication is in copyright. Subject to statutory exception
and to the provisions of relevant collective licensing agreements,
no reproduction of any part may take place without the written
permission of Cambridge University Press.

First published 2006
First paperback printing 2010

A catalogue record for this publication is available from the British Library

ISBN 978-0-521-84242-6 Hardback
ISBN 978-0-521-14359-2 Paperback

Cambridge University Press has no responsibility for the persistence or accuracy of
URLs for external or third-party Internet Web sites referred to in this publication, and
does not guarantee that any content on such Web sites is, or will remain, accurate or
appropriate.

Contents

Preface

The past 50 years have seen much progress in our understanding of the behaviour of complex physical systems, made up of large numbers of strongly interacting particles. This includes a rather detailed, if not complete, understanding of such phenomena as phase transitions of various kinds, macroscopic quantum phenomena such as magnetic order and superconductivity, and the response of such systems to external probes, vital for the interpretation of experimental results. Such unifying concepts as scaling and universality, long-range order (including off-diagonal, long-range order), and spontaneous symmetry breaking have led to a unified understanding of diverse and complex phenomena.

Central to this endeavour has been the detailed and systematic study of lattice models of various genera; models which are precisely defined mathematically, which are believed to embody the *essential physics* of interest, and which are, to a greater or lesser extent, mathematically tractable. Exact analytic treatment of these models is rarely possible. Series expansion techniques, the subject of this book, provide one of the main systematic and powerful approximate methods to treat such lattice models.

Our decision to write this book arose from a request from a journal editor to write a review of our group's work over the last decade on series studies of quantum lattice models. On reflection, we came to the view that it would be more useful to write a book covering the entire field, at a level which would be accessible to graduate students and other researchers wishing to learn about these methods. We were also strongly influenced by the appearance of another book, *A Guide to Monte Carlo Simulations in Statistical Physics* by Landau and Binder (Cambridge University Press, 2000), which provides an excellent introduction to Monte Carlo methods. We felt there was a need for a book, with the same general approach, for the series expansion field. The response of the physics community will show whether we were justified in this belief.

Our archetypal reader, then, is a graduate student or young researcher who wishes to use series expansion methods for some particular project and who is not overly familiar with the subject and does not have access to a local expert. As with any technical area, there are many skills to learn and pitfalls to be avoided. Many of the computational algorithms needed are heavily combinatorial in nature and wise or unwise programming can make orders of magnitude difference in both time and memory demands. In this book we provide examples of efficient computer programs to do the most commonly needed tasks – there is no need for every worker in the field to reinvent these things from scratch.

Our approach is to demonstrate results, wherever possible, by means of specific calculations for simple models, which are worked out in some detail. We hope that these will provide useful examples and checks for researchers building their own programs in these areas. We do not attempt to give formal proofs of results, although an outline proof is sketched in a few basic cases: in general, the reader must consult the listed references if a more rigorous approach is required.

The choice of content is, as in any book, partly a reflection of our interests. The later chapters, in particular, are influenced by our own work, over the last 10–15 years, on series studies of quantum spin models and electronic models. However we wish to explain rather than to expound and the early chapters, in particular, are intended to be largely pedagogical. While much of this material is well known (to those who know it well) and well documented in volume 3 of the Domb and Green series (Domb, 1974) and in a number of books (e.g. Baker, 1990), it has not, to our knowledge, been presented in a unified *hands-on* way with supporting computer programs.

Inevitably some interesting and important topics have had to be curtailed: for example the large field of series analysis, where other good sources exist, and the area of disordered systems, which is treated very briefly. Very little has been included on those lattice models which are primarily of interest in mathematical physics, such as Potts models, random walks, and lattice polygons. We have concentrated primarily on models with more direct physical applications.

The chapters are relatively self contained and some may be skipped at first reading. In particular Chapter 3 (the free graph method) and Chapter 5 (quantum antiferromagnets at $T = 0$) are not prerequisites for chapters which come after them. A comment on referencing is appropriate. While we have tried to be true to history, and to acknowledge the pioneers on particular topics by name, we have not attempted to cite every source. It seemed appropriate to cite in the book only those sources which, in our view, readers may benefit from following up, or more recent articles, which give references to earlier work. We can only apologize to any of our fellow workers whose work has not been fully referenced.

Our knowledge of this field has been acquired over many years and through interactions with colleagues too numerous to mention individually. We thank all of them, but particularly George Baker, Michael Barber, Donald Betts, Conrad Burden, Chuck Elliott, Shuohong Guo, Tony Guttmann, Hong-Xing He, Alan Irving, and Rajiv Singh. We also gratefully acknowledge support, over many years, from the Australian Research Council.

We are grateful for permission to reproduce the following figures: Fig. 11, T.M.R. Byrnes *et al.*, *Phys. Rev.* D69, 074509 (2004); Fig. 1, M. Creutz, *Phys. Rev. Letts.* 43, 553 (1979); Fig. 4, M. Creutz and K.J.M. Moriarty, *Phys. Rev.* D26, 2166 (1982); Figs. 3, 5, C.J. Hamer *et al.*, *Phys. Rev.* D56, 55 (1997); Figs. 3, 15, C.J. Hamer *et al.*, *Phys. Rev.* B68, 214408 (2003); Fig. 2, D. Horn and G. Lana, *Phys. Rev.* D44, 2864 (1991); Figs. 4, 5, D. Horn and D. Schreiber, *Phys. Rev.* D47, 2081 (1993); Fig. 2, J. Oitmaa and W. Zheng, *Phys. Rev.* B54, 3022 (1996); Fig. 3, J.D. Stack, *Phys. Rev.* D29, 1213 (1984); Figs. 3, 5, O.P. Sushkov *et al.*, *Phys. Rev.* B63, 104420 (2001); Fig. 4, S. Trebst *et al.*, *Phys. Rev. Letts.* 85, 4373 (2000); Fig. 4, C.P. Van en Doel and D. Horn, *Phys. Rev.* D33, 3011 (1986); Figs. 1, 6, W-H. Zheng, *Phys. Rev.* B55, 12267 (1997); Figs. 2, 3, W-H. Zheng *et al.*, *Phys. Rev. Letts.* 91, 037206 (2003); Figs. 1, 2, 5, W-H. Zheng *et al.*, *Phys. Rev.* B63, 144411 (2001); Fig. 2, W-H. Zheng and J. Oitmaa, *Phys. Rev.* B63, 064425 (2001); Figs. 3, 4, W-H. Zheng *et al.*, *Phys. Rev.* B58, 14147 (1998); Fig. 6, W-H. Zheng *et al.*, *Phys. Rev.* B65, 014408 (2001); all copyright by the American Physical Society; Fig. 2, G.T. Rado, *Solid State Commun.*, 8, 1349 (1970); Fig. 1, C.J. Hamer, *Phys. Letts.* B224, 339 (1989); Fig. 12, C.J. Hamer and A.C. Irving, *Nucl. Phys.* B230, 336 (1984); Figs. 2, 5, J. Smit, *Nucl. Phys.* B206, 309 (1982); copyright by Elsevier; and Fig. 3, G. Aeppli *et al.*, *Physics World* 10, 33 (December 1997), copyright by Institute of Physics (UK).

1

Introduction

1.1 Lattice models in theoretical physics

A major part of theoretical physics involves the construction and systematic analysis of mathematical models as a description of the physical world. For microscopic phenomena, in particular, the reference to models becomes quite explicit. The phenomena are often complex, and any theory that attempts to include every detail soon becomes intractable. It is more fruitful to develop models which ignore irrelevant details but, hopefully, capture the essential physics of the phenomena of interest. Thus we have the Heisenberg model of magnetic order, the Bardeen–Cooper–Schrieffer (BCS) model of superconductivity, and so on. It is important to note, at the outset, that the models we shall be discussing describe strongly interacting, and therefore highly correlated, systems of particles. These are difficult and interesting problems. Where interactions are absent, or weak, elementary treatments are possible and the resulting phenomena are generally unspectacular.

Why lattice models? In solid-state phenomena there is usually an underlying lattice structure, and the symmetry properties of this lattice play an important role in the analysis. Even the process of electrical conduction in metals or semiconductors can be equally well described in terms of localized quantum states or in terms of the more usual continuum picture. In quantum field theory, which is formulated in a space–time continuum, a lattice is often introduced for computational purposes. One can think of this in two ways: as an approximation to the continuum, which is then recovered as a limit at the end of the calculations, or as a necessary means of regulating the theory (i.e. controlling divergences) in calculating the Feynman path integral.

1.2 Examples and applications

To provide the reader with an overview of the subsequent material, which is often rather technical, we will introduce some of the models in a qualitative way,

1

explain why they are interesting, and outline some of the important questions to be addressed.

1.2.1 The Ising model

As is well known, this model was proposed by Lenz to his student Ising in 1925 or so, as a simple model for ferromagnetism. It is defined in terms of a set of two-valued classical variables (usually termed 'spins') $\sigma_i = \pm 1$, at the N sites of a lattice, with an interaction energy

$$E(\{\sigma\}) = -J \sum_{\langle ij \rangle} \sigma_i \sigma_j - h \sum_i \sigma_i \tag{1.1}$$

Here J (> 0) is an interaction constant, the sum is over nearest-neighbour pairs of sites $\langle ij \rangle$ on the lattice, and h is an external magnetic field. The interaction favours neighbouring spins being in the same state (either $++$ or $--$) rather than in different states ($+-$ or $-+$) because this minimizes the energy.

It is clear then that the most probable state(s), or the macroscopic thermodynamic state, will change with temperature. At high temperatures, provided that h is zero or small, the spins will be more-or-less randomly arranged due to thermal fluctuations, while at low temperatures they will tend to be aligned. Thus it is at least plausible that this simple model might reasonably describe the onset of magnetic order below the Curie temperature of a ferromagnet.

As is also well known, Ising solved the model exactly for a one-dimensional lattice and found no spontaneous magnetization and no phase transition. Thermal fluctuations in one dimension destroy the possibility of a non-zero magnetic moment at any non-zero temperature. In the two-dimensional case, however, Onsager's exact solution of the Ising model in zero field, which is regarded as one of the most important developments in the whole field, demonstrated the existence of a critical point with a divergent specific heat, and a spontaneous magnetization below the critical temperature. This showed, for the first time, that a careful statistical mechanical calculation could yield the kind of spectacular non-analytic behaviour characteristic of phase transitions in general.

Much of our current understanding of the physics of second-order phase transitions/critical points was first obtained from studies of the Ising model: the existence of critical exponents and of universality, the concept of a divergent correlation length at criticality, and the importance of thermal fluctuations at all length scales, leading to the ideas of renormalization (see e.g. Yeomans, 1992) Figure 1.1 shows typical configurations of Ising spins on a 64×64 square lattice at temperatures $T/T_c = 1.25, 1.0$, obtained from a Monte Carlo simulation. The existence of clusters of all sizes at the critical point is evident.

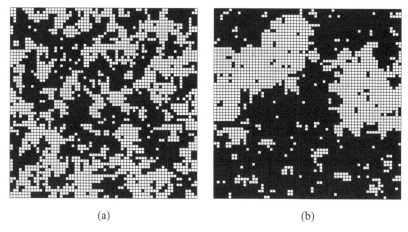

Fig. 1.1. A typical configuration of Ising spins on a 64×64 square lattice at temperatures $T/T_c = 1.25$ (a) and $T/T_c = 1$ (b), obtained from a Monte Carlo simulation.

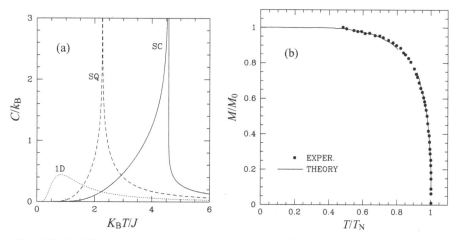

Fig. 1.2. (a) The temperature dependence of the specific heat of the Ising model in one (1D), two (SQ) and three-dimensions (SC). (b) The measured magnetization versus temperature of $DyPO_4$, compared with low-temperature series for the Ising model. From Rado (1970).

Although exact solutions exist for Ising models in two dimensions, there are no exact results for any three-dimensional lattice. Very precise and reliable results have been obtained, however, from two numerical methods: series expansions, the subject of this book, and Monte Carlo simulations. Figure 1.2(a) shows the temperature dependence of the specific heat of the Ising model in one, two and three dimensions. The one- and two-dimensional results are exact, while the three-dimensional result is from series expansions (simple cubic lattice). Figure 1.2(b) shows the measured magnetization versus temperature of $DyPO_4$, compared with low-temperature series for the Ising model. The agreement is impressive.

As well as being a model for (uniaxial) magnets, the Ising model can be used to model many other kinds of system: binary alloys, lattice fluids and systems outside physics. We will describe the series expansion approach to the Ising model in Chapter 2.

1.2.2 The Ising model in a transverse field

In this model the Ising spins σ_i are replaced by Pauli operators σ_i^z, and an external field couples to the x-component, σ_i^x. The resulting Hamiltonian is

$$H = -J \sum_{\langle ij \rangle} \sigma_i^z \sigma_j^z - \Gamma \sum_i \sigma_i^x \qquad (1.2)$$

This is now a fully quantum mechanical model, as the terms in H do not commute. We remind the reader of the usual matrix representation

$$\sigma^x = \begin{pmatrix} 0 & 1 \\ 1 & 0 \end{pmatrix}, \quad \sigma^y = \begin{pmatrix} 0 & -i \\ i & 0 \end{pmatrix}, \quad \sigma^z = \begin{pmatrix} 1 & 0 \\ 0 & -1 \end{pmatrix} \qquad (1.3)$$

and the commutation rule

$$[\sigma_j^\alpha, \sigma_k^\beta] = 2i \delta_{jk} \epsilon_{\alpha\beta\gamma} \sigma_j^\gamma \qquad (1.4)$$

where $\epsilon_{\alpha\beta\gamma}$ is the usual Levi–Civita symbol. We also note that many authors use the spin-$\frac{1}{2}$ operators $S_j^\alpha = \frac{1}{2}\hbar\sigma_j^\alpha$.

Historically this model appears to have been first used by de Gennes, to model the order–disorder transition in hydrogen bonded ferroelectrics, such as KH_2PO_4. Each H atom sits in a double-well potential and the two states can be regarded as the eigenstates of a 'pseudospin' operator σ^z. The transverse field term then describes tunnelling between the two states, and the parameter Γ is related to the tunnelling frequency. This is illustrated in Figure 1.3. The model has subsequently been used to describe a large variety of systems (see e.g. Chakrabarti *et al.*, 1996).

Apart from its role in describing real physical systems, this model is of interest for a number of fundamental theoretical reasons.

Unlike the classical Ising model discussed in the previous section, which has no intrinsic dynamics, the Ising model in a transverse field (for brevity the 'transverse field Ising model') is described by a proper Hamiltonian and does have intrinsic collective excitation modes ('spin waves') with an energy–momentum relation $\omega = \omega(\mathbf{k})$.

The transverse field Ising model can be solved exactly in one dimension at zero temperature (Pfeuty, 1970), and has a quantum phase transition at $\Gamma/J = 1$.

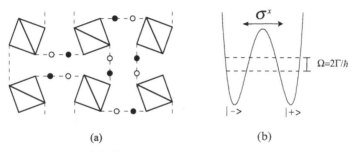

(a)   (b)

Fig. 1.3. (a) Structure of the ferroelectric material KH_2PO_4 (schematic). The tilted squares represent PO_4 tetrahedra, connected via hydrogen bonds (dashed lines). Each bond has a H atom in one of two equilibrium positions (filled and empty circles); (b) Double-well potential for a H atom. The lowest two eigenstates are shown as dashed lines, with tunnelling frequency $\Omega = 2\Gamma/\hbar$.

It is the simplest model to show such a transition, which results from strong quantum fluctuations rather than thermal fluctuations. The subject of quantum phase transitions is a very active one at present (Sachdev, 1999), and we shall investigate a number of quantum phase transitions in succeeding chapters. Here the transition is in fact a quantum critical point, with associated critical exponents which are found to be the same as the thermal critical exponents of the classical Ising model in two dimensions. This is an example of a rather general correspondence between ground state properties of a quantum model in d dimensions and thermodynamic properties of a corresponding classical model in $(d + 1)$ dimensions (Appendix 8).

The transverse field Ising model cannot be solved exactly in dimensions $d > 1$. However series expansions can be used to locate the quantum critical point to high precision, and to study other properties. We will describe this approach in Chapter 4.

1.2.3 The Heisenberg model

Very soon after the development of quantum mechanics, Heisenberg and Dirac independently proposed that the phenomenon of magnetic order in solids might be understood on the basis of a model of exchange coupled quantum angular momenta ('spins'), with a Hamiltonian of the form

$$H = -J \sum_{\langle ij \rangle} \mathbf{S}_i \cdot \mathbf{S}_j \tag{1.5}$$

where the $\mathbf{S}_i$ are spin-S operators, the sum is over nearest-neighbour pairs, and the dot product is the usual

$$\mathbf{S}_i \cdot \mathbf{S}_j = S_i^x S_j^x + S_i^y S_j^y + S_i^z S_j^z \tag{1.6}$$

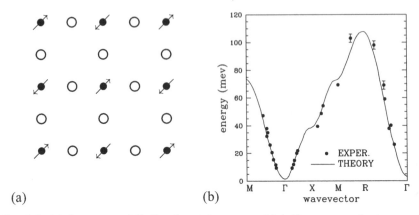

Fig. 1.4. (a) Structure of CuO_2 planes in cuprate high-T_c superconductors, and the antiferromagnetic state. The solid and empty circles represent Cu^{++} and O^{--} ions, respectively; (b) Measured spin-wave dispersion curve for a cubic manganite compound, compared with a model calculation. From Aeppli *et al.* (1997).

The model describes a ferromagnet (antiferromagnet) for $J > 0 (< 0)$ respectively. Note that the Hamiltonian (1.5) is fully symmetric under rotations in spin space, with symmetry group O(3). This model has been intensively studied as the generic prototype for magnetic materials. The first systematic use of the model to describe real magnetic materials appears to be due to Van Vleck (1965). Figure 1.4 shows examples of two magnetic systems of current interest that are well described by the Heisenberg model.

The Heisenberg model (1.5) can be generalized to

$$H = -\sum_{\langle ij \rangle} \left[J_x S_i^x S_j^x + J_y S_i^y S_j^y + J_z S_i^z S_j^z \right] \tag{1.7}$$

when it is known as the XYZ model (or the XXZ model when $J_x = J_y \neq J_z$). The special case $J_z = 0$, $J_x = J_y$ is known as the XY model, and has been used as a lattice model of superfluid helium (for $S = \frac{1}{2}$). Obviously, with the freedom to vary the spin value, the degree of exchange anisotropy, the sign of the exchange constant, and the inclusion of further neighbour interactions we have a multi-dimensional space of possible Hamiltonians, with a resulting variety of properties and applications.

The isotropic Hamiltonian (1.5) has no long range order at finite temperatures in less than three dimensions. In three dimensions there is a critical point for both ferromagnetic (Curie point) and antiferromagnetic (Néel point) interactions. Unlike the case of the classical Ising model, the Curie and Néel temperatures differ slightly. The ground state of the ferromagnet is simple, with all spins aligned in an arbitrary direction, but for the anti-ferromagnet the situation is considerably more complex. In one dimension the famous solution of Bethe gives the wavefunction, and leads to an exact result for the ground state energy (see e.g. Takahashi, 1999). Very limited results are known, however, for general correlations or for thermodynamic properties. In two dimensions there are rigorous results for the existence of long-range order in the ground state for $S \geq 1$, and strong numerical evidence that this also the case for $S = \frac{1}{2}$.

The Heisenberg model also has interesting dynamical properties, including collective spin-wave excitations, which determine the low-temperature thermodynamics, higher energy bound states and time-dependent correlations which can be compared with neutron scattering studies of real materials.

Interesting recent developments, which have kept the Heisenberg model at the forefront of research, include the following.

- The discovery that, for the $S = \frac{1}{2}$ antiferromagnetic chain, the true elementary excitations are $S = \frac{1}{2}$ objects, termed 'spinons', rather than $S = 1$ magnons as previously supposed.
- The discovery that integer-spin antiferromagnetic chains differ in an essential way from half-integer spin chains, the former having an energy gap between the ground state and the lowest excitations ('Haldane gap') while the latter are gapless. This leads, *inter alia*, to fundamentally different low-temperature thermodynamics for the two cases, which is confirmed experimentally.
- The discovery of various compounds, usually low-dimensional, which have energy gaps, and attempts to understand these through Heisenberg models with competing interactions.

Series expansion methods have played an important role in understanding the Heisenberg model. This work will be described in Chapters 5, 6.

1.2.4 The Hubbard model

While models with localized spins do successfully describe an important class of real magnetic materials, it is nevertheless a restricted class since there are no charge degrees of freedom, i.e. no conduction electrons. A number of relatively recent discoveries, such as 'heavy fermion' behaviour in actinides, quantized Hall effects in high-mobility two-dimensional electronic systems, and the high-temperature cuprate superconductors, have pointed to the need for a better understanding of strongly correlated electron systems.

Perhaps the simplest generic model for strongly correlated electron systems is the Hubbard model, which arose from work of Hubbard and others in the 1960s on narrow-band metals. For a single band, the Hamiltonian is

$$H = -t \sum_{\langle ij \rangle, \sigma} (c_{i\sigma}^{\dagger} c_{j\sigma} + c_{j\sigma}^{\dagger} c_{i\sigma}) + U \sum_{i} n_{i\uparrow} n_{i\downarrow} \qquad (1.8)$$

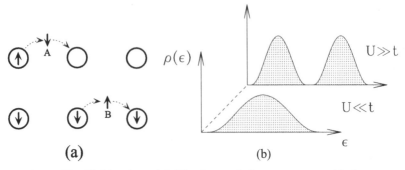

Fig. 1.5. (a) The Hubbard model. The large circles represent orbitals at atomic sites, which may hold up to two electrons of opposite spin. Process A, in which a spin down electron hops from a doubly occupied site to an empty site decreases the energy by U; process B results in no energy change. (b) Electron density of states (schematic) showing the evolution of a 'Mott–Hubbard' gap for large U, and the origin of a metal-insulator transition.

where $c_{i\sigma}^{\dagger}$, $c_{i\sigma}$ are creation and destruction operators, respectively, for electrons of spin σ on site i, and $n_{i\sigma} = c_{i\sigma}^{\dagger}c_{i\sigma}$ is the number of electrons of spin σ on site i. The first term represents non-interacting electrons of spin σ ($\uparrow, \downarrow$) 'hopping' between localized Wannier orbitals on nearest-neighbour sites of a lattice, the parameter t being determined by an overlap integral. This is just a one-electron theory, giving rise to a band of width $W \propto t$. The second term mimics the most important part of the Coulomb repulsion between electrons, giving a positive energy U for two electrons of opposite spin on the same site. Figure 1.5 shows the model pictorially and the development of a gap in the density of states arising from the Coulomb repulsion term. Many generalizations of this model have been proposed including multi-band models, the case of negative U, and the so-called 'extended Hubbard model', which includes a nearest-neighbour Coulomb repulsion. We shall not consider such cases here.

There is another important parameter in the model, the total electron density n = number of electrons per site. When $n = 1$, an average of one electron per site, the band is half-filled. For large U at half-filling the motion of the electrons is strongly suppressed and it can be shown that the lowest energy state has electrons of opposite spin at alternating sites, effectively a Heisenberg antiferromagnet, with exchange constant $J = 4t^2/U$. The removal of a fraction of the electrons from half-filling (generally referred to as 'doping' since this is precisely what happens in many real materials) allows the charge degrees of freedom to become active and leads to major changes in the physics of the model.

It is the presence of the U-term which introduces strong electron correlations, and makes the model much more difficult to analyse. Apart from some exact solutions in one dimension we have no really comprehensive knowledge of the physics of the model in the space of U/t, electron density n and temperature T. Much work has been done, using a variety of techniques including series expansions. Each method has its own strengths and weaknesses, and it is important to compare results from different approaches. We will describe the series approach to the Hubbard model and other models of strongly correlated electrons in Chapter 8.

1.2.5 Lattice gauge models

The lattice approach to gauge field theories was pioneered by Wilson (1974) as a non-perturbative approach to Quantum Chromodynamics (QCD), the standard theory of the strong interactions in particle physics. Subsequently, Monte Carlo simulations of lattice gauge models have become the preferred method for *a priori* calculations of the properties of QCD at low energies or large distances. Spin versions of these lattice gauge models were discussed even earlier, by Wegner (1971).

A lattice gauge model is primarily defined in terms of *link* variables or 'parallel transporters' U_{ij}, defined on the links between neighbouring sites i and j of the lattice. One may also introduce site variables or 'matter fields' ϕ_i residing on sites of the lattice. The link variables are taken to be elements of some gauge group G, and under gauge transformations the variables transform as

$$U_{ij} \rightarrow U'_{ij} = V_i U_{ij} V_j^\dagger \qquad (1.9a)$$

$$\phi_i \rightarrow \phi'_i = V_i \phi_i \qquad (1.9b)$$

where $U_{ji} = U_{ij}^\dagger$, and V_i is a unitary transformation belonging to the group G, which may vary according to the site i. This is an example of a 'local' or gauge transformation which may vary from one site to another, as opposed to a 'global' transformation which is the same at all sites.

We are interested in models where the terms in the lattice Hamiltonian (or action) are *invariant* under gauge transformations. Terms involving only link variables ('pure gauge' terms) will be invariant if they consist of ordered strings of link variables around a closed path on the lattice. The simplest example is a 'plaquette' interaction

$$P = \text{Tr}\{U_{ij} U_{jk} U_{kl} U_{li}\} \qquad (1.10)$$

where i, j, k, l form the corners of an elementary square of the lattice [Figure 1.6(a)]. The trace is taken because for a general group the U_{ij} will be matrix variables. It is easy to check that a term such as (1.10) is invariant under the transformation (1.9a). Invariant interactions can also be constructed from open strings of link variables

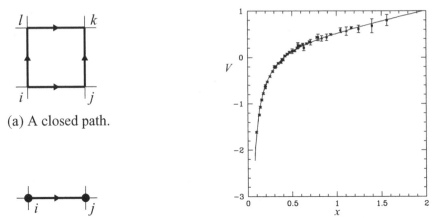

(a) A closed path.

(b) An open path.   (c) The static quark potential.

Fig. 1.6. (a), (b): Interaction terms corresponding to different paths on the lattice.
Link variables are denoted by directed arrows, and site variables by filled points.
(c) The static quark potential for SU(3) Yang–Mills theory. From Stack (1984).

terminated by a matter field at each end, such as the nearest-neighbour interaction
[Figure 1.6(b)]

$$\phi_i^\dagger U_{ij} \phi_j \tag{1.11}$$

The invariance of this term may also be verified by inspection.

The simplest example of a lattice gauge model is the Ising or Z_2 gauge model
introduced by Wegner (1971)

$$H = \beta \sum_{\text{plaquettes}} \sigma_{ij} \sigma_{jk} \sigma_{kl} \sigma_{li} \tag{1.12}$$

where σ_{ij} is an Ising spin variable taking values ± 1 on each link (ij). This
Hamiltonian is invariant under the Z_2 gauge group, where the transformation V_i
either reverses the spin or leaves it unchanged on every link connected to site i.

A primary object of lattice gauge theory is to explain phenomena such as *confinement* of quark and gluons in QCD. Figure 1.6(c) shows a Monte Carlo calculation of the potential between static quarks by Stack (1984). The linear rise of the
potential at large radii provides convincing evidence that the quarks cannot escape
from one another. We shall postpone further discussion of the physics of lattice
gauge models, and their correspondence with continuum gauge field theories, until
Chapter 9.

1.3 The important questions

In the previous section we introduced a number of generic models which will be the
subjects of study in later chapters. In this section we will briefly discuss some of the

physics involved, and the quantities one would like to calculate. These questions transcend particular models and are best discussed in an unified way. Another task of this section is to summarize many of the basic statistical mechanics results needed. We assume a degree of prior familiarity with this material.

The properties of any system at finite temperature T can, at least in principle, be obtained from a knowledge of the canonical partition function

$$\mathcal{Z}_N(T) = \text{Tr}\{e^{-\beta H}\} \tag{1.13}$$

where $\beta = 1/k_B T$ and the trace is taken over a complete set of states for a system of N sites. For a system with a variable number, N_p, of particles (such as the Hubbard model), the grand partition function

$$\mathcal{Z}_g(\mu, T) = \text{Tr}\{e^{-\beta(H - \mu N_p)}\} \tag{1.14}$$

with μ the chemical potential, plays the same role, as 'generating functional' for the theory. The thermodynamic potentials (Helmholtz free energy F or grand potential Ω) are obtained via

$$-\beta F(T) = \ln \mathcal{Z}_N \tag{1.15}$$

and

$$-\beta \Omega(\mu, T) = \ln \mathcal{Z}_g \tag{1.16}$$

Usually one is interested in a large system ($N \to \infty$) in which case both F and Ω are extensive quantities, and we consider the corresponding potential per site, or 'thermodynamic limit'

$$-\beta f(T) = \lim_{N \to \infty} \frac{1}{N} \ln \mathcal{Z}_N \tag{1.17a}$$

$$-\beta \omega(\mu, T) = \lim_{N \to \infty} \frac{1}{N} \ln \mathcal{Z}_g \tag{1.17b}$$

From the corresponding thermodynamic potential, or the logarithm of the partition function, one obtains the usual thermodynamic quantities, per site.

Internal energy:

$$u = -\frac{\partial}{\partial \beta} \left(\frac{1}{N} \ln \mathcal{Z} \right) \tag{1.18}$$

(for variable particle number the derivative must be taken at constant fugacity $\zeta = e^{\beta \mu}$).

Particle number:

$$n = -\zeta \frac{\partial}{\partial \zeta} \left(\frac{1}{N} \ln \mathcal{Z}_g \right) \tag{1.19}$$

Specific heat:

$$C = \frac{du}{dT} = -k_B\beta^2\frac{du}{d\beta} \tag{1.20}$$

Entropy:

$$S/k_B = \frac{1}{N}\ln\mathcal{Z} + \beta u - \beta\mu n \tag{1.21}$$

Magnetization or order parameter:

$$m = -\frac{1}{\beta}\frac{\partial}{\partial h}\left(\frac{1}{N}\ln\mathcal{Z}\right) \tag{1.22}$$

(h is an appropriate field which couples to the order parameter operator in the Hamiltonian).

Susceptibility:

$$\chi = \frac{\partial m}{\partial h} = -\frac{1}{\beta}\frac{\partial^2}{\partial h^2}\left(\frac{1}{N}\ln\mathcal{Z}\right) \tag{1.23}$$

We will be concerned with models which exhibit finite-temperature phase transitions, particularly critical points, where the free energy develops a mathematical singularity at some temperature T_c (for this the thermodynamic limit is crucial). Not only is the determination of T_c important but, even more so, the asymptotic behaviour of thermodynamic quantities in the vicinity of T_c

$$\begin{aligned}C(T) &\sim A_\pm|T_c - T|^{-\alpha} & T &\to T_c\pm\\ m(T) &\sim B(T_c - T)^\beta & T &\to T_c-\\ \chi(T) &\sim C_\pm|T_c - T|^{-\gamma} & T &\to T_c\pm\end{aligned} \tag{1.24}$$

where α, β, γ are termed critical exponents and are found to be universal for systems with the same order parameter symmetry, and in the same number of dimensions, while A, B, C are non-universal amplitudes. We have made the assumption, now well established, that the exponents above and below T_c are identical. For a thorough discussion of the statistical mechanics of phase transitions we refer the reader to Yeomans (1992).

We will also be concerned with quantum phase transitions (see e.g. Sachdev, 1999), where the nature of the ground state changes as some parameter λ in the Hamiltonian is tuned through a critical value λ_c. A quantum critical point of this type occurs in the transverse field Ising model. The ground state energy per site $e_0(\lambda)$ now plays a similar role to the free energy previously. A quantity analogous to the specific heat is

$$C(\lambda) = -\frac{d^2e_0}{d\lambda^2} \tag{1.25}$$

and, by adding a field term to the Hamiltonian, one can define an order parameter and susceptibility

$$m(\lambda) = -\left(\frac{de_0}{dh}\right)_{h=0} \tag{1.26a}$$

$$\chi(\lambda) = -\left(\frac{d^2 e_0}{dh^2}\right)_{h=0} \tag{1.26b}$$

These quantities will be singular at λ_c, and critical exponents can be defined in an analogous way.

In addition to locating singular points, corresponding to classical or quantum phase transitions, we will also wish to compute many other kinds of physical quantities. These include correlation functions and their Fourier transforms, dispersion relations for elementary excitations, energy gaps, and the possibility of higher energy bound states of two or more excitations. In a spin system, for example, the spin correlation function

$$S^{\alpha\beta}(\mathbf{r}, t) = \langle S_0^\alpha S_{\mathbf{r}}^\beta(t)\rangle \tag{1.27}$$

describes both the nature of any static ordering (for $t = 0$) and the dynamical response to an external agent coupling to the spins. Here $\alpha, \beta = x, y, z$ and the angular brackets denote a ground state average or a thermodynamic average at finite temperature. The dynamical structure function

$$S(\mathbf{k}, \omega) = \int dt \int d^3 \mathbf{r} e^{i(\mathbf{k}\cdot\mathbf{r} - \omega t)} S(\mathbf{r}, t) \tag{1.28}$$

can be measured in scattering experiments, and thus provides a link between real systems and model calculations.

It is often possible to identify, in strongly interacting systems, low-energy excited states which are 'particle like'. The idea goes back to Landau's seminal work on interacting Fermi systems leading to the idea of 'quasiparticles' with a definite energy–momentum relation or dispersion law $\epsilon(\mathbf{k})$. It is well known that the form of this relation at small $\mathbf{k}$ (e.g. linear, quadratic) determines the low-temperature behaviour of thermodynamic quantities. We will show how one can compute dispersion relations for elementary excitations of various kinds using series methods. This provides another link between model calculations and real materials.

Elementary excitations are not usually independent and, following the particle analogy, can scatter from each other, or form bound states. Both cases lead to important and observable physical effects. The computation of bound states and their energies is an area of considerable current interest.

All of these issues will be taken up in subsequent sections of this book.

1.4 Series expansion methods

In the absence of exact solutions, which exist for only a few problems of interest, there is a branching set of approaches which can be used. The first division is between analytic approximations and numerical methods. The former can be very powerful, but are often uncontrolled and difficult to assess for accuracy. The latter can also be very powerful, and are usually systematically improvable with increased effort. The word 'numerical' is, by the way, potentially misleading – computers are sometimes used simply because the algebra would be impossibly lengthy otherwise. Among such methods we mention:

- exact diagonalizations for small systems;
- classical and quantum Monte Carlo simulations;
- renormalization methods, in particular the Density Matrix Renormalization Group (DMRG) approach; and
- series expansion methods.

Each of these has its strengths and weaknesses and it can be important to study a given problem using a number of different approaches.

This book is concerned with series expansion methods, of which there are again a number of different kinds. Their common feature is that one computes a number of coefficients in a power series expansion for some quantity represented by a function $f(x)$

$$f(x) = a_0 + a_1 x + a_2 x^2 + \cdots + a_N x^N + \cdots \qquad (1.29)$$

where the coefficients (a_i) are computed **exactly** for a finite number of terms (typically 10–20, although much longer series are available in some special cases). The problem then is to use the information contained in these coefficients to construct (approximately) the function $f(x)$ or some of its properties.

In one sense this is of course a nonsensical goal, as it is obvious that the early terms of an infinite power series cannot, in general, provide information about the function as a whole. However, in physics one assumes, as a necessity, that the functions that occur are not pathological and that useful information can be extracted from the first N coefficients. This is confirmed in actual calculations.

The calculation of the coefficients is a computational problem of considerable complexity with, usually, an exponentially increasing computer time and memory requirement. As a rule of thumb, the effort to compute another term is at least as great as that used to compute all of the preceding terms. This limits the length of series that can be obtained.

It is usually not possible to know or prove that a series is convergent for a given value of x. Clearly for small enough x one can evaluate $f(x)$ to high precision simply by summing the known terms. Usually this is not enough. One needs to extrapolate to larger x values, or to estimate the location of the closest physical singularity on the positive real axis. There are a number of numerical analysis methods for doing this, which will be discussed in the following section. Before doing so, we will introduce various kinds of expansions which have been used.

1.4.1 High-temperature series

For any model with Hamiltonian H one can expand the partition function

$$\mathcal{Z} = \text{Tr}\{e^{-\beta H}\} \tag{1.30}$$

in powers of $\beta = 1/k_B T$ yielding

$$\mathcal{Z} = \sum_{n=0}^{\infty} \frac{(-\beta)^n}{n!} \text{Tr}(H^n) \tag{1.31}$$

If we extract from H some energy parameter J, then from (1.31) one can derive a high-temperature series for the free energy per site, in the form

$$-\beta f(K) = \sum_{n=0}^{\infty} a_n K^n, \quad (K \equiv J/k_B T) \tag{1.32}$$

where the a_n are numerical coefficients. This expansion will be convergent at high temperature and can be used to compute any thermodynamic property of the model at sufficiently high temperature. By including auxiliary fields in the Hamiltonian and taking appropriate derivatives one can obtain corresponding high-temperature series for susceptibilities and for correlation functions.

If there are no interfering singularities off the positive real axis then the series will converge up to some value K_c, which corresponds to the critical temperature $k_B T_c/J = K_c^{-1}$. We will see, in the next section, how this estimate can be made.

It should be noted that K is not the only possible expansion variable, nor is it necessarily the optimal one. We will see in Chapter 2 that $v = \tanh K$ is a natural high-temperature expansion variable for the Ising model.

Another point worth making is that, in many cases of interest, the Hamiltonian may include more than one coupling constant. For the case of two, say J_1 and J_2, proceeding exactly as above yields a free energy expansion

$$-\beta f(K_1, K_2) = \sum_{n=0}^{\infty} a_n(\alpha) K_1^n, \quad (K_i \equiv J_i/k_B T) \tag{1.33}$$

where the coefficients are polynomials in $\alpha = J_2/J_1$. Analysis of the series can then yield the locus of a critical line in the (K_1, K_2) plane.

1.4.2 Low-temperature series

For systems in which the excitations above the ground state have a discrete and regular spectrum it is also possible to derive expansions which are convergent at low temperature. This is the case for the Ising model and other classical models such as Potts and vertex models. In such cases the partition function can be written as

$$\mathcal{Z} = \sum_n e^{-\beta E_n} = e^{-\beta E_0}\left(1 + \sum_{n \neq 0} e^{-\beta \Delta E_n}\right), \quad \Delta E_n \equiv E_n - E_0 \quad (1.34)$$

Now, if it is possible to find an energy increment ϵ such that, for all n, $\Delta E_n = p\epsilon$ with p an integer, then an expansion for the free energy of the form

$$-\beta f(u) = -\beta \epsilon_0 + \sum_{n=1}^{\infty} a_n u^n \quad (1.35)$$

can be obtained, with $\epsilon_0 = E_0/N$ and $u = e^{-\beta \epsilon}$. Since $u \to 0$ as $T \to 0$, this expansion will converge at low temperatures.

In the following chapter we will show how to derive low-temperature expansions for the Ising model. Such an expansion can be particularly valuable if a first-order transition is known, or suspected, to exist. By matching the high- and low-temperature free energies one can locate the transition from their crossing, and identify the order from whether the two branches meet smoothly (second order) or cross with different slopes (first order). This approach has been successfully used in a number of studies.

Low-temperature expansions are only possible where the spectrum is discrete. Most of the models of interest studied in this book have a continuous energy spectrum and, in such cases, low-temperature expansions are not possible.

1.4.3 Perturbation expansions at zero temperature

Every undergraduate student of quantum mechanics learns the techniques of low-order perturbation theory for quantum Hamiltonians of the form

$$H = H_0 + \lambda V \quad (1.36)$$

where the reference or unperturbed system H_0 is exactly solvable, V represents the remaining part or 'perturbation', and λ is a parameter. The energy of any non-degenerate level can be expressed as a power series in λ; in particular the ground-state energy is

$$E_g = E_0 + \lambda \langle 0|V|0 \rangle + \lambda^2 \sum_{n \neq 0} \frac{\langle 0|V|n\rangle\langle n|V|0\rangle}{E_0 - E_n} + \cdots \quad (1.37)$$

where $|n\rangle$, E_n denote the eigenstates and eigenenergies of H_0. This is known as 'Rayleigh–Schrödinger' perturbation theory, and in fact predates quantum mechanics.

Second-order perturbation theory is fine if successive terms fall rapidly in magnitude (say by a factor of 10 or so), but questionable if the decrease is less rapid or if, for reasons of symmetry, the first-order term is zero! By judicious choice of technique it is, in fact, possible to extend such perturbation expansions to many terms (typically 10–20), when it becomes possible to treat 'large' perturbations, obtain results to high numerical precision, and to study analytic properties of the perturbation series.

Such methods have been developed and used extensively over the last 20 years, by our group and by others, to investigate not only the ground-state energy but many other ground-state properties of many quantum models, and much of this work will be described in later chapters. In these calculations there is often no naturally small term, so the division into H_0 and V is made in any convenient way. This leads to 'Ising expansions', 'dimer expansions', 'plaquette expansions', etc, which will be explained later. In each case an expansion is obtained for some ground-state property f in the form of a power series

$$f(\lambda) = \sum_n a_n \lambda^n \tag{1.38}$$

and the series must be evaluated at $\lambda = 1$, corresponding to the full Hamiltonian. This is not simply a matter of summing the known terms, as the series may be only slowly convergent or $\lambda = 1$ may be a singular point!

We defer further discussion until Chapter 4, when we will explain the procedure in some detail within the context of a particular model.

1.4.4 Thermodynamic perturbation theory

Another type of perturbation expansion can be developed for systems in which the Hamiltonian is written as $H = H_0 + \lambda V$, but where the interest is not in ground-state properties, but rather in thermodynamic properties at finite temperature. Thus we start with the canonical partition function

$$\mathcal{Z}(T) = \mathrm{Tr}\{e^{-\beta H}\} = \mathrm{Tr}\{e^{-\beta(H_0 + \lambda V)}\} \tag{1.39}$$

and seek to expand this (or more precisely, its logarithm) in powers of λ. If H_0 and V commute this is relatively straightforward, but usually they do not. The following approach can then be used: we write

$$e^{-\beta(H_0 + \lambda V)} = e^{-\beta H_0} f(\beta) \tag{1.40}$$

and obtain a differential equation for $f(\beta)$

$$\frac{d}{d\beta}f(\beta) = -\lambda\tilde{V}(\beta)f(\beta) \tag{1.41}$$

where

$$\tilde{V}(\beta) \equiv e^{\beta H_0}Ve^{-\beta H_0} \tag{1.42}$$

The differential equation can then be re-cast as an integral equation

$$f(\beta) = 1 - \lambda\int_0^\beta \tilde{V}(\tau)f(\tau)d\tau \tag{1.43}$$

and solved by iteration, yielding

$$f(\beta) = 1 + \sum_{n=1}^\infty (-\lambda)^n \int_0^\beta d\tau_1 \int_0^{\tau_1} d\tau_2 \cdots \int_0^{\tau_{n-1}} d\tau_n \tilde{V}(\tau_1)\tilde{V}(\tau_2)\cdots\tilde{V}(\tau_n) \tag{1.44}$$

Inserting this expansion into Eq.(1.39) gives

$$\mathcal{Z}(T) = Z_0\left\{1 + \sum_{n=1}^\infty (-\lambda)^n \int_0^\beta d\tau_1 \int_0^{\tau_1} d\tau_2 \cdots \int_0^{\tau_{n-1}} d\tau_n \langle\tilde{V}(\tau_1)\tilde{V}(\tau_2)\cdots\tilde{V}(\tau_n)\rangle_0\right\} \tag{1.45}$$

where $\langle\cdots\rangle_0$ denotes the unperturbed expectation value

$$\langle A\rangle_0 \equiv \text{Tr}\{e^{-\beta H_0}A\}/\text{Tr}\{e^{-\beta H_0}\} \tag{1.46}$$

for any operator A.

To evaluate the nth term of this expansion one then needs to compute the expectation value of a product of 'time-dependent' V operators (we refer to the τ variables as 'times'), and then compute the multiple-time integral. Many readers will recognize this as analogous to the expansion for the evolution operator in the interaction representation of quantum mechanics. In that context it is normal to introduce a time-ordering operator, extend the range of all integrals from 0 to β and go to a frequency representation. The same could be done here, but does not appear to confer any significant advantage in calculations. The upshot is an expansion, for the free energy or for other thermodynamic quantities, of the form

$$A(\beta) = \sum_{n=0}^\infty a_n(\beta)\lambda^n \tag{1.47}$$

where the $a_n(\beta)$ are complete functions of inverse temperature β and other parameters of the problem. The series can then be analysed by standard methods.

We will describe particular calculations of this type in later chapters. It is worth noting that if we take $H_0 = 0$, and $V = H$, then we immediately recover the standard high-temperature expansion

$$\mathcal{Z} = 1 + \sum_{n=1}^{\infty} \frac{(-\beta)^n}{n!} \langle H^n \rangle \tag{1.48}$$

as in Eq. (1.31). However, the thermodynamic perturbation theory is more than simply a high-temperature expansion. As the coefficients are complete functions of β the series can be successfully used down to quite low temperatures, yielding information about the ordered phase in quantum systems. In cases where the ground state is non-degenerate the coefficients are regular at $\beta \to \infty$ and the zero-temperature perturbation result can be recovered.

1.5 Analysis of series

How can we analyse series expansions to obtain the information we want? This is a large field, about which much is known. Rather than reproduce much of what is available elsewhere, we will outline the main methods only and refer the reader to existing work (Guttmann, 1989; Baker, 1990), not only for the important mathematical analysis which underlies the methods, but also for discussions of many other, more specific, methods which are available for particular cases.

However, the short answer to our initial question is that there is no single method which can be guaranteed to give reliable results for any particular series. The art of series analysis lies in systematically using a battery of methods, and of making an educated guess as to the type of behaviour expected and choosing the method of analysis to suit. It is from the analysis that the uncertainty (and the error bars in our subsequent figures in later chapters) arises. These uncertainties are hardly ever true statistical errors, but rather subjective 'confidence limits' estimated from each particular analysis. They can be overly optimistic!

The three main methods, which we will describe here, and use in much of the book, are:

- the ratio method;
- Padé approximants; and
- integrated differential approximants.

We will illustrate their use, and their effectiveness, on three different test functions, each expanded to 15 terms. The advantage of using test functions in this way is, firstly, that we known what the correct answer is and, secondly, that we can compare and contrast the effectiveness of each method for functions which have different analyticity properties.

The *ratio method* is based on the following result. If a function has a power law singularity of the form

$$f(x) \simeq A \left(1 - \frac{x}{x_c} \right)^{-\theta} + B \tag{1.49}$$

where $A(x)$ and $B(x)$ are regular at x_c, then the ratios $r_n = a_n/a_{n-1}$ of successive coefficients in the power series expansion

$$f(x) = \sum_{n=0}^{\infty} a_n x^n \tag{1.50}$$

satisfy the relation

$$r_n = \frac{1}{x_c} \left[1 + \frac{\theta - 1}{n} + 0(1/n^2) \right] \tag{1.51}$$

Hence a plot of r_n versus $1/n$ should become linear for large enough n, and the intercept and slope yield estimates of x_c and θ. This method actually works well for Ising model series, but in many cases the analytic structure of the function is more complex and ratios become too erratic.

A *Padé approximant* (PA) is a representation of the first N terms of a power series as a ratio of two polynomials. Thus

$$f(x) = \sum_{n=0}^{N} a_n x^n = P_L(x)/Q_M(x) \tag{1.52}$$

where $P_L(x)$, $Q_M(x)$ are polynomials of degree L, M and $L + M \leq N$. This is denoted as the $[L, M]$ (or [L/M]) approximant. The polynomial coefficients can be determined by solving a set of linear equations. For a given number of known coefficients of the original series, a set of PAs with different (L, M) can be constructed, forming a *Padé table*. Padé approximants allow the evaluation of series and provide a way of estimating their analytic structure. Baker (1990) gives a nice example of the evaluation of $f(\infty) = \sqrt{2}$ where $f(z) = \sqrt{(1 + 2z)/(1 + z)}$ is defined by its Taylor series expansion. The [5, 5] approximant, which uses 11 terms, is accurate to nine significant figures.

A Padé approximant can represent functions with simple poles *exactly*. Thus in the field of critical phenomena it is usual to take the logarithmic derivative of the series, which converts an algebraic singularity into a simple pole. For example, if $f(x) = A(x_c - x)^{-\theta}$ then

$$\text{Dlog} f(x) \equiv \frac{f'(x)}{f(x)} = \frac{\theta}{x_c - x} \tag{1.53}$$

Hence the positions of the singular points can be estimated from roots of the denominator polynomial $Q_M(x)$, and the exponents from the corresponding residues.

Integrated differential approximants (IDAs) (also referred to as 'integral approximants' or 'differential approximants') fit the known coefficients of the series to a homogeneous or inhomogeneous differential equation, usually of first or second order, of the general form

$$P_K(x)\frac{d^2 f}{dx^2} + Q_L(x)\frac{df}{dx} + R_M(x)f + S_T(x) = 0 \qquad (1.54)$$

where P_K, Q_L, R_M, S_T are polynomials of degree K, L, M, T respectively. We use the notation $[T/M, L, K]$ to denote such an approximant. The homogeneous case, with $S_T(x) = 0$, is denoted $[-1/M, L, K]$. These can again be obtained by solving a set of linear equations. In general longer series are needed since $K + L + M + T + 4 \leq N$. IDAs have the advantage that they can handle additive analytic or non-analytic terms which cause difficulties for PAs. Some known exact solutions satisfy equations of this sort. For example, the spontaneous magnetization $M(u)$ of the square lattice Ising model satisfies

$$(1 - u^2)(1 - 6u + u^2)\frac{dM}{du} + 4uM = 0, \qquad u = e^{-4K} \qquad (1.55)$$

from which the exact result can be obtained immediately. Likewise the internal energy can be shown to satisfy a second-order inhomogeneous differential equation.

Singularities in $f(x)$ are represented by the roots of the polynomial multiplying the highest order derivative, and the exponents are determined by the solution of the indicial equation

$$\theta = m - 1 - \frac{P_{m-1}(x_c)}{P'_m(x_c)} \qquad (1.56)$$

where m is the order of the highest derivative. The function $f(x)$ can be evaluated at any point within the circle of convergence by numerical integration of the differential equation.

To see how these methods work we consider the three test functions, each expanded up to x^{14}

$$f_1(x) = x(1 - x)^{-3/2} + e^{-x} \qquad (1.57a)$$
$$f_2(x) = (1 + 2x)^{1/2} + (1 - x)^{3/2} \qquad (1.57b)$$
$$f_3(x) = (1 - x)^{-5/4} + 2(1 - x)^{-3/4} \qquad (1.57c)$$

Function f_1 has a simple algebraic singularity plus an additive analytic term. Such a function is well suited to ratio analysis, and a ratio plot is shown in Figure 1.7(a).

Table 1.1. *Estimates of critical point and exponent (in brackets) from $[L, M]$ Padé approximants to the series $\frac{d}{dx} \ln f_1(x)$.*

	$L = 5$	$L = 6$	$L = 7$	$L = 8$
$M = 5$	1.000593	0.999985	1.000087	1.000085
	(1.51336)	(1.50130)	(1.50408)	(1.50402)
$M = 6$	1.000004	1.000101	1.000084	
	(1.50185)	(1.50447)	(1.50402)	
$M = 7$	1.000093	1.000085		
	(1.50424)	(1.50403)		
$M = 8$	1.000083			
	(1.50398)			

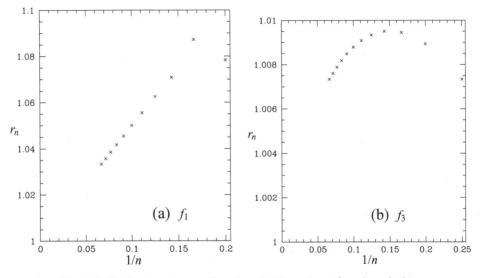

Fig. 1.7. Ratio plots for test function f_1 (a), and test function f_3 (b).

Pairwise linear extrapolation from the last five points recovers the critical point and exponent to six significant figures or better.

In Table 1.1 we show estimates of the critical point and exponent from Dlog Padé approximants $[L, M]$, as defined above. The presence of additive terms will, in general, cause difficulties for Dlog PAs, because taking the logarithmic derivative will no longer yield a simple pole. Nevertheless for this function the method works well, as can be judged from the results of Table 1.1. One should, however, be cautious about uncertainty estimates. An optimist might conclude from the highest order estimates that the exponent $\theta \simeq 1.504 \pm 0.001$, which does not encompass the

Table 1.2. *Estimates of critical point and exponent (in brackets) from* [T/L, L] *first-order IDAs to the series* $f_1(x)$.

	$T = 2$	$T = 3$	$T = 4$	$T = 5$
$L = 3$	0.99659	0.99699	1.00060	1.00221
	(1.4401)	(1.4301)	(1.5188)	(1.5659)
$L = 4$	0.99976	1.00014	0.99984	
	(1.4934)	(1.50516)	(1.49313)	
$L = 5$	1.00001	1.00000		
	(1.5005)	(1.4999)		
$L = 6$	1.00001			
	(1.5008)			

correct value. On the other hand if we look at a sequence of diagonal approximants one sees convergence towards the correct values.

	[5,5]	[6,6]	[7,7]
Critical point	1.000593	1.000101	1.000052
Exponent	1.51336	1.50447	1.50300

Integrated differential approximants, on the other hand, can deal effectively with additive terms. The function f_1 satisfies the differential equation

$$2(1 - x)\frac{df}{dx} - 3f(x) + (5 - 2x)e^{-x} = 0 \tag{1.58}$$

and is thus not represented exactly as a [T/M, L] approximant. However we would expect good convergence as the degree of the approximant increases. This is borne out by the results of Table 1.2.

It is left as an exercise for the reader to show that f_1 exactly satisfies a homogeneous second-order differential equation with polynomial coefficients, and is represented exactly by an approximant [$-1/1,2,2$].

Turning now to test function f_2, the task is to evaluate the series at the singular point $x = 1$, from a knowledge of the first 15 terms. As can be seen by expanding the function, the coefficients increase in magnitude and alternate in sign. If we form PAs to the series (not to the logarithmic derivative here!) and evaluate these at $x = 1$ we obtain a reasonable estimate of $f(1) = \sqrt{3}$, accurate to better than 0.5% (Table 1.3). Note again the increase in estimates with increasing order suggesting that the true value is somewhat greater than 1.728.

Table 1.3. *Estimates of* $f_2(1) = \sqrt{3}$ *from* $[L, M]$ *PAs to the series for* $f_2(x)$ *evaluated at* $x = 1$.

	$L = 5$	$L = 6$	$L = 7$	$L = 8$	$L = 9$
$M = 5$	1.7208	1.7260	1.7260	1.7164	1.7275
$M = 6$	1.7194	1.7260	1.7260	1.7280	
$M = 7$	1.7228	1.7293	1.7278		
$M = 8$	1.7208	1.7274			

Table 1.4. *Estimates of critical point and exponent (in brackets) from* $[L, M]$ *Padé approximants to the series* $\frac{d}{dx} \ln f_3(x)$.

	$L = 5$	$L = 6$	$L = 7$	$L = 8$
$M = 5$	0.99879	0.99900	0.99916	0.99927
	(−1.115)	(−1.123)	(−1.129)	(−1.135)
$M = 6$	0.99905	0.99922	0.99935	
	(−1.125)	(−1.132)	(−1.138)	
$M = 7$	0.99922	0.99937	0.99947	
	(−1.132)	(−1.140)	(−1.146)	
$M = 8$	0.99935			
	(−1.138)			

First-order IDAs do not yield better results in this case. However, as for the first test function, a second-order homogeneous IDA $[-1/1,2,3]$ can represent this function exactly.

Finally we turn to test function $f_3(x)$, a more difficult case. This is a function with a so-called 'confluent singularity'. There are strong reasons to believe that the analytic structure of many lattice model quantities has this form. Figure 1.7(b) shows a ratio plot for this series. It is evident that, even discarding the first few points, there is strong residual curvature left and that it would be difficult, if not impossible, to obtain reliable results.

Turning to Padé approximants, we show in Table 1.4 estimates of the critical point and leading exponent obtained from the logarithmic derivative.

As is evident, while the critical point is given quite accurately to 0.1% or better, the leading exponent is seriously underestimated. The PAs show a consistent second singularity at ~ 1.10 with a residue of ~ -0.11: an attempt to represent the confluent term.

Table 1.5. *Estimates of critical point and leading exponent (in brackets) from $[T/L, L]$ first-order IDAs to the series $f_3(x)$.*

	$T = 1$	$T = 2$	$T = 3$	$T = 4$
$L = 3$	0.9991	0.9994	0.9995	0.9996
	(-1.142)	(-1.159)	(-1.170)	(-1.178)
$L = 4$	0.9995	0.9997	0.9998	0.9998
	(-1.171)	(-1.185)	(-1.194)	(-1.200)
$L = 5$	0.9997	0.9998	0.9999	
	(-1.189)	(-1.200)	(-1.208)	
$L = 6$	0.9998			
	(-1.2001)			

Differential approximants do much better with this function. Table 1.5 gives estimates of the critical point and leading exponent from first-order IDAs. While the correct exponent is not achieved, there is a clear convergence towards the correct value, both across and down the table.

Second-order IDAs with $K = 1$ do not provide an improvement. However on trying second-order IDAs with $K = 2$ one quickly finds an exact representation given by $[-1/0,1,2]$ and higher approximants of this type. The reader may wish to verify that f_3 satisfies the second-order equation

$$(1 - x)^2 f''(x) - 3(1 - x) f'(x) + \frac{15}{16} f(x) = 0 \qquad (1.59)$$

Of course real series with confluent singularities will be more difficult to analyse. However, the existence of an exact or near double root in $P_K(x)$ is a strong indicator of a confluent term. This example also illustrates the usefulness of IDAs as another tool for analysis of power series.

2

High- and low-temperature expansions for the Ising model

2.1 Introduction

The Ising model, introduced in Chapter 1, provides the simplest lattice model for which long series expansions can be derived. The material to be presented here goes back to the pioneering work of Domb and colleagues at King's College, London, half a century ago, and is rather well known. It is nevertheless a nice pedagogical example with which to introduce many of the ideas and techniques to the unfamiliar reader.

We start with the familiar 'Hamiltonian' (1.2)

$$H = -J \sum_{\langle ij \rangle} \sigma_i \sigma_j - h \sum_i \sigma_i \tag{2.1}$$

where σ_i is the classical dimensionless Ising variable at site i, taking values ± 1, and J, h are constant parameters with dimensions of energy. We assume a regular lattice of N sites, with nearest-neighbour interactions. The thermodynamic properties are derivable from the partition function

$$\mathcal{Z}(K, h) = \sum_{\{\sigma\}} e^{-\beta H} = \sum_{\{\sigma\}} \exp \left(K \sum_{\langle ij \rangle} \sigma_i \sigma_j + \beta h \sum_i \sigma_i \right) \tag{2.2}$$

where the first sum is over all spin configurations, $K = \beta J$ is a temperature-dependent coupling constant, and $\beta = 1/k_B T$ as usual. We note that at high temperatures K is small.

The basic idea is to expand the partition function in a power series in K. Setting $h = 0$, for the moment, we obtain

$$\mathcal{Z}(K) = \sum_{\{\sigma\}} \prod_{\langle ij \rangle} e^{K \sigma_i \sigma_j} = \sum_{\{\sigma\}} \prod_{\langle ij \rangle} \sum_{l=0}^{\infty} \frac{K^l}{l!} (\sigma_i \sigma_j)^l \tag{2.3}$$

A graphical or diagrammatic representation helps in the book-keeping. The factor $(\sigma_i \sigma_j)^l$ is associated with an l-fold line joining sites i and j on the lattice. A general term in the sum (2.3) is then represented by a diagram of the entire lattice with each bond $\langle ij \rangle$ having some multiplicity l_{ij}. At each site i there will be a factor σ^p, where p is the sum of multiplicities of all bonds connecting to site i. We refer to this as the degree of site i. The simple result

$$\sum_{\sigma=\pm 1} \sigma^p = \begin{cases} 2, & p \text{ even} \\ 0, & p \text{ odd} \end{cases} \qquad (2.4)$$

immediately shows that the only non-zero terms come from graphs in which every vertex is of even degree (including zero). This leads to the result

$$\mathcal{Z}_N(K) = 2^N \sum_{\{g_0\}} \frac{C(g)}{w(g)} K^{l_g} \qquad (2.5)$$

where the sum is over all possible graphs with all even vertices, l_g is the number of lines, including multiplicities, $w(g)$ is a combinatorial factor for multiple lines, and $C(g)$ is the number of ways in which the graph can be located on the lattice of N sites (the embedding factor).

Table 2.1 gives a list of graphs and the corresponding multiplicity and embedding factors up to order six, on the triangular lattice. Adding together the contributions from these 25 graphs, gives

$$\mathcal{Z}_N(K) = 2^N \left[1 + \tfrac{3}{2}NK^2 + 2NK^3 + (\tfrac{9}{8}N^2 + \tfrac{11}{4}N)K^4 + (3N^2 + 4N)K^5 \right.$$
$$\left. + (\tfrac{9}{16}N^3 + \tfrac{49}{8}N^2 + \tfrac{106}{15}N)K^6 + \cdots \right] \qquad (2.6)$$

Taking the logarithm gives

$$\ln \mathcal{Z}_N(K) = N \ln 2 + \left[\tfrac{3}{2}NK^2 + 2NK^3 + (\tfrac{9}{8}N^2 + \tfrac{11}{4}N)K^4 + \cdots \right]$$
$$- \tfrac{1}{2}\left[\tfrac{3}{2}NK^2 + 2NK^3 + \cdots \right]^2 + \tfrac{1}{3}\left[\tfrac{3}{2}NK^2 + \cdots \right]^3 + \cdots$$
$$= N\left(\ln 2 + \tfrac{3}{2}K^2 + 2K^3 + \tfrac{11}{4}K^4 + 4K^5 + \tfrac{106}{15}K^6 + \cdots \right) \qquad (2.7)$$

from which we see that all of the terms proportional to N^2, N^3, $\cdots$, can-cel on taking the logarithm. This must be so because the thermodynamic free energy, $-k_B T \ln \mathcal{Z}_N$, is an extensive quantity, proportional to the system size. We are usually interested in the bulk properties of large systems, or the thermodynamic limit, which gives

$$\tfrac{1}{N} \ln \mathcal{Z}_N(K) = \ln 2 + \tfrac{3}{2}K^2 + 2K^3 + \tfrac{11}{4}K^4 + 4K^5 + \tfrac{106}{15}K^6 + \cdots \qquad (2.8)$$

Before proceeding further, it is worth noting the following.

Table 2.1. *List of graphs for the partition function on the triangular lattice, to order* K^6.

g	(graph)	(graph)	(graph)	(graph)	(graph)
$w(g)$	2	1	24	4	1
$C(g)$	$3N$	$2N$	$3N$	$15N$	$3N$

g	(graph)	(graph)	(graph)	(graph)	(graph)
$w(g)$	4	6	2	1	2
$C(g)$	$\frac{9}{2}N^2 - \frac{33}{2}N$	$6N$	$24N$	$6N$	$6N^2 - 30N$

g	(graph)	(graph)	(graph)	(graph)	(graph)
$w(g)$	720	48	8	8	8
$C(g)$	$3N$	$30N$	$69N$	$20N$	$2N$

g	(graph)	(graph)	(graph)	(graph)	(graph)
$w(g)$	2	6	2	1	1
$C(g)$	$3N$	$12N$	$42N$	$15N$	$9N$

g	(graph)	(graph)	(graph)	(graph)	(graph)
$w(g)$	48	8	1	2	8
$C(g)$	$9N^2 - 33N$	$45N^2 - 234N$	$2N^2 - 13N$	$9N^2 - 57N$	$\frac{27}{6}N^3 - \frac{99}{2}N^2 + 144N$

- The number of graphs grows rather rapidly, and if one wishes to proceed much further it is necessary to streamline the process and use a computer.
- Likewise the computation of the embedding constants rapidly becomes too tedious to do without an automated procedure.
- Within this formulation both connected and disconnected graphs occur. The latter can be a nuisance, not only because their numbers proliferate rapidly but also because calculation of embedding factors becomes tricky.

An important goal, then, is to endeavour to reduce the number of graphs which need to be considered explicitly.

In the case of the Ising model, an immediate simplification is possible by use of the identity

$$e^{K\sigma_i\sigma_j} = (\cosh K)(1 + v\sigma_i\sigma_j) \tag{2.9}$$

Table 2.2. *List of graphs for the partition function to order v^6, for the triangular lattice.*

which is valid for $\sigma_i, \sigma_j = \pm 1$, with $v = \tanh K$. The zero-field partition function can then be written as

$$\mathcal{Z}_N(K) = (\cosh K)^{Nq/2} \sum_{\{\sigma\}} \prod_{\langle ij \rangle} (1 + v\sigma_i\sigma_j)$$

$$= 2^N (\cosh K)^{Nq/2} \sum_{\{g_0\}} C(g)v^{l_g} \tag{2.10}$$

where q is the coordination number of the lattice, i.e. the number of neighbours of any site ($\frac{1}{2}Nq$ is the number of nearest-neighbour pairs), and the sum is again over a set of even-vertex graphs. However, only single-bonded graphs occur. Table 2.2 lists these up to order 6.

Taking the embedding constant data from Table 2.1, and taking the logarithm as before, yields

$$\frac{1}{N} \ln \mathcal{Z}_N = \ln 2 + 3 \ln \cosh K + 2v^3 + 3v^4 + 6v^5 + 11v^6 + \cdots \tag{2.11}$$

When re-expanded in terms of K we recover the previous result (2.8). There is usually no need to do this, as v can itself serve as a high-temperature expansion variable. We note that the number of graphs (to sixth order) has been reduced from 25 to 6 – this is a simple example of *renormalization*, which is an idea that will recur later.

Let us now return to the full Hamiltonian (2.1), with the field term, and derive a high temperature series for the zero-field magnetic susceptibility

$$\chi(v) = \beta^{-1} \lim_{h \to 0} \frac{\partial^2}{\partial h^2} \left(\frac{1}{N} \ln \mathcal{Z}_N \right) \tag{2.12}$$

Using the identity (2.9), and a similar one for the field term, yields

$$\mathcal{Z}_N = (\cosh K)^{Nq/2}(\cosh \beta h)^N \sum_{\{\sigma\}} \prod_{\langle ij \rangle} (1 + v\sigma_i\sigma_j) \prod_k (1 + \tau\sigma_k)$$

$$\equiv (\cosh K)^{Nq/2}(\cosh \beta h)^N \Lambda_N, \quad (\tau = \tanh \beta h) \tag{2.13}$$

and hence

$$\frac{1}{N} \ln \mathcal{Z} = \frac{q}{2} \ln \cosh K + \ln \cosh \beta h + \frac{1}{N} \ln \Lambda_N \qquad (2.14)$$

The quantity $\frac{1}{N} \ln \Lambda_N$ can be expanded graphically, as before. Each bond carries a factor of $v \sigma_i \sigma_j$ and, in addition, each site has a factor either 1 or $\tau \sigma_k$. Only those graphs with precisely two τ factors contribute to (2.12). As a result of (2.4) then the graphs which contribute are those with precisely two vertices of odd degree, those to be compensated by the two $\tau \sigma_k$ factors. We obtain the following result

$$\beta^{-1} \chi(v) \equiv \bar{\chi}(v) = 1 + 2 \sum_{\{g_2\}} c(g) v^{l_g} \qquad (2.15)$$

where the sum is over the set of graphs $\{g_2\}$, and $c(g)$ denotes the coefficient of N in the embedding factor. This is termed the *lattice constant* of the graph.

This result (2.15) can also be derived in a slightly different way.

$$\chi = \lim_{h \to 0} \frac{1}{N \beta \mathcal{Z}} \frac{\partial^2 \mathcal{Z}}{\partial h^2} \qquad \text{(in the paramagnetic phase)}$$

$$= \frac{\beta}{N \mathcal{Z}} \sum_{i,j} \sum_{\{\sigma\}} \sigma_i \sigma_j e^{-\beta H} \qquad (2.16)$$

$$\beta^{-1} \chi(v) = \frac{1}{N} \sum_{i,j} \langle \sigma_i \sigma_j \rangle = 1 + 2 \sum_{\langle ij \rangle} \langle \sigma_i \sigma_j \rangle \qquad (2.17)$$

where $\langle \sigma_i \sigma_j \rangle$ is a correlation function. Expanding the correlation function graphically leads immediately to (2.15).

Let us consider again the triangular lattice. Table 2.3 lists the graphs which contribute to the zero-field susceptibility up to sixth order, together with their lattice constants. Collecting these results gives

$$\bar{\chi}(v) = 1 + 6v + 30v^2 + 138v^3 + 606v^4 + 2586v^5 + 10818v^6 + \cdots \qquad (2.18)$$

Again the number of graphs is growing rapidly, and it is not feasible to proceed much further by hand.

A final point: if we wanted to study the full field dependence of $\ln \mathcal{Z}$ then graphs with all numbers of odd vertices would be needed.

2.2 Graph generation and computation of lattice constants

The previous section led us to the point where we need efficient automated procedures for generating graphs, usually of a restricted class, and for computing the lattice constants. These are combinatorial problems, of some complexity, in the field

Table 2.3. *List of graphs and lattice constants for the zero-field susceptibility to order v^6, for the triangular lattice.*

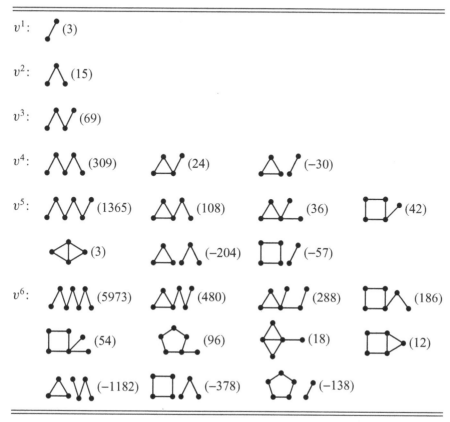

v^1: (3)

v^2: (15)

v^3: (69)

v^4: (309)   (24)   (−30)

v^5: (1365)   (108)   (36)   (42)

(3)   (−204)   (−57)

v^6: (5973)   (480)   (288)   (186)

(54)   (96)   (18)   (12)

(−1182)   (−378)   (−138)

of mathematics known as Graph Theory. There are many books on the subject. A recommended one is Chartrand (1977).

A graph is basically a collection of points ('sites' or 'vertices') and lines ('edges' or 'bonds'). Appendix 1 gives a brief summary of nomenclature and examples, and should be read prior to proceeding further.

The first issue is how to identify/represent a graph in a computer program. As an example we consider the labelled graph in Figure 2.1(a). One could simply list the bonds (1,2), (1,3), (1,4), (2,3), (4,5) but it is more useful to define an *adjacency matrix* by

$$m_{ij} = \begin{cases} 1, & \text{if vertices } i, j \text{ are connected by a bond} \\ 0, & \text{otherwise} \end{cases}$$

The diagonal elements can be set to zero or, usefully, to the order of each vertex. Figure 2.1(b) shows the adjacency matrix for this graph, which is symmetric provided no 'direction' is imposed on the bonds.

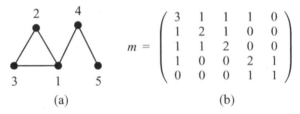

(a)   (b)

Fig. 2.1. (a) A labelled connected graph of five bonds and five vertices. (b) The corresponding adjacency matrix.

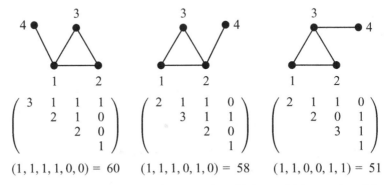

$(1, 1, 1, 1, 0, 0) = 60$ $(1, 1, 1, 0, 1, 0) = 58$ $(1, 1, 0, 0, 1, 1) = 51$

Fig. 2.2. Three equivalents of a bare graph, which would be produced by the generating algorithm, together with their adjacency matrices and keys. The first labelling is canonical.

An efficient way to generate a complete set of graphs with up to n vertices and l bonds is to start with a set of n isolated points, and to place bonds on all possible pairs of points in all possible ways. This is an example of a powerful enumeration algorithm, which we will call the 'Pegs in Holes' (PIH) algorithm, and which forms the basis of many of our combinatorial computer programs. We describe this principle in Appendix 2, in sufficient detail that the reader can follow the structure and logic of our various computer programs. In the present case the 'pegs' are the l bonds and the 'holes' are the $\frac{1}{2}n(n-1)$ possible links between pairs of vertices.

Most applications require a list of unlabelled or 'bare' graphs, or a subset of these. The generating program yields graphs in which the vertices are numbered and yields, in general, a number of labelled realisations of the same bare graph. It would be possible to ignore this, and to eliminate equivalents subsequently; however, it is far more efficient to avoid duplicates altogether, at as early a stage in the generating program as possible. This can be achieved using the concept of a *canonical labelling* and the associated *graph key*. This very powerful idea is illustrated in Figure 2.2.

As an example let us consider the bare graph ⚠. The generating program, if unmodified, would generate the three labellings shown in Figure 2.2. Each of these has its own unique adjacency matrix, also shown. It we take the off-diagonal elements (0 or 1) in the order $m_{12}, m_{13}, m_{23}, m_{14}, \cdots, m_{n-1,n}$ as the binary bits of an integer (by convention we take the right-most bit as the least significant) then we obtain for each labelled graph a unique identifier, or key. The labelling which maximizes the key is termed canonical (in Figure 2.2 the leftmost case), and the corresponding key is then a unique identifier for the bare graph.

A computer program *gen.f*, that can be obtained at www.cambridge.org/ 9780521842426, generates all bare graphs up to an input maximum number of bonds and vertices. In this program, whenever a partial graph is complete, a subroutine is called to check for canonical labelling. If the labelling is canonical the program continues by adding further bonds. If it is not then adding further bonds would be futile and wasteful, since the resulting graph would never be canonical. Instead the program backtracks and changes the previous bond before continuing.

Another important consideration is to restrict, as far as possible, the number of graphs produced because, in most applications, only a small subset of graphs is required. This can be achieved by including a set of constraints into the generating program. Examples of possible constraints might be:

- limiting the maximum vertex order;
- limiting the number of odd vertices to 0 or 2; or
- requiring the graphs to be embeddable in a particular lattice.

A way to do this is discussed in the application in Section 2.3.

The other part of computing series expansions for lattice models is the determination of embedding factors or lattice constants. In the following pages we describe a systematic automated way of doing this, firstly for connected graphs and secondly for disconnected graphs. We will use as an example the bare graph in Figure 2.2, embedded into the triangular lattice.

The counting algorithm is again an application of the PIH procedure, the pegs being the vertices of the graph and holes the sites of the lattice. A program *count.f* is supplied at www.cambridge.org/9780521842426, and the reader is encouraged to study its logical structure and to experiment with different graphs and lattices. The lattice is, of course, infinite (or toroidal) but we only need a finite segment. The lattice sites could be specified by a pair or triple of coordinates but it is more convenient to map these onto a line. The procedure is illustrated in Figure 2.3.

Without loss of generality the graph vertices 1 and 2 are fixed at lattice sites 0 and 1. The other vertices are then embedded in turn, subject to the constraints:

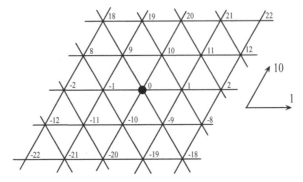

Fig. 2.3. Mapping the vertices of a section of the triangular lattice onto a segment of a line. The two basic vectors ($\mathbf{e}_1$, $\mathbf{e}_2$) are given integer values $(1, N)$, and every lattice site then has a unique index with reference to an origin 0. The integer N (here $N = 10$) must be large enough to avoid wrap-around effects.

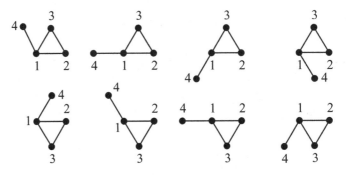

Fig. 2.4. Embedding of the example graph on the triangular lattice.

- no lattice site can be occupied by more than one graph vertex; and
- all graph bonds must correspond to nearest-neighbour links on the lattice.

For the present example this yields the embeddings shown in Figure 2.4. The number of successful counts is 8. This must be multiplied by the coordination number $q = 6$ (since the first bond (1,2) could have been placed in any of q positions) and divided by the symmetry number of the graph nsym $= 2$, to give a lattice constant $8 \times 6/2 = 24$. Of course for this simple example the result is obvious by direct inspection; however, the computer program is perfectly general.

There are a few words of caution. The procedure outlined above and used in *count.f* is essentially 'brute force'. While the program is very fast on modern computers, the CPU time needed is roughly proportional to the lattice constant, and eventually limits the length of series which can be calculated. The most time consuming graphs to count directly are the chain and the polygon. For example, for the face-centred cubic lattice with coordination number $q = 12$, the lattice constant for

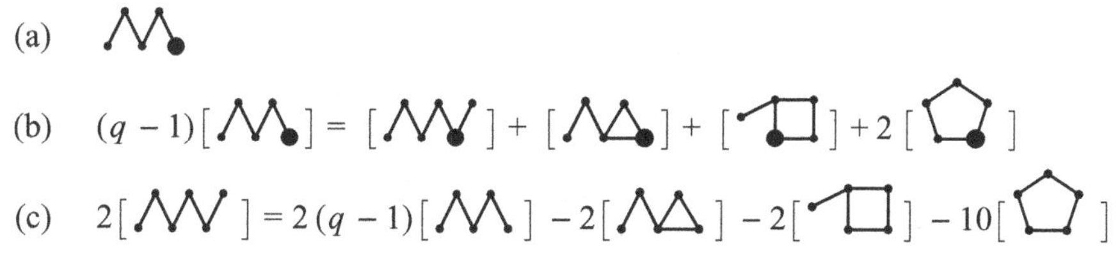

(a)

(b) $(q-1)$ [] = [] + [] + [] + 2 []

(c)   2[] = $2(q-1)$[] − 2[] − 2[] − 10[]

Fig. 2.5. (a) Initial graph with root vertex ●. (b) Symbolic representation of possible outcomes. The square brackets designate the set of all such graphs, or equivalently the lattice constant. (c) Symbolic equation for lattice constant of resulting tree.

a 14-sided polygon is 8 798 329 080. The program *count.f* takes about 2 hours on an AlphaServer SC supercomputer to count this. The lattice constant for a 14-bond chain on the same lattice is of order 10^{14} and would take some 200 hours to count directly. Fortunately this is not necessary as there is a very fast algebraic procedure for all tree graphs, as discussed next.

The lattice constants of tree graphs, including chains, can be obtained algebraically in term of lattice constants of graphs with fewer vertices which, it is assumed, have already been computed. We illustrate this idea, with a simple example, in Figure 2.5. Consider the possible results of adding a bond to the root vertex ● in Figure 2.5(a), which can be done in $(q-1)$ ways (in general $q-o$ ways where o is the order of the root vertex). The result is shown schematically in Figure 2.5(b). This can be rearranged to give the lattice constant of the resulting tree graph [Fig. 2.5(c)]. Note the additional symmetry factors which arise when rooted graphs are replaced by bare graphs.

We can check this for the triangular lattice, using data from Table 2.3. The result is

$$2\left[\,\text{}\,\right] = 10 \times 309 - 2 \times 108 - 2 \times 42 - 10 \times 6$$

$$\left[\,\text{}\,\right] = 1365$$

A computer program *treecnt.f*, which computes the lattice constants of tree graphs sequentially for an ordered list is supplied at www.cambridge.org/9780521842426, and will be employed in the following section.

In the most straightforward form of high-temperature expansion for the Ising model, disconnected graphs must also be included. Generation of disconnected graphs is most conveniently done from a starting list of connected graphs. A program *disc.f* that does this is provided at www.cambridge.org/9780521842426, and will be employed in the following section. Computing the lattice constants of disconnected graphs can be tedious; however, an algebraic procedure, somewhat similar to that for tree graphs, can be developed. Let us start with a simple example. Suppose we

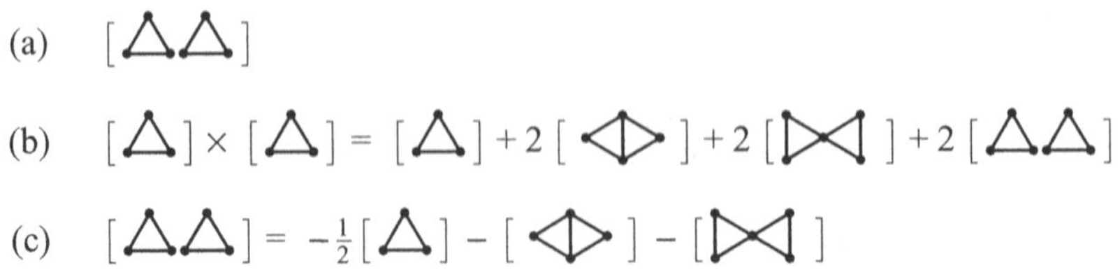

(a) $[\triangle\triangle]$

(b) $[\triangle]\times[\triangle]=[\triangle]+2[\diamondsuit]+2[\bowtie]+2[\triangle\triangle]$

(c) $[\triangle\triangle]=-\tfrac{1}{2}[\triangle]-[\diamondsuit]-[\bowtie]$

Fig. 2.6. (a) Disconnected graph; (b) symbolic representation of possible outcomes when the set of all triangles is multiplied by itself; (c) symbolic equation for the lattice constant.

want the lattice constant (remember this is the coefficient of N in the number of embeddings) for two separated triangles (Fig. 2.6(a)). If we denote by $[\triangle]$ the set of all embeddings of a single triangle, and if we multiply this set by itself, we will generate the list of sets shown in Figure 2.6(b). From this we obtain a symbolic equation for the lattice constant, given in Figure 2.6(c). A check using the data for the triangular lattice in Table 2.1 confirms the lattice constant as $-(\tfrac{1}{2}\times 2+3+9)$ $=-13$.

The general principle is that lattice constants of disconnected graphs can be obtained recursively from graphs with fewer vertices and fewer components, by isolating one component and enumerating all possible overlaps with the graph formed from the remaining components.

2.3 A case study: high-temperature susceptibility for the Ising model on the simple cubic lattice

It is our view that the only way to become proficient in this area, as in any technical area, is through hands-on practice. This section is written somewhat as a tutorial, based on the ideas and computer programs discussed previously. The reader who follows through the steps will derive a 15-term series for the zero-field susceptibility for the Ising model on the simple cubic lattice. This would have been near state-of-the-art some 30 years ago, and is still a respectably long series, far longer than could be achieved by hand.

We will need a list of graphs, of up to $nb = 15$ bonds and $nv = 16$ vertices, which are embeddable on the simple cubic lattice. If we simply try to run the basic graph-generating program with this input we will rapidly be overwhelmed with output. The subset of graphs actually needed is a tiny fraction of the total number at this order. We need to include in the program a judicious set of constraints or tests to eliminate most or all of the unwanted graphs. Furthermore it is of little use simply to filter out unwanted graphs after they have been generated. This will reduce the size of the output file but will not reduce (in fact will increase) the running time, which will prove excessive. As graphs are generated recursively from smaller graphs we want to impose the constraints at the earliest possible point, to eliminate whole branches of graphs being generated at all!

There are three types of constraint that we might choose to impose:

(i) For the simple cubic lattice, with coordination number $q = 6$, no graph with a vertex of order greater than 6 can be embedded. The generating program can be modified to keep track of vertex orders, and to backtrack if any vertex of order greater than 6 is produced.

(ii) If we are concerned only with the zero-field free energy (graphs with no odd vertices) or the initial susceptibility (graphs with two odd vertices) we can keep track of the number of odd vertices and backtrack if this number is too large at any stage. This constraint is a bit more subtle because each extra bond added to the graph can potentially reduce the number of odd vertices by 2. At the same time we filter out all graphs with more than two vertices of degree 1.

(iii) We can attempt to eliminate all graphs which are not embeddable in the simple cubic lattice. This could be done by including a simplified version of the counting program which returns a *yes* flag as soon as a single embedding is found. This is somewhat inefficient. Most of the non-embeddable graphs can be eliminated at the earliest stage by checking for embeddability of one or more small subgraphs, such as $\triangle$, $\bigcirc$, $\diamondslash$, which are known not to be embeddable on the simple cubic lattice. This is much faster. Other non-embeddable graphs can be eliminated subsequently by the counting program.

These constraints are incorporated in the program *gensc.f* (at www.cambridge.org/9780521842426), which the reader is encouraged to experiment with. With input parameters (16,15) the program generates a list of 7901 connected graphs with 15 or fewer bonds and 16 or fewer vertices. Subsequent filtering through a modified version of the counting program leaves 5250 embeddable graphs. Many of these have zero or more than two odd vertices, and thus do not contribute directly to the susceptibility, but are needed for the indirect counting of disconnected graphs. For later use this list is sorted in order of increasing number of bonds and vertices.

The next step is to obtain the lattice constants of all graphs in this list. We use the basic counting program *count.f* to count all stars and, arbitrarily, all trees with $nb \leq 6$. The program *treecnt.f* is then used to count the remaining trees, using the ideas of the previous section.

Disconnected graphs are generated using the program *disc.f*. Constraints include the number of odd vertices being less than $2(17 - nb)$, and no more than two vertices of degree 1. We obtain 641 disconnected graphs with 15 bonds or less. These are added to the graph list, which is then re-sorted in terms of increasing number of bonds, vertices and components. Finally we need to obtain the lattice constants of disconnected graphs recursively. A program *sepcnt.f*

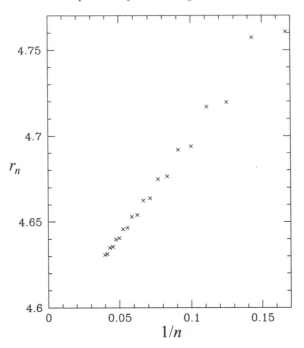

Fig. 2.7. Ratio plot for the susceptibility series of the Ising model on the simple cubic lattice.

(available at www.cambridge.org/9780521842426) was developed to do this. It is based on the principle of embedding a multi-component graph into all graphs with fewer vertices and components. This program is relatively inefficient, taking some 12 minutes of CPU time for this calculation . A faster, but more complex, program could be developed, based on the ideas of the previous section.

At the conclusion we have a list of 5891 graphs, connected and disconnected, with lattice constants. Sifting out those with two odd vertices, of which there are 1718, and accumulating the data gives the susceptibility series

$$\begin{aligned}
\chi = {} & 1 + 6v + 30v^2 + 150v^3 + 726v^4 + 3510v^5 + 16710v^6 \\
& + 79494v^7 + 375174v^8 + 1769686v^9 + 8306862v^{10} \\
& + 38975286v^{11} + 182265822v^{12} + 852063558v^{13} \\
& + 3973784886v^{14} + 18527532310v^{15} + \cdots
\end{aligned} \tag{2.19}$$

One could easily extend the series further using our approach, but for our purposes this is sufficient. The reader may like to know that this series is presently known up to v^{25}! The substantial recent extension has been made using the 'free graph expansion' method, which will be described in Chapter 3.

To conclude this section we present a brief analysis of this series. The ratio plot, shown in Figure 2.7, shows good convergence but with a strong odd-even

Table 2.4. *Estimates of critical point v_c and exponent γ*
from $[M, M]$ Dlog Padé approximants to the Ising model
susceptibility for the simple cubic lattice.

M	v_c	γ
5	0.218165	−1.2511
6	0.218135	−1.2496
7	0.218138	−1.2498
8	0.218142	−1.2501
9	0.218091	−1.2418
10	0.218077	−1.2375
11	0.218107	−1.2458
12	0.218107	−1.2458

oscillation. This results from the antiferromagnetic singularity on the negative real axis. A simple visual extrapolation gives $v_c^{-1} \simeq 4.585$ or $v_c \simeq 0.2180$. Table 2.4 give estimates of the critical point v_c and exponent γ from diagonal $[M, M]$ Padé approximants to $d[\ln \chi(v)]/dv$.

As can be noted, these estimates are well converged, with a trend to slightly lower values with increasing number of terms. There is an interesting historical tale here. In the 1960s, when the only evidence was from the ~ 15 term series available at the time, it was generally believed that γ might well be exactly 5/4. When the first renormalization group calculations in the 1970s indicated a lower value $\gamma \simeq 1.24$ a flurry of activity took place to reconcile the two estimates, assuming, for example, a confluent term (e.g. Nickel, 1982). It is certainly true that additive terms to the asymptotic susceptibility expression will affect the rate of convergence of Dlog Padé approximants. The current best series estimates (Butera and Comi, 2000) are $v_c = 0.218094(1)$ and $\gamma = 1.2375(6)$, the latter result coming from the series for the body-centred cubic lattice.

2.4 Low-temperature expansion

As mentioned in the Introduction (1.4.2) it is possible to derive another kind of expansion for the Ising model, based on enumeration of low-energy excitations above the ground state. The partition function is written as

$$\mathcal{Z} = e^{-\beta E_0} \left\{ 1 + \sum_{\{c\}} e^{-\beta(E_c - E_0)} \right\} \tag{2.20}$$

where the sum is over all configurations $\{c\}$ consisting of $1, 2, 3, \cdots$ overturned spins from the ground state. If we consider a regular lattice with coordination number q

Table 2.5. *List of low-temperature graphs with up to four overturned spins, together with their embedding constants and Boltzmann weights, for the triangular lattice.*

g	•	⟋	$(• \; •)$	△
$C(g)$	N	$3N$	$\frac{1}{2}N^2 - \frac{7}{2}N$	$2N$
wt	$u^3\mu$	$u^5\mu^2$	$u^6\mu^2$	$u^6\mu^3$

g	⋀	$(⟋ \, •)$	$(• \; • \; •)$	⊠
$C(g)$	$9N$	$3N^2 - 30N$	$\frac{1}{6}N^3 - \frac{7}{2}N^2 + \frac{58}{3}N$	$3N$
wt	$u^7\mu^3$	$u^8\mu^3$	$u^9\mu^3$	$u^7\mu^4$

g	◺	⋈	⌁	$(△ \, •)$
$C(g)$	$12N$	$27N$	$2N$	$2N^2 - 24N$
wt	$u^8\mu^4$	$u^9\mu^4$	$u^9\mu^4$	$u^9\mu^4$

g	$(⟋ \, ⟋)$	$(⋀ \, •)$	$(⟋ \, • \, •)$	$(• \; • \; • \; •)$
$C(g)$	$\frac{9}{2}N^2 - \frac{123}{2}N$	$9N^2 - 117N$	$\frac{3}{2}N^3 - \frac{81}{2}N^2 + 288N$	$\frac{N}{24}(N^3 - 42N^2 + 611N - 3114)$
wt	$u^{10}\mu^4$	$u^{10}\mu^4$	$u^{11}\mu^4$	$u^{12}\mu^4$

then it is easily seen that

$$E_0/N = -qJ/2 - h \tag{2.21}$$

and

$$E_c - E_0 = 2(qn - 2r)J + 2nh \tag{2.22}$$

where n is the number of overturned spins and r is the number of nearest-neighbour bonds between them. If we introduce the variables

$$u = e^{-4\beta J}, \quad \mu = e^{-2\beta h} \tag{2.23}$$

then the Boltzmann factor associated with an excited configuration of class (n, r) is

$$u^{(qn-2r)/2}\mu^n \tag{2.24}$$

To evaluate the sum in (2.20) we need to identify all possible configurations of overturned spins (which will again be represented by graphs), evaluate the appropriate embedding factors and multiply by the corresponding Boltzmann weight, and accumulate the data.

In Table 2.5 we display all configurations with up to four overturned spins, together with their embedding factors and Boltzmann weights, for the triangular lattice. The vertices of the graphs denote overturned spins and the bonds represent nearest-neighbour links. As for the high-temperature case, contributions come from both connected and disconnected graphs, and the embedding factors of the latter are polynomials in N. However, just as for the high-temperature case, these higher powers of N cancel on taking the logarithm, yielding a free energy which is properly extensive.

In the present case we obtain the result

$$\frac{1}{N} \ln \mathcal{Z} = 3\beta J + \beta h + \sum_{n=1}^{4} L_n(u)\mu^n \tag{2.25}$$

where

$$
\begin{aligned}
L_1(u) &= u^3 \\
L_2(u) &= 3u^5 - \tfrac{7}{2}u^6 \\
L_3(u) &= 2u^6 + 9u^7 - 30u^8 + \tfrac{58}{3}u^9 \\
L_4(u) &= 3u^7 + 12u^8 + 5u^9 - \tfrac{357}{2}u^{10} + 288u^{11} - \tfrac{519}{4}u^{12}
\end{aligned} \tag{2.26}
$$

The form (2.25) is termed the 'field grouping'. Alternatively we can write the logarithm of the partition function as a 'temperature grouping'

$$\frac{1}{N} \ln \mathcal{Z} = 3\beta J + \beta h + \sum_{s} \psi_s(\mu)u^s \tag{2.27}$$

where the ψ_s are polynomials in μ. For the triangular lattice

$$
\begin{aligned}
\psi_3(\mu) &= \mu \\
\psi_4(\mu) &= 0 \\
\psi_5(\mu) &= 3\mu^2 \\
\psi_6(\mu) &= -\tfrac{7}{2}\mu^2 + 2\mu^3 \\
\psi_7(\mu) &= 9\mu^3 + 3\mu^4 \\
\psi_8(\mu) &= -30\mu^3 + 12\mu^4 + \cdots \\
\psi_9(\mu) &= \tfrac{58}{3}\mu^3 + 5\mu^4 + \cdots
\end{aligned} \tag{2.28}
$$

etc., where the ellipses $\cdots$ denote terms arising from configurations with five or more overturned spins.

From the general expression (2.25) or (2.27) it is straightforward to derive expansions in powers of u and in zero field ($\mu = 1$) for the internal energy

$$U/J = -3 + 12u^3 + 60u^5 - 36u^6 + 336u^7 + \cdots \tag{2.29}$$

the specific heat

$$C/k_B(\ln u)^2 = 9u^3 + 75u^5 - 54u^6 + 588u^7 + \cdots \tag{2.30}$$

the spontaneous magnetization

$$M = 1 - 2u^3 - 12u^5 + 2u^6 - 78u^7 + \cdots \tag{2.31}$$

and the low-temperature susceptibility

$$\beta^{-1}\chi = u^3 + 12u^5 + 4u^6 + 129u^7 + \cdots \tag{2.32}$$

It is possible to extend the series much further, by using special techniques, based on partial generating functions, and by extensive use of computers. We will not pursue this here, but refer the reader to standard works (Domb, 1974, and references therein).

Before leaving this topic we draw attention to two points. Firstly, it is important to note the difference between embeddings and lattice constants for high- and low-temperature expansions. In both cases, sites connected by bonds must be nearest neighbours on the lattice. However, if two sites are not connected by a bond there is an *important difference*. In the high-temperature case the two sites may or may not be embedded onto nearest-neighbour lattice sites, whereas in the low-temperature case they *cannot* be. The high-temperature rules lead to so-called 'weak' embeddings/'weak' lattice constants, whereas the low-temperature rules lead to 'strong' embeddings/'strong' lattice constants. We will return to this distinction in later chapters. As a simple example, consider the graph ⋀ on the triangular lattice. The weak/strong lattice constants are 69, 27 respectively (Tables 2.3 and 2.5). Another example is the graph ⊔ which has a weak lattice constant of 3 but is not strongly embeddable at all on the triangular lattice! Any graph which is strongly embeddable will certainly be weakly embeddable on the same lattice. Conversely, if a graph is not weakly embeddable then it will not be strongly embeddable in the same lattice. This result is of practical use in efficient graph generation.

The second point to note is that low-temperature Ising model series are irregular. The coefficients are typically of both signs and their magnitude varies erratically. This behaviour is symptomatic of functions having their nearest singularity to the origin lying off the positive real axis. The ratio method will be quite unable to deal with such series, but Padé or differential approximants can be used. Historically, the first use of Padé approximants in the field of critical phenomena was to analyse low-temperature Ising series (Baker, 1961).

2.5 Reducing the number of graphs

In a previous section we derived the high temperature susceptibility series for the simple cubic Ising model using the 'primitive' method, in which both connected

Table 2.6. *(a) Multi-graphs and weights for* ln $\mathcal{Z}$ *expansion to order* v^6*;*
(b) multi-graphs and weights for χ *expansion to order* v^5*.*

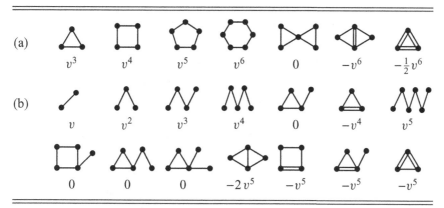

and disconnected graphs occur. At higher orders the number of disconnected graphs increases more rapidly than the number of connected graphs, and approaches have been developed which avoid disconnected graphs altogether.

2.5.1 Connected graph expansion

It is clear from the previous section that the lattice constants of disconnected graphs can be algebraically expressed in terms of lattice constants of connected graphs, by considering all possible overlaps of the separate components. Thus we can write expressions for the logarithm of the partition function and for the susceptibility in the form

$$\frac{1}{N}\ln \mathcal{Z} = \ln 2 + \frac{1}{2}q\ln\cosh K + \sum_{\{g\}}c(g)w_g^{(0)}(v) \tag{2.33}$$

$$\beta^{-1}\chi = 1 + 2\sum_{\{g\}}c(g)w_g^{(2)}(v) \tag{2.34}$$

where the sum in each case is over a certain set of connected graphs (to be specified shortly), $c(g)$ is the lattice constant as before, and $w_g^{(0)}(v)$, $w_g^{(2)}(v)$ are 'graph weights', to be determined. An important point is that these weights are independent of the underlying lattice and, once calculated, can be used for any Ising problem.

The most straightforward connected graph expansion is in terms of multi-graphs (graphs with multiple bonds) which are generated directly from the overlaps. Table 2.6 shows the graphs to order v^6 for ln $\mathcal{Z}$ and to order v^5 for χ, together with their weights. It should be noted that the symmetry of multi-graphs is lower, in general, and the lattice constants are therefore increased by the appropriate factors.

It is seen that, in this formalism, the $\ln \mathcal{Z}$ graphs still have no odd vertices, and the χ graphs have two odd vertices (allowing for multiple bonds), and the weight is proportional to v^l, where l is the *total* number of lines in the graph. It is also noted that some of the graphs, which have articulation points, have zero weight (and hence could have been omitted from the list – we include them to make the point!). This occurs when a graph arises twice, directly with weight $+1$ and as a simple overlap with weight -1. This approach has been developed and used by one of the authors (Oitmaa, 1981; Soehianie and Oitmaa, 1997).

It is more common, however, to write the expansions in terms of simple graphs only, by collapsing multiple bonds into single bonds. This reduces the number of graphs, but introduces two complications. Firstly the restriction of allowed graphs in terms of the number of odd vertices (zero or two) needs to be relaxed, and secondly, the graph weights become functions of v. To illustrate this, consider the bare graph which now represents a set of multigraphs,

$$\ln(1 + v^3) \qquad v^3 \qquad -\tfrac{1}{2}v^6 \qquad \tfrac{1}{3}v^9$$

and has weight given by $\ln(1 + v^3)$.

In Table 2.7 we give the graphs which contribute to order v^8 to the $\ln \mathcal{Z}$ series, together with their weights and lattice constants, for the triangular lattice.

Combining the data gives the $\ln \mathcal{Z}$ series

$$\frac{1}{N} \ln \mathcal{Z} = \ln 2 + \frac{1}{2}q \ln \cosh K + 2v^3 + 3v^4$$
$$+ 6v^5 + 11v^6 + 24v^7 + 55\tfrac{1}{2}v^8 + \cdots \qquad (2.35)$$

As can be seen from Table 2.7 the weights are now more complicated but still depend only on the graph topology. They can be computed in a number of (equivalent) ways, each of which can be computerized:

- By direct enumeration of overlaps of disconnected bare graphs.
- By a 'finite cluster method', whereby the partition function for the graph is computed and the weight of proper sub-graphs subtracted. This is the technique illustrated in Table 2.7.
- By the so-called 'complete-term method' (Domb, 1970; Sykes and Hunter, 1974).

A similar procedure can be followed for the susceptibility, and in Table 2.8 we give the graphs which contribute to order v^6 with weights and lattice constants, again for the triangular lattice.

Combining these data gives the χ series

$$\beta^{-1}\chi = 1 + 6v + 30v^2 + 138v^3 + 606v^4 + 2586v^5 + 10818v^6 + \cdots \qquad (2.36)$$

in agreement with the earlier calculation (Eq. 2.18).

Table 2.7. *Graphs, lattice constants (LC) and weights for connected graph expansion to order v^8 for the $\ln \mathcal{Z}$ series.*

	Graph	LC	Weight
1.		2	$w_1 = \ln(1 + v^3) = v^3 - \frac{1}{2}v^6 + \cdots$
2.		3	$w_2 = \ln(1 + v^4) = v^4 - \frac{1}{2}v^8 + \cdots$
3.		6	$w_3 = \ln(1 + v^5) = v^5 + \cdots$
4.		3	$w_4 = \ln(1 + 2v^3 + v^4) - 2w_1 - w_2 = -v^6 - 2v^7 + \cdots$
5.		15	$w_5 = \ln(1 + v^6) = v^6 + \cdots$
6.		12	$w_6 = \ln(1 + v^3 + v^4 + v^5) - w_1 - w_2 - w_3 = -v^7 - v^8 + \cdots$
7.		42	$w_7 = \ln(1 + v^7) = v^7 + \cdots$
8.		30	$w_8 = \ln(1 + v^3 + v^5 + v^6) - w_1 - w_3 - w_5 = -v^8 + \cdots$
9.		12	$w_9 = \ln(1 + 2v^4 + v^6) - 2w_2 - w_5 = -v^8 + \cdots$
10.		6	$w_{10} = \ln(1 + 3v^3 + 2v^4 + v^5 + v^6) - 3w_1 - 2w_2 - w_3 - 2w_4 - 2w_6 = -2v^8 + \cdots$
11.		123	$w_{11} = \ln(1 + v^8) = v^8 + \cdots$

2.5.2 *Star graph expansion*

The strategy of the last section, to reduce the number of graphs by reducing lattice constants in terms of more compact graphs, can be carried one step further. The lattice constant of any graph with an articulation point can be so reduced. This was discussed in Section 2.2 for tree graphs, but the same ideas can be used for any articulated graph. This process will then lead to expansions in terms of multiply-connected or star graphs.

The expansion for $\ln \mathcal{Z}$ developed in the previous section is, in fact, already a star graph expansion. The expansion for the susceptibility χ is not. Of the 16 graphs in Table 2.8, 10 have articulation points and could, in principle, be eliminated. It turns out to be more convenient (see e.g. McKenzie, 1982) to consider the inverse susceptibility, for which a star graph expansion has been developed. Table 2.9 gives

Table 2.8. *Graphs, lattice constants (LC) and weights for connected graph expansion to order v^6 for the χ series.*

Graph	LC	Weight	Graph	LC	Weight
1.	3	v	2.	15	v^2
3.	69	v^3	4.	2	$-3v^4 - 3v^5 + \cdots$
5.	309	v^4	6.	3	$-4v^5 - 4v^6 + \cdots$
7.	24	$-2v^5 - 2v^6 + \cdots$	8.	1365	v^5
9.	6	$-5v^6 + \cdots$	10.	42	$-2v^6 + \cdots$
11.	108	$-2v^6 + \cdots$	12.	90	$-v^6 + \cdots$
13.	3	$-2v^5 - 8v^6 + \cdots$	14.	5973	v^6
15.	18	$-2v^6 + \cdots$	16.	12	$-2v^6 + \cdots$

Table 2.9. *Graphs and weights for star graph expansions for the inverse susceptibility χ^{-1}.*

	Graph	Lattice constant	Weight
1.		1	1
2.		3	$-2v + 2v^2 - 2v^3 + 2v^4 - 2v^5 + 2v^6 + \cdots$
3.		2	$6v^3 - 12v^4 + 12v^5 - 6v^6 + \cdots$
4.		3	$8v^4 - 16v^5 + 16v^6 + \cdots$
5.		6	$10v^5 - 20v^6 + \cdots$
6.		15	$12v^6 + \cdots$
7.		3	$-8v^5 + 4v^6 + \cdots$
8.		12	$-8v^6 + \cdots$

the graphs to order v^6, together with their weights and lattice constants for the triangular lattice.

Collecting these data gives, for the triangular lattice,

$$\beta\chi^{-1}(v) = 1 - 6v + 6v^2 + 6v^3 + 6v^4 + 6v^5 + 18v^6 + \cdots \qquad (2.37)$$

Inverting this series recovers the previous result to order v^6. Historically, the star graph expansion for the inverse susceptibility was used by McKenzie (1975), in extending the face-centred-cubic susceptibility series to order v^{15}.

2.6 More on Ising models

This chapter, so far, has considered only the spin-$\frac{1}{2}$ Ising model with ferromagnetic nearest-neighbour interactions. We have considered explicitly only the triangular lattice and the simple cubic lattice (Section 2.3). Evidently the same techniques can be used, and have been used, to study the model on any other regular lattice in any spatial dimensions.

There are, also, many directions in which this model has been generalized and much of our knowledge of these generalized models has been obtained using series methods. In the present section we provide an overview of this work, without trying to be exhaustive.

2.6.1 General spin-S

The spin-S Ising model is defined by

$$H = -\frac{J}{S^2} \sum_{\langle ij \rangle} S_i S_j - \frac{h}{S} \sum_{\langle i \rangle} S_i \qquad (2.38)$$

where each Ising variable now is allowed $(2S + 1)$ values $S_i = -S, -S + 1, \cdots, S - 1, S$, and the exchange and field parameters are normalized appropriately. The case $S = \frac{1}{2}$ corresponds to the original Ising model. There are several reasons why one might want to study such spin-S models. One is to study questions of universality where the critical exponents and certain amplitude ratios are expected to be independent of S. Indeed one of the first hints of universality as a general principle came from series work on this problem by the King's College group in the 1960s.

While the primitive method of Section 2.1 can be easily adapted to this case, and was used in early work (Domb, 1974), the 'free graph' method, which we will describe in Chapter 3, has proved to be more efficient and has been used in recent work. The Milan group have computed high-temperature expansions for the nearest-neighbour correlator, the zero-field susceptibility and the second correlator moment to order K^{25} and the second field derivative of the susceptibility to order K^{23}, for the square lattice in $d = 2$ and the simple and body-centred cubic lattice in $d = 3$ (Butera and Comi, 2002). Analysis of these long series has confirmed universality and the so-called 'hyperscaling relation' $d\nu = 2 - \alpha$ to high accuracy.

Equally impressive is the recent work of the Melbourne group (Jensen and Guttmann, 1996; Jensen *et al.*, 1996) who have obtained long low-temperature series (over 100 terms) for the square lattice with $S = 1, 3/2, 2, 5/2, 3$. This work has used the so-called 'finite lattice' method, which is particularly suited to two-dimensional lattices, and which we describe in Section 11.2.1.

Comparable extensions have not been made, as yet, for the close-packed triangular and face-centred cubic lattices but there is no reason to anticipate any surprises here.

Ising models with $S > \frac{1}{2}$ have also been proposed for specific systems. The best known is perhaps the $S = 1$ Blume–Capel model

$$H = -J \sum_{\langle ij \rangle} S_i S_j + \Delta \sum_{\langle i \rangle} S_i^2 \qquad (2.39)$$

which has been used as a lattice model of ^{4}He–^{3}He mixtures and alloys of magnetic ($S = \pm 1$) and non-magnetic ($S = 0$) atoms. This model has a tricritical-point, where a critical line ($\Delta < \Delta_t$) meets with a first-order phase boundary ($\Delta > \Delta_t$). The most comprehensive series study of this model was made some time ago (Saul *et al.*, 1974) and an extension of this work seems feasible and worthwhile.

2.6.2 Antiferromagnets and ferrimagnets

Antiferromagnets and ferrimagnets are materials in which, at low temperatures, moments on neighbouring sites point in opposite directions, resulting in a complete (antiferromagnets) or partial (ferrimagnets) cancellation. An Ising model for an antiferromagnet is obtained by simply changing the sign of the exchange constant

$$H_{AF} = J \sum_{\langle ij \rangle} \sigma_i \sigma_j - h \sum_{\langle i \rangle} \sigma_i \qquad (2.40)$$

with $J > 0$.

The behaviour of antiferromagnets depends on the structure of the lattice. For 'bipartite lattices', which consist of two equivalent sublattices A,B, a simple anti-ferromagnetic state is possible where, at $T = 0, \sigma = +1$ on sub-lattice A (say) and $\sigma = -1$ on sublattice B. This order will be reduced by thermal fluctuations but will persist to some finite critical temperature T_N, known as the Néel point. The square, simple cubic and body-centred cubic lattices are bipartite structures: the triangular and face-centred cubic lattices are not.

On bipartite lattices the zero-field free energy is unchanged on changing the sign of J, as can be easily seen by transforming the spins $\sigma \to -\sigma$ on one of the sub-lattices. Hence the Néel point and Curie point occur at the same temperature. The same transformation shows that the divergent susceptibility for an antiferromagnet is the 'staggered' susceptibility which is the response to a non-physical staggered field which takes alternating signs on the two sub-lattices. The physical susceptibility does not diverge at the Néel point, but does have a weak singularity. Antiferromagnets have a finite region of long-range order in the (T, h) plane, bounded by a critical line.

For non-bipartite lattices the picture is more complex. The triangular lattice antiferromagnet has an infinitely degenerate ground state and a finite entropy at $T = 0$. There is no transition in zero field. The FCC antiferromagnet also has an infinitely degenerate ground state but has zero entropy, and a complicated form of order.

The early series work on Ising antiferromagnets is described by Domb (1974). More recent work includes a combined high- and low-temperature series study of the FCC antiferromagnet (Styer, 1985), which identifies a first-order transition, systems of ferromagnetic planes coupled antiferromagnetically, or vice versa, and systems with competing ferromagnetic and antiferromagnetic interactions as discussed in the following section. This is an area where further study may be profitably directed.

Ising models of ferrimagnets can consist of bipartite lattices with different spin values $S_A \neq S_B$, or more complex structures where the sub-lattices have different numbers of spins. Series work to date on ferrimagnetic Ising systems is relatively sparse.

2.6.3 Further neighbour interactions

Including interactions between further neighbours, particularly if these are 'competing', opens up a huge new field, with much rich physics and many potential applications. We make no attempt to summarize this whole body of work here but rather discuss two examples: the square and body-centred cubic lattices with ferromagnetic nearest and antiferromagnetic next-nearest-neighbour interactions, of strength J_1 and J_2 respectively. In either case, we write the Hamiltonian as

$$H = -J_1 \sum_{\langle ij \rangle} \sigma_i \sigma_j + J_2 \sum_{[kl]} \sigma_k \sigma_l \tag{2.41}$$

where $\langle ij \rangle$ and $[kl]$ denote nearest-neighbour and next-nearest-neighbour pairs, and $J_1, J_2 > 0$. Each system has two possible ground states: a ferromagnetic state for small J_2 and a doubly degenerate antiferromagnetic state for large J_2. In this

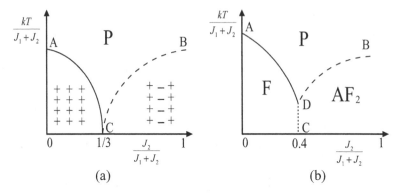

Fig. 2.8. Schematic phase diagrams for the $J_1 - J_2$ Ising model on (a) the square lattice, and (b) the body-centred cubic lattice. Note the choice of variables to allow the full range of (J_1, J_2) to be included. The various features are discussed in the text.

latter state each of the two sub-lattices orders antiferromagnetically, driven by the dominant J_2 interaction. In Figure 2.8 we show qualitative phase diagrams for these models.

The square lattice is exactly solvable at points A $(J_2 = 0)$ and B $(J_1 = 0)$ with, in each case $kT/J = 2.269$ (the Onsager value). Starting at point A and increasing J_2 introduces frustration and reduces the stability of the ferromagnetic phase, and hence the critical temperature, which becomes zero at the point $J_2/J_1 = \frac{1}{2}$ where the two competing ground states have equal energy. In fact at this point there are many other states which are also degenerate in energy. It is believed that the whole line AC lies in the normal Ising universality class, and hence the critical exponents have constant values along the entire line. Line CB separates the low-temperature 'columnar' phase (as shown) from the high-temperature disordered phase. There is strong numerical evidence that this is also a line of critical points, but one which has exponents varying with the ratio J_2/J_1. This was surprising when first proposed, and is still not fully understood. The longest current high-temperature series (Soehianie and Oitmaa, 1997) are only 13 terms. It should be possible to obtain much longer series at both high and low temperatures, perhaps using the finite lattice method, and this would be an interesting project.

Turning to the body-centred cubic lattice, the points A and B correspond to nearest-neighbour body-centred and simple cubic cases respectively with accurately known critical temperatures $kT/J_1 = 6.353$ and $kT/J_2 = 4.511$. At $J_2/J_1 = 2/3$ (point C) the two competing ground states, ferromagnetic and antiferromagnetic type 2 (AF$_2$ – see e.g. Smart, 1966) are degenerate but it is believed that the transition temperature remains non-zero. The best current evidence is that the line DB is a line of first-order transitions, in which case the point D is a critical end point. The most comprehensive series work on this model is that of Velgakis and Ferer (1983) who used a combination of high- and low-temperature series to match the free energies and locate the apparent first-order transition. Again this scenario is far from conclusive and a further study using longer series would be valuable.

2.6.4 Continuous spin Ising models

It is possible to generalize the Ising model with discrete spin states to a continuous scalar spin model, with energy

$$-\beta E(\{\phi\}) = K \sum_{\langle ij \rangle} \phi_i \phi_j - \sum_i V(\phi_i) \qquad (2.42)$$

where the ϕ_i are continuous variables $-\infty < \phi_i < \infty$ and $V(\phi_i)$ is a single-site potential. The partition function can then be obtained as

$$\mathcal{Z}(K) = M^{-1} \int_{-\infty}^{\infty} \cdots \int_{-\infty}^{\infty} d\phi_1 \cdots d\phi_N e^{-\beta E(\{\phi\})} \qquad (2.43)$$

where M is a normalization constant. Clearly if we choose

$$V(\phi) = \delta(\phi + 1) + \delta(\phi - 1) \qquad (2.44)$$

we recover the usual two-state Ising model.

A common choice has been to take $V(\phi)$ of the generic form

$$V(\phi) = a_2 \phi^2 + a_4 \phi^4 \qquad a_4 > 0 \qquad (2.45)$$

which has a single quadratic minimum for $a_2 > 0$ and a double-well form if $a_2 < 0$. The model then is a discrete lattice version of scalar ϕ^4 field theory.

This model is of interest for a number of reasons. It is the starting point for renormalization group and other field theoretic approaches to critical phenomena (e.g. Zinn-Justin, 1996). Systematic numerical Monte Carlo and series studies are thus of fundamental interest. If one assumes, as is commonly done, that the model lies in the Ising universality class, one can use numerical methods to obtain high-precision estimates of Ising critical exponents. This is particularly true if one 'tunes' the model to a point where the leading corrections to scaling vanish (Hasenbusch *et al.*, 1999; Campostrini *et al.*, 1999), the so-called 'improved action'.

Impressively long (25-term) high-temperature series have been derived for this model by the Pisa/Rome group (Campostrini *et al.*, 2002), using the free graph expansion method, to be described in the following chapter. Analysis yields very precise exponent estimates $\alpha = 0.1096(5)$, $\beta = 0.32653(10)$, $\gamma = 1.2373(2)$, $\nu = 0.63012(16)$, $\cdots$. However, the assumption of Ising universality for this whole class of models has been questioned by Baker (Baker and Johnson, 1985; Baker *et al.*, 1994). In particular, the so-called 'border model' with $a_2 = 0$ in two dimensions appears to have non-Ising exponent values $\gamma = 1.85(5)$, $\nu = 1.07(1)$. These results are from quite short series and it would be valuable to check them using the longer series now available.

3

Models with continuous symmetry and the free graph expansion

3.1 Introduction

According to the principle of universality, the symmetry of the order parameter is one of the factors determining the universality class of a system and, *inter alia*, the values of critical exponents. In this chapter we will study models in which the order parameter is a two-component classical vector which can point in any direction on a circle (O(2) symmetry) or a three-component vector which can point in any direction on a sphere (O(3) symmetry). To be specific we take a model with Hamiltonian

$$H = -J \sum_{\langle ij \rangle} \boldsymbol{\mu}_i \cdot \boldsymbol{\mu}_j - \boldsymbol{h} \cdot \sum_i \boldsymbol{\mu}_i \tag{3.1}$$

where the $\boldsymbol{\mu}_i$ are N-component classical unit vectors and the dot product is defined in the usual way. From a mathematical point of view one can analytically continue results to arbitrary N, including the case $N = 0$, which, surprisingly, is related to the problem of self-avoiding random walks. The case $N = 1$ is just the Ising model of the previous chapter.

However we will be thinking here in terms of $N = 2$, which is variously known as the plane rotator model and the classical X–Y model (although this latter terminology is potentially misleading since the quantum X–Y model has three-component vector spins), and $N = 3$ which is usually called the classical Heisenberg model. Unless otherwise stated we take $J > 0$ (ferromagnetic) so the ground state consists of all spins aligned in some arbitrary direction.

There are two (related) ways in which the $N = 2, 3$ models differ significantly from the Ising model considered in the previous chapter. The first is that these models possess intrinsic dynamics (the $\boldsymbol{\mu}$'s are angular momenta) and thus, if the system is disturbed in some way, the spins will respond by changing their configuration. This leads to well-defined collective modes (spin waves and more complex excitations). The second is that the Hamiltonian has a continuous global symmetry. Rotating all of the spins in any configuration through a constant angle ϕ costs no energy, and this is manifested through the excitation energy at zero wavevector ($\mathbf{k} = 0$) becoming zero, i.e. gapless. This is known as a Goldstone mode.

The presence of a Goldstone mode leads directly to a very important distinction between such models in two and in three spatial dimensions. Remember that we can define such a model on any lattice in any spatial dimension – the spin and lattice dimensionalities are completely separate quantities. The Mermin–Wagner theorem (Mermin and Wagner, 1966; Hohenberg, 1967) shows that there can be no state of long-range order in dimension $d = 2$ at any non-zero temperature, if there is a continuous symmetry. This implies that in $d = 2$ thermal fluctuations are strong enough to disorder the system at any $T > 0$, and there is no phase transition of the usual kind. In $d = 3$, on the other hand, these models have finite-temperature critical points, which depend on the lattice, and universal exponents.

Although the $d = 2$ plane rotator model does not have a finite temperature ordered phase it does have a phase transition at finite $K_c \equiv J/k_B T_c$, albeit of a rather unusual kind, as first predicted by Berezinski (1971) and Kosterlitz and Thouless (1973). At K_c the correlation length ξ and susceptibility χ diverge with a weak 'essential singularity'

$$\xi(K) \sim \exp[b(K_c - K)^{-\sigma}] \qquad K \to K_c - \tag{3.2}$$

$$\chi(K) \sim [\xi(K)]^{2-\eta} \tag{3.3}$$

where σ, η are universal exponents with values $1/2$, $1/4$ respectively and b and K_c are non-universal parameters. Both ξ and χ remain infinite for all $K > K_c$. Physically the *Kosterlitz–Thouless transition* represents the break-up of bound 'vortex–antivortex' pairs, which are present at low temperatures. There is, by now, strong and convincing verification of these predictions from model studies, including careful analysis of long high-temperature series (Butera and Comi, 1993, 1994).

The aim of this chapter is to explain how one can derive high-temperature series for such vector spin models. Needless to say, there are several ways of doing this; however, the most efficient method, and the one used in all recent work is one which was developed in the early 1970s by Wortis and co-workers (Wortis, 1974). This approach, which we will describe in some detail in the following section, is often termed the 'linked cluster expansion'. Unfortunately this phrase has been used in many different contexts to refer to different things and so we will refer to the method as the 'free graph expansion', for reasons which will become clear.

Some readers may detect a logical inconsistency in treating this method here, since it has also been used very successfully in treating scalar or Ising systems. However, we believe there is pedagogic value in describing general methods, such as this, in close association with particular problems.

3.2 The free graph expansion

The free graph expansion method has its historical roots in the classical treatment of imperfect gases by Mayer and others, and has also been influenced by developments in perturbative quantum field theory. The overarching philosophy is to take as the unperturbed system free spins interacting with external fields, which of course is easily solved, and to treat the interactions perturbatively by means of a graphical expansion. To this end we write

$$-\beta H = \sum_{i\alpha} h_{i\alpha}\mu_{i\alpha} + \frac{1}{2}\sum_{i,j,\alpha} v^{\alpha}_{i,j}\mu_{i\alpha}\mu_{j\alpha} \qquad (3.4)$$

where $\mu_{i\alpha}$ ($\alpha = x, y, z$) are the Cartesian components of the vector spins and we have distinguished, for mathematical reasons, between different local fields and different pair interactions. We have subsumed the inverse temperature β into the fields and into the interactions $v_{ij} = \beta J_{ij}$.

The logarithm of the partition function is

$$W(h, v) = \ln \text{Tr}(e^{-\beta H}) \qquad (3.5)$$

and we seek to expand this perturbatively in powers of the v_{ij}. From a mathematical point of view we can regard this as a multi-variable Taylor series expansion about $W(h, 0)$ in which the coefficients are partial derivatives of various orders, evaluated at $v = 0$. Alternatively we can write

$$\begin{aligned}
W(h, v) &= \ln \text{Tr}(e^{-\beta H_0}e^{-\beta V}) \\
&= \ln \text{Tr}(e^{-\beta H_0}) + \ln\langle e^{-\beta V}\rangle_0 \\
&= W(h, 0) + \ln\left[1 + \sum_{r=1}^{\infty} \frac{(-\beta)^r}{r!}\langle V^r\rangle_0\right] \\
&= W(h, 0) + \sum_{r=1}^{\infty} \frac{(-\beta)^r}{r!}\langle V^r\rangle_{0c} \qquad (3.6)
\end{aligned}$$

where the angular bracket denotes an average in the *non-interacting system*, i.e.

$$\langle A\rangle_0 = \text{Tr}(e^{-\beta H_0}A)/\text{Tr}(e^{-\beta H_0}) \qquad (3.7)$$

and the subscript c denotes the usual *cumulant average* – see Appendix 6. Thus we are doing thermodynamic perturbation theory (Section 1.4.4) but in a classical system where all variables commute. If we keep these ideas in mind we will avoid getting lost in the formalism.

Let us first evaluate the unperturbed free energy. This is, straightforwardly,

$$W(h, 0) \equiv W_0(h) = \sum_i M_0^0(h_i) \tag{3.8}$$

where

$$M_0^0(h_i) = \ln \mathrm{Tr}_i(e^{h_i \cdot \mu_i}) \tag{3.9}$$

We can evaluate this for different N.

$$N = 1: \quad M_0^0(h) = \ln \sum_{\mu=\pm 1} e^{h\mu} = \ln(2 \cosh h) \tag{3.10}$$

$$N = 2: \quad M_0^0(h) = \ln \int_0^{2\pi} d\theta e^{h \cos \theta} = \ln(2\pi I_0(h)) \tag{3.11}$$

where $I_0(x)$ is a modified Bessel function

$$I_0(x) = 1 + \frac{x^2}{2^2} + \frac{x^4}{2^2 \cdot 4^2} + \frac{x^6}{2^2 \cdot 4^2 \cdot 6^2} + \cdots \tag{3.12}$$

and

$$N = 3: \quad M_0^0(h) = \ln \int_0^{2\pi} d\phi \int_0^{\pi} d\theta e^{h \cos \theta} \sin \theta$$

$$= \ln[4\pi \sinh(h)/h] \tag{3.13}$$

We note that the $M_0^0(h)$ are even functions of h in all cases.

Before we can explicitly carry out the expansion in (3.6) we need to look at the form of single site spin averages. Let us define

$$M_1^0(i, \alpha) = \langle \mu_{i\alpha} \rangle_{0c} = \langle \mu_{i\alpha} \rangle_0 \tag{3.14}$$

$$M_2^0(i, \alpha\beta) = \langle \mu_{i\alpha} \mu_{i\beta} \rangle_{0c} = \langle \mu_{i\alpha} \mu_{i\beta} \rangle_0 - \langle \mu_{i\alpha} \rangle_0 \langle \mu_{i\beta} \rangle_0 \tag{3.15}$$

and, in general,

$$M_n^0(i, \alpha\beta \cdots \nu) = \langle \mu_{i\alpha} \mu_{i\beta} \cdots \mu_{i\nu} \rangle_{0c} \tag{3.16}$$

It is straightforward to establish the result

$$M_n^0(i, \alpha\beta \cdots \nu) = \frac{\partial^n}{\partial h_{i\alpha} \partial h_{i\beta} \cdots \partial h_{i\nu}} M_0^0(h_i) \tag{3.17}$$

We will refer to these quantities as *bare vertex functions* (in the original work they were termed 'bare semi-invariants', but we prefer to avoid this terminology).

Turning now to (3.6), the first term in the sum is

$$\langle -\beta V \rangle_{0c} = \frac{1}{2} \sum_{ij\alpha} v_{ij}^\alpha \langle \mu_{i\alpha} \mu_{j\alpha} \rangle_{0c}$$

$$= \frac{1}{2} \sum_{ij\alpha} v_{ij}^\alpha \langle \mu_{i\alpha} \rangle_{0c} \langle \mu_{j\alpha} \rangle_{0c}$$

$$= \frac{1}{2} \sum_{ij\alpha} v_{ij}^\alpha M_1^0(i, \alpha) M_1^0(j, \alpha) \tag{3.18}$$

This term is represented graphically as $i \overset{\alpha}{\bullet\!\!-\!\!\bullet} j$, where the v_{ij}^α is represented by a line, carrying an index α, joining sites i, j. The site symbols $\bullet$ carry the bare vertex functions, of order equal to the number of lines incident on the vertex (in this case 1). The factor of $1/2$ accounts for the symmetry, since i and j are summed independently.

The next term requires a bit more work:

$$\frac{1}{2!} \langle (-\beta V)^2 \rangle_{0c} = \frac{1}{8} \sum_{ijkl\alpha\beta} v_{ij}^\alpha v_{kl}^\beta \langle \mu_{i\alpha} \mu_{j\alpha} \mu_{k\beta} \mu_{l\beta} \rangle_{0c} \tag{3.19}$$

We first note that if the pairs (ij), (kl) are completely disjoint the cumulant average is zero. There are two other possibilities: either the pair coincides ($i = k$, $j = l$ or $i = l$, $j = k$) or one of the sites is common ($i = k$, $j \neq l$ or $i = l$, $j \neq k$ or $j = k$, $i \neq l$ or $j = l$, $i \neq k$). The former case leads to

$$\frac{1}{4} \sum_{ij\alpha\beta} v_{ij}^\alpha v_{ij}^\beta \langle \mu_{i\alpha} \mu_{i\beta} \rangle_{0c} \langle \mu_{j\alpha} \mu_{j\beta} \rangle_{0c}$$

$$= \frac{1}{4} \sum_{ij\alpha\beta} v_{ij}^\alpha v_{ij}^\beta M_2^0(i, \alpha\beta) M_2^0(j, \alpha\beta) \tag{3.20}$$

while the latter case gives

$$\frac{1}{2} \sum_{ijk\alpha\beta} v_{ij}^\alpha v_{jk}^\beta \langle \mu_{i\alpha} \rangle_{0c} \langle \mu_{j\alpha} \mu_{j\beta} \rangle_{0c} \langle \mu_{k\beta} \rangle_{0c}$$

$$= \frac{1}{2} \sum_{ijk\alpha\beta} v_{ij}^\alpha v_{jk}^\beta M_1^0(i, \alpha) M_2^0(j, \alpha\beta) M_1^0(k, \beta) \tag{3.21}$$

The graphical representations are, respectively

$$i \bullet \overset{\alpha}{\underset{\beta}{\smile}} \bullet j \qquad \qquad i \bullet \overset{\alpha \quad \overset{j}{\bullet} \quad \beta}{\diagup \diagdown} \bullet k \qquad (3.22)$$

If we also include the zeroth-order term, represented by the graph $\bullet$, then formally

$W(h, v) =$ sum of all unrooted connected graphs, with both

single and multiple bonds, with the contribution

of each graph given by the following rules.

Rules for unrenormalized free energy

(1) Label each vertex with a dummy label, and each line with a Cartesian index α.
(2) Write a factor $M_l^0(i, \alpha\beta \cdots)$ for each vertex of order l with incident lines $\alpha, \beta, \cdots$.
(3) Write a factor v_{ij}^α for each line.
(4) Sum each vertex over the whole lattice. i.e. multiply by the so-called *free lattice constant*.
(5) Sum over all line indices.
(6) Divide by the symmetry factor of the graph.

In the early work the factor in rule (4) was termed the 'free multiplicity', but we believe a more descriptive term is 'free lattice constant'. It is a numerical factor which arises from summing over lattice sites, and is thus like the lattice constants defined in the previous chapter. However in this free graph method different vertices of the graph are not constrained to lie on different lattice sites. Hence the free lattice constant of a graph is numerically greater than the corresponding weak lattice constant, but is easier to evaluate. Further details are given in Appendix 3. The symmetry factor in rule (6) above is a product of the basic spatial symmetry with permutation factors $l!$ for each multiple edge.

To illustrate the ideas so far let us consider a simple example, the Ising model on the triangular lattice in zero field. In zero field the vertex functions M_l^0 vanish for l odd, i.e. only graphs with even order vertices contribute. For the Ising case the Cartesian indices α are redundant, and the only vertex functions we will need are $M_0^0 = \ln 2$, $M_2^0 = 1$, $M_4^0 = -2$, $M_6^0 = 16$ (see Appendix 3). To aid comparison with the previous chapter we set all $v_{ij} \equiv K$. Table 3.1 lists the graphs and the various factors needed for the expansion to order K^4. Adding these terms gives

$$W(K) = \ln 2 + \tfrac{3}{2}K^2 + 2K^3 + \tfrac{11}{4}K^4 + \cdots \qquad (3.23)$$

which agrees with the previous result (2.8).

Table 3.1. *The zero-field free energy for the Ising model on the triangular lattice, to order K^4.*

Graph	$\prod M_l^0$	Symmetry	Free lattice constant	Contribution
(graph)	$\ln 2$	1	1	$\ln 2$
(graph)	1	$2 \times 2! = 4$	6	$\frac{3}{2}K^2$
(graph)	1	6	12	$2K^3$
(graph)	4	48	6	$\frac{1}{2}K^4$
(graph)	-2	8	36	$-9K^4$
(graph)	1	8	90	$\frac{45}{4}K^4$

The formalism of the free graph expansion for lattice models was developed by Wortis and co-workers in the late 1960s (see Wortis 1974), and is sometimes called the 'Wortis method'. The method of derivation was somewhat different to that given above but is, of course, equivalent. The reader may wonder why one would use this more complex approach rather than the more direct methods of Chapter 2. There are (at least) three important reasons:

- the free lattice constants are easier to compute than the restricted lattice constants;
- a program of successive renormalizations can be implemented, leading to a reduction in the number of graphs; and
- the method can be used, with little extra work, for more complex models, such as higher-spin Ising, continuous-spin and vector spin models.

Let us now return to the general formalism. The expansion for $W(h, v)$ reads

$$W(h, v) = W(h, 0) + \frac{1}{2} \sum_{ij\alpha} v_{ij}^\alpha M_1^0(i, \alpha) M_1^0(j, \alpha)$$

$$+ \frac{1}{4} \sum_{ij\alpha\beta} v_{ij}^\alpha v_{ij}^\beta M_2^0(i, \alpha\beta) M_2^0(j, \alpha\beta)$$

$$+ \frac{1}{2} \sum_{ijk\alpha\beta} v_{ij}^\alpha v_{jk}^\beta M_1^0(i, \alpha) M_2^0(j, \alpha\beta) M_1^0(k, \beta) + \cdots \quad (3.24)$$

The magnetization is obtained by differentiation. Using (3.17) we obtain

$$\langle \mu_{i\alpha} \rangle = \frac{\partial W}{\partial h_{i\alpha}}$$

$$= M_1^0(i, \alpha) + \sum_{j\beta} v_{ij}^\beta M_2^0(i, \alpha\beta) M_1^0(j, \beta)$$

$$+ \frac{1}{2} \sum_{j\beta\gamma} v_{ij}^\beta v_{ij}^\gamma M_3^0(i, \alpha\beta\gamma) M_2^0(j, \beta\gamma)$$

$$+ \sum_{jk\beta\gamma} v_{ij}^\beta v_{jk}^\gamma M_2^0(i, \alpha\beta) M_2^0(j, \beta\gamma) M_1^0(k, \beta\gamma)$$

$$+ \frac{1}{2} \sum_{jk\beta\gamma} v_{ij}^\beta v_{jk}^\gamma M_1^0(j, \beta) M_3^0(i, \alpha\beta\gamma) M_1^0(k, \gamma) + \cdots \tag{3.25}$$

This can be represented graphically as

$$M_1(i, \alpha) = \circ + \ \vcenter{\hbox{}} + \ \vcenter{\hbox{}} + \ \vcenter{\hbox{}} + \ \vcenter{\hbox{}} + \cdots \tag{3.26}$$

where the open circle is a *root point* which is fixed. A similar set of rules apply to the evaluation of these graphs, except that the root vertex carries a factor M_{l+1}^0 where l is its order, and the symmetry factor may be reduced. This quantity $M_1(i, \alpha)$ is also termed the order-1 *renormalized vertex function*. In the same way we can define renormalized vertex functions of higher order

$$M_n(i, \alpha\beta \cdots \nu) = \langle \mu_{i\alpha} \mu_{i\beta} \cdots \mu_{i\nu} \rangle$$

$$= M_n^0(i, \alpha\beta \cdots \nu) + \sum_{j\omega} v_{ij}^\omega M_{n+1}^0(i, \alpha\beta \cdots \nu\omega) M_1^0(j, \omega)$$

$$+ \cdots \tag{3.27}$$

with the same graphical expansion as for M_1, but with the root vertex carrying a factor M_{l+n}^0. These renormalized vertex functions will play an important role a little later.

The other main quantities of interest are the pair correlators or two-point functions

$$C_{\alpha\beta}(i, j) \equiv \langle \mu_{i\alpha} \mu_{j\beta} \rangle - \langle \mu_{i\alpha} \rangle \langle \mu_{j\beta} \rangle$$

$$= \frac{\partial W}{\partial h_{i\alpha} \partial h_{j\beta}}$$

$$= \delta_{ij} M_2^0(i, \alpha\beta) + \sum_\gamma v_{ij}^\gamma M_2^0(i, \alpha\gamma) M_2^0(j, \beta\gamma)$$

$$+\frac{1}{2}\sum_{\gamma\delta}v_{ij}^{\gamma}v_{ij}^{\delta}M_3^0(i,\alpha\gamma\delta)M_3^0(j,\beta\gamma\delta)$$

$$+\sum_{k\gamma\delta}v_{ik}^{\gamma}v_{kj}^{\delta}M_2^0(i,\alpha\gamma)M_2^0(k,\gamma\delta)M_2^0(j,\beta\delta)$$

$$+\sum_{k\gamma\delta}v_{ij}^{\gamma}v_{ik}^{\delta}M_3^0(i,\alpha\gamma\delta)M_2^0(j,\beta\gamma)M_1^0(k,\delta)$$

$$+\sum_{k\gamma\delta}v_{ij}^{\gamma}v_{jk}^{\delta}M_2^0(i,\alpha\gamma)M_3^0(j,\beta\gamma\delta)M_1^0(k,\delta)+\cdots \tag{3.28}$$

These terms can be represented by graphs with two root vertices (i, j), as

$$\tag{3.29}$$

The rules are basically unchanged, except that the root vertices carry Cartesian indices α, β and vertex factors M_{l+1}^0 and are not summed over. The symmetry factors and free lattice constants have to be modified appropriately.

The zero-field susceptibility is the sum of all pair correlators. For this case all internal vertices must be of even order while the root or 'field' vertices have odd order. Let us again illustrate this for the Ising model on the triangular lattice, where the complication of Cartesian indices is absent. Tables 3.2 and 3.3 contain the necessary data to obtain the series up to v^4.

Combining the data in Table 3.2 gives

$$M_2 = 1 - 6K^2 - 12K^3 + 10K^4 + \cdots \tag{3.30}$$

and, similarly, from Table 3.3

$$\chi - M_2 = 6K + 36K^2 + 148K^3 + 576K^4 + \cdots \tag{3.31}$$

whence

$$\chi = 1 + 6K + 30K^2 + 136K^3 + 586K^4 + \cdots \tag{3.32}$$

This agrees with the earlier result (2.18) when the appropriate change of variables is made.

As mentioned previously one of the advantages of this approach is that it allows renormalizations. What does this mean? Let us consider the following expansion

$$M_2(i)M_2(j)\sum_{k}v_{ik}v_{jk}M_2(k) \tag{3.33}$$

Table 3.2. *Data for the vertex function M_2 for the Ising model on the triangular lattice, to order K^4.*

Graph	$\prod M_l$	Symmetry	Free lattice constant	Contribution
○	1	1	1	1
	-2	2	6	$-6K^2$
	-2	2	12	$-12K^3$
	-32	24	6	$-8K^4$
	16	8	36	$72K^4$
	4	4	36	$36K^4$
	-2	2	90	$-90K^4$

This is just the second graph in Table 3.3, but with the bare vertex functions M_n^0 replaced by the renormalized quantities M_n, which we might represent as ,
using squares to distinguish the M_n from the M_n^0 (we have dropped the Cartesian indices to simplify things). Since the M_n are themselves composed of diagrams of increasing order and complexity, we see that this renormalized diagram represents an infinite subset of bare diagrams. Schematically

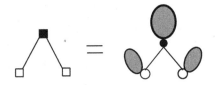

where the blobs represent all possible attachments to the bare vertices. The process in which bare vertex functions, in our previous expansions, are replaced by renormalized vertex functions is called *vertex renormalization*. We can then omit from our list of graph all graphs with factorizable vertex attachments. For example, in Table 3.3 the graphs

should be omitted. The saving becomes much greater at higher orders.

Table 3.3. *Data for* $(\chi - M_2)$ *for the Ising model on the triangular lattice, to order* K^4.

Graph	$\prod M_l$	Symmetry	Free lattice constant	Contribution
(graph)	1	1	6	$6K$
(graph)	1	1	36	$36K^2$
(graph)	4	6	6	$4K^3$
(graph)	1	1	216	$216K^3$
(graph)	−2	2	72	$-72K^3$
(graph)	4	6	72	$48K^4$
(graph)	4	2	12	$24K^4$
(graph)	−2	2	144	$-144K^4$
(graph)	−2	2	216	$-216K^4$
(graph)	−2	2	432	$-432K^4$
(graph)	1	1	1296	$1296K^4$

To proceed along this path we need to streamline the calculation of the vertex functions M_n. We could obtain them via the unrenormalized expansion, as in (3.25) and Table 3.2. However one might as well renormalize this expansion also. For example, the expansion (3.25) would become

$$M_n\,(i,\,\alpha\beta\ldots v) = \text{O} + \;\text{(graph)} + \;\text{(graph)} + \;\text{(graph)} + \cdots$$

$$(3.34)$$

or, algebraically

$$M_n(i, \alpha\beta \cdots v) = M_n^0(i, \alpha\beta \cdots v) + \sum_{\omega_1} M_{n+1}^0(i, \alpha\beta \cdots v\omega_1)G_1(i, \omega_1)$$

$$+ \sum_{\omega_1\omega_2} M_{n+2}^0(i, \alpha\beta \cdots v\omega_1\omega_2)G_2(i, \omega_1\omega_2)$$

$$+ \frac{1}{2} \sum_{\omega_1\omega_2} M_{n+2}^0(i, \alpha\beta \cdots v\omega_1\omega_2)G_1(i, \omega_1)G_1(i, \omega_2)$$

$$+ \cdots \tag{3.35}$$

We have introduced a set of new quantities $G_n(i, \omega_1 \cdots \omega_n)$, known as *self-fields*, to represent all of the possible attachments in the above diagrammatic expansion. In particular

$$G_1(i, \omega) = \; \rule{0pt}{1em} \!\! = \sum_j v_{ij}^\omega M_1(j, \omega)$$

$$G_2(i, \omega_1\omega_2) = \rule{0pt}{1em} + \; \rule{0pt}{1em} + \; \rule{0pt}{1em} + \; \rule{0pt}{1em} + \cdots$$

$$= \sum_j v_{ij}^{\omega_1} v_{ij}^{\omega_2} M_2(j, \omega_1\omega_2)$$

$$+ \sum_{jk\gamma} v_{ij}^{\omega_1} v_{ik}^{\omega_2} v_{jk}^\gamma M_2(j, \omega_1\gamma)M_2(k, \omega_2\gamma) + \cdots$$

$$G_3(i, \omega_1\omega_2\omega_3) = \rule{0pt}{1em} + \cdots \tag{3.36}$$

Note that the so-called *external vertex* which is not marked by a circle or square carries a factor 1.

The two sets of equations (3.35) and (3.36) implicitly determine the vertex functions M_n. While they look complicated, they can be solved by iteration, yielding expansions in v for the M_n. In doing this we note that the leading term in G_n is proportional to v^n. The procedure is best illustrated by example, and we again look at the Ising model in zero field on the triangular lattice. The self-fields G_n and vertex functions M_n will be non-zero only for n even and, again, the Cartesian indices will be redundant. To order K^4 equations (3.35) and (3.36) give

$$M_2 = M_2^0 + M_4^0 G_2 + M_6^0 G_4 + \cdots + \frac{1}{2} M_6^0 G_2^2 + \cdots$$

$$= 1 - 2G_2 + 16G_4 + 8G_2^2 + \cdots$$

$$M_4 = M_4^0 + M_6^0 G_2 + M_8^0 G_4 + \cdots + \frac{1}{2} M_8^0 G_2^2 + \cdots$$

$$= -2 + 16G_2 - 272G_4 + 136G_2^2 + \cdots \tag{3.37}$$

Table 3.4. *Data for* $(\chi - M_2)$ *for the Ising model on the triangular lattice, to order* K^4, *from the vertex-renormalized expansion.*

Graph	Symmetry	Free lattice constant	Contribution
	1	6	$6\,KM_2^2 = 6K - 72K^3 - 144K^4 + \cdots$
	1	36	$36K^2 M_2^3 = 36K^2 - 648K^4 + \cdots$
	6	6	$K^3 M_4^2 = 4K^3 + \cdots$
	1	216	$216K^3 M_2^4 = 216K^3 + \cdots$
	6	72	$12K^4 M_2 M_4^2 = 48K^4 + \cdots$
	2	12	$6K^4 M_2 M_4^2 = 6K^4 + \cdots$
	1	1296	$1296K^4 M_2^5 = 1296K^4 + \cdots$

and

$$G_2 = 3K^2 M_2 + 6K^3 M_2^2 + 45K^4 M_2^3 + \cdots$$
$$G_4 = \frac{1}{4} K^4 M_4 + \cdots \tag{3.38}$$

The reader can easily verify that these yield

$$M_2 = 1 - 6K^2 - 12K^3 + 10K^4 + \cdots$$
$$M_4 = -2 + 48K^2 + 96K^3 - 656K^4 + \cdots \tag{3.39}$$

Knowing the renormalized vertex functions allows us to eliminate many graphs from the expansion for the pair correlators and susceptibility. Let us illustrate this for our simple example. The graphs and their contributions, to order K^4, are given in Table 3.4.

Collecting the various contributions gives

$$\chi = M_2 + 6K + 36K^2 + 148K^3 + 576K^4 + \cdots$$
$$= 1 + 6K + 30K^2 + 136K^3 + 586K^4 + \cdots \tag{3.40}$$

as obtained previously [Eq. (3.32)] from the unrenormalized expansion.

This is as far as we will go with the formalism of the free graph expansion method. In concluding we note two further points.

- It is possible to push the idea of renormalization further by replacing the edge factors v_{ij} by corresponding correlators. This is called *bond renormalization*, and eliminates many more graphs from the expansion. We refer to Wortis (1974) for details.
- In practice the free graph method is usually used for correlators, rather than for the free energy. However, from the nearest neighbour correlator we can directly get the series for the internal energy and hence the specific heat.

3.3 The plane rotator ($N = 2$) model

As a case study we will use the 'free graph expansion' method, in the vertex-renormalized form, to derive the high-temperature susceptibility series for the plane rotator model. We consider both the square and simple-cubic lattices, and aim to derive the series to order K^{12}. It is clear from the discussion of the previous section that this can only be done by computerizing the process.

We will need the renormalized vertex functions to the orders shown: $M_2(K^{12})$, $M_4(K^8)$, $M_6(K^6)$, $M_8(K^4)$, $M_{10}(K^2)$, $M_{12}(K^0)$. In turn we will need the self-fields G_2, G_4, $\cdots$, G_{12}, all to order K^{12}. The first step will be to obtain these. The G-graphs can be obtained from a set of unrooted single edge bare graphs consisting of the single bond, plus all star graphs with number of bonds ≤ 12, and no odd-length loops. There are 48 of these. They can be generated by our standard generating program with constraints to eliminate graphs with articulation points and/or odd-length loops. In Table 3.5 we show the first 14 of these, together with their free lattice constants.

A point to note here is the occurrence of some graphs (no. 3 is the first) which cannot be embedded (in the weak- or strong-embedding sense) but have non-zero free lattice constants because more than one graph vertex can be on the same lattice point.

Two steps are needed to obtain the G graph list from this list of bare graphs:

- 'decorate' the bonds with multiplicities in all possible inequivalent ways, with the constraint that in the final graph all vertices must be of even degree; and
- choose the external vertex in all possible ways.

We illustrate this in Figure 3.1 by showing all the allowed G graphs, to order 10, which arise from the third graph in Table 3.5.

We have written a computer program *ggrafs.f* to carry out this procedure, and it can be obtained at: www.cambridge.org/9780521842426. From the list of 48 bare graphs a total of 615 G graphs of order ≤ 12 are produced. Note that the process of decoration does not change the free lattice constant but will, in general, reduce the symmetry number.

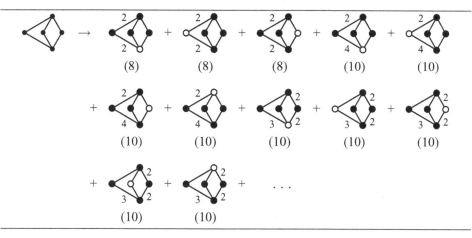

Fig. 3.1. Generation of *G* graphs from a bare graph. The external vertex is represented by an open circle. The number on a bond is the multiplicity, single bonds not being numbered. The number in parentheses is the total number of bonds, or the order of the graph.

Table 3.5. *Bare graph data needed to obtain G graphs up to ninth order, with free lattice constants.*

	Graph	SQ	SC		Graph	SQ	SC
1.		4	6	8.		1024	5712
2.		36	90	9.		4900	44730
3.		100	318	10.		196	642
4.		400	1860	11.		900	4770
5.		324	1350	12.		784	3804
6.		256	960	13.		3600	27900
7.		324	1496	14.		3136	22632

Table 3.6. *Self-fields G_n and renormalized vertex functions*
M_n for the square (SQ) lattice $N = 2$ model. Each Cartesian
index represents a pair, e.g. $(xxy) \equiv (xxxxyy)$ or any
permutation thereof. The functions are unchanged under any
permutation of the indices, or under the interchange $x \leftrightarrow y$.

$$G_2(x) = K^2 + \tfrac{5}{4}K^4 + \tfrac{9}{16}K^6 + \cdots$$

$$G_4(xx) = -\tfrac{1}{16}K^4 - \tfrac{5}{16}K^6 + \cdots$$

$$G_6(xxx) = \tfrac{1}{144}K^6 + \cdots$$

$$G_6(xxy) = \tfrac{1}{720}K^6 + \cdots$$

$$M_2(x) = \tfrac{1}{2} - \tfrac{1}{2}K^2 + \tfrac{1}{4}K^4 - \tfrac{23}{96}K^6 + \cdots$$

$$M_4(xx) = -\tfrac{3}{8} + \tfrac{3}{2}K^2 - \tfrac{453}{128}K^4 + \cdots$$

$$M_4(xy) = -\tfrac{1}{8} + \tfrac{1}{2}K^2 - \tfrac{151}{128}K^4 + \cdots$$

$$M_6(xxx) = \tfrac{5}{4} - \tfrac{165}{16}K^2 + \cdots$$

$$M_6(xxy) = \tfrac{1}{4} - \tfrac{333}{16}K^2 + \cdots$$

The next step is to solve the pair of equations (3.35) and (3.36) and iteratively obtain expansions for the renormalized vertex functions to the order needed. This is done in the middle part of the computer program *free2.f*, for the plane rotator model ($N = 2$). We need to remember that for this case each line of a graph also carries a Cartesian index $\alpha = x, y$. The computer program generates each of these separate configurations, again using the 'pegs in holes' algorithm, and sums over them. The generalized form of Eqs. (3.37) is read as input.

In Table 3.6 we display results to order K^6 for the SQ lattice. To this order the expansion can be carried out by hand and the reader is encouraged to do this. A technical point in connection with this step is that the G functions at higher order do not automatically preserve the symmetry with respect to permutations of Cartesian indices, but they can be suitably symmetrized.

The final step is to compute the susceptibility series. For this we need a new set of graphs: two-rooted multibond graphs, with no reducible parts and all vertices of even degree except the root vertices, which are of odd degree. The generation of these follows parallel lines to the generation of G graphs. We first generate a list of bare graphs, the only constraint being that graphs with more than two vertices of degree one can be discarded. This yields a list of 703 bare graphs with 12 bonds or less. A computer program *xgrafs.f* then generates the list of irreducible χ graphs – there

Table 3.7. *Irreducible χ graphs up to order 6, with their free lattice constants. Multiple bonds are denoted by a number on the bond.*

	Graph	SQ	SC		Graph	SQ	SC
1.		4	6	11.		1024	7776
2.		16	36	12.		16	36
3.		4	6	13.		16	36
4.		64	216	14.		36	90
5.		16	36	15.		256	1296
6.		256	1296	16.		256	1296
7.		4	6	17.		144	540
8.		64	216	18.		100	318
9.		64	216	19.		4096	46656
10.		36	90				

are a total of 2384 of these, with 12 bonds or less. In Table 3.7 we display the first 19 of these (up to order 6) with their free lattice constants.

We can then use our program *free2.f* to compute the susceptibility series for the $N = 2$ model to order K^{12} on both the SQ and SC lattices. The series coefficients are given in Table 3.8, and are in agreement with known results. The reader is encouraged to use the program and data to repeat this calculation. To this order the computer memory and CPU requirements are minimal. Of course one could go much further (the series are known up to K^{21}), but this is sufficient to understand the method. We will analyse these series in the following section.

3.4 Analysis of the $N = 2$ susceptibility

The series derived in the previous section appear regular and so we will first attempt a ratio analysis. In Figure 3.2 we show a ratio plot based on the known coefficients up to order K^{21}.

Table 3.8. *Series coefficients of the high temperature susceptibility
of the plane rotator model on the square (SQ) and simple cubic
(SC) lattices (to 16 significant figures).*

Order	SQ	SC
0	$0.1000000000000000 \times 10^1$	$0.1000000000000000 \times 10^1$
1	$0.2000000000000000 \times 10^1$	$0.3000000000000000 \times 10^1$
2	$0.3000000000000000 \times 10^1$	$0.7500000000000000 \times 10^1$
3	$0.4250000000000000 \times 10^1$	$0.1837500000000000 \times 10^2$
4	$0.5500000000000000 \times 10^1$	$0.4350000000000000 \times 10^2$
5	$0.6854166666666667 \times 10^1$	$0.1023437500000000 \times 10^3$
6	$0.8265625000000000 \times 10^1$	$0.2370546875000000 \times 10^3$
7	$0.9722005208333333 \times 10^1$	$0.5469462890625000 \times 10^3$
8	$0.1120507812500000 \times 10^2$	$0.1252004882812500 \times 10^4$
9	$0.1267555338541667 \times 10^2$	$0.2858817529296875 \times 10^4$
10	$0.1415201280381944 \times 10^2$	$0.6496151407877604 \times 10^4$
11	$0.1560190022786458 \times 10^2$	$0.1473537464124891 \times 10^5$
12	$0.1701930067274306 \times 10^2$	$0.3331475377468533 \times 10^5$

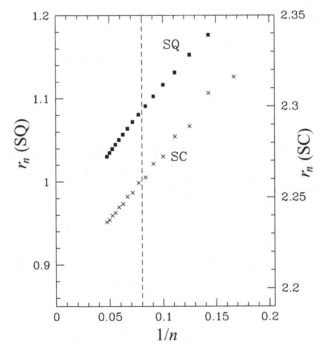

Fig. 3.2. Ratio plots for the $N = 2$ susceptibility series for both the square (SQ)
and simple cubic (SC) lattices. The data to the right of the vertical dashed line are
from the 12th order series derived in the previous section.

Table 3.9. *Poles and residues from diagonal [M,M]*
Padé approximants to d[ln χ(K)]/dK for square (SQ)
and simple cubic (SC) lattices.

M	SQ		SC	
3	1.0097	2.42	complex	
4	0.9799	2.18	0.45404	1.325
5	1.0349	2.71	0.45410	1.326
6	1.0712	3.58	0.45410	1.326
7	1.0710	3.57	0.45421	1.329
8	1.1169	10.41	0.45420	1.329
9	complex		0.45422	1.330
10	1.0977	5.53	0.45423	1.330

What can we conclude from this? The SC plot shows an odd–even oscillation, similar to the Ising case in Section 2.3, due to an antiferromagnetic singularity. One could reasonably conjecture a critical point at around $K_c^{-1} \simeq 2.205$, even from the 12-term series. The additional terms confirm this trend and provide greater precision. The SQ ratio plot differs in two ways. Firstly the odd–even oscillation is not present, indicating that any antiferromagnetic singularity must be weak. More importantly the plot shows a definite residual curvature indicating either that longer series are needed to see the true asymptotic behaviour or, more probably, that the singularity has a more complex form. This is not so clear from the twelfth-order series, and early work (Ferer and Velgakis, 1983) was unable to convincingly conclude whether the singularity was of the predicted Kosterlitz–Thouless form or not. Here the additional terms have clarified the situation and the most recent analysis supports the Kosterlitz–Thouless form quite convincingly (Butera and Comi, 1993; Campostrini *et al.*, 1996).

We now turn to a Padé approximant analysis. Table 3.9 shows estimates of the critical point and exponent from diagonal Dlog Padé approximants, for both SQ and SC lattices.

Looking first at the SC case, we observe good convergence in the last four entries at least. The results are strongly suggestive of a power law singularity of the standard form, with the critical point at $K_c \simeq 0.4542$ and an exponent $\gamma \simeq 1.33$. The results are consistent with the best current estimates (Butera and Comi, 1997), although the value of γ is a little higher than the latest field theory estimates $\gamma = 1.318$ (Guida and Zinn-Justin, 1998).

On the other hand the SQ series is clearly not giving converged results. From 12 terms there is, surprisingly, an indication of a pole at around 1.0 with an anomalously

Table 3.10. *Estimates of the critical point K_c and exponent σ for the plane rotator model on the square lattice from $[M, M]$ Padé approximants to the series $\frac{d}{dK} \ln(\frac{d}{dK} \ln \chi)$.*

M	5	6	7	8	9
K_c	1.1275	–	1.1006	1.1138	1.1119
σ	0.477	–	0.270	0.408	0.384

large exponent around 2.5, but the inclusion of more terms shows that this is false. Clearly there is something very different between the two lattices.

If the SQ lattice susceptibility has the Kosterlitz–Thouless form

$$\chi(K) = A \exp[b(K_c - K)^{-\sigma}] \qquad (3.41)$$

with A, b constants then the logarithmic derivative will have a power law singularity. Taking a second logarithmic derivative will yield a simple pole, viz.

$$\frac{d}{dK} \ln \left(\frac{d}{dK} \ln \chi \right) = \frac{\sigma + 1}{K_c - K} \qquad (3.42)$$

Thus one might expect Padé approximants to this series to give consistent results for both K_c and σ. In Table 3.10 we show results of this analysis. These results are reasonably stable and consistent with the best current estimates. We refer the reader to the original literature for more comprehensive analyses (Pelissetto and Vicari (2002), and references therein).

3.5 Discussion

The free graph expansion method, described in this chapter, appears to be the most powerful and efficient for deriving high-temperature expansions for classical lattice models. The longest existing series for most models have been derived by this method: the only exceptions are where special features may exist, such as in two dimensions or for special Hamiltonians.

We have illustrated this method for the case of the plane rotator ($N = 2$) spin model on the square and simple cubic lattices. For the general N-vector model series are known up to order K^{21} for the nearest-neighbour correlator, the susceptibility, and other quantities for the bipartite lattices (Butera and Comi, 1997). For the Ising ($N = 1$) case the same group has obtained series to order K^{25}, for general spin (Butera and Comi, 2002) (cf. Section 2.6.1). More general Hamiltonians, motivated by lattice field theory, have been investigated by Lüscher and Weisz (1988), Nickel and Rehr (1990), and most recently, by the Pisa/Rome group (Pelissetto and Vicari, 2002).

In principle, as already remarked by Wortis (1974), the method can also be applied to quantum spin models. Some attempts have been made to do this (Wang and Wentworth, 1985), but, in practice, other approaches have yielded longer series, as will be described in Chapters 7 and 8.

4

Quantum spin models at $T = 0$

4.1 Introduction

In the limit of zero temperature, the partition function is dominated by the ground state of the system, the free energy becomes equivalent to the ground state energy, and the central object of interest is the Hamiltonian of the system $H(\lambda)$, dependent in general on one or more parameters λ. A 'quantum' Hamiltonian, as noted previously, is one which contains non-commuting operators. Here we concentrate on quantum spin Hamiltonians, of which the prime example is the Heisenberg model. Of particular interest is the possibility of a *quantum phase transition*, i.e. non-analytic behaviour as a function of coupling λ. A quantum phase transition may exert a dominating influence on the system over a range of low temperatures above $T = 0$. Such phenomena have figured prominently in theoretical discussions of experiments on the cuprate superconductors, the heavy fermion materials, organic conductors, and related materials (e.g. Sachdev, 1999).

In some cases, a correspondence can be found between a quantum system at zero temperature in 1 time and $(d - 1)$ space dimensions, and an equivalent classical system at finite temperature in d space dimensions, based on the Feynman path integral formalism – see Chapter 9 and Appendix 8 for further discussions. Under this mapping, the temperature kT in the classical system corresponds to a coupling λ in the equivalent quantum system. If a phase transition occurs, the same universal critical exponents are expected to apply in both cases. An example is provided by the transverse Ising chain, discussed below.

4.2 Linked cluster expansions

This chapter begins our discussion of quantum models at $T = 0$ via long perturbation expansions. Our aim is to develop standard Rayleigh–Schrödinger perturbation theory in a form where we can express the quantities of interest as sums of contributions from a sequence of finite clusters of sites. At the same time we will develop efficient algorithms to calculate the contribution of each cluster, and to manage the bookkeeping. As we shall see, only connected or *linked clusters* occur in the formalism.

The standard approach to many-body perturbation theory, as described in many texts (e.g. Negele and Orland, 1988), associates the different physical processes with diagrams, analogous to Feynman diagrams in quantum field theory. Using appropriate rules it is possible to develop the expansions in terms of *connected* or *linked* diagrams only. The *linked cluster expansion* we develop and use here is rather different. The *clusters* are real physical groups of sites on the lattice (we use the words *cluster* and *graph* more or less interchangeably), and in evaluating the contribution of each cluster to the quantity of interest we include all possible physical processes. In this sense each cluster includes many different diagrams. At least for the problems we study here, this new formalism is more powerful and effective.

To the best of our knowledge the first application of the method was by Marland (1981), based on unpublished work of Nickel. Hamer and coworkers began to use these methods extensively in the mid-1980s (Irving and Hamer, 1984c), and this led directly to our ongoing collaborative work since the late 1980s. At about the same time, Singh *et al.* (1988) rediscovered the method and began applying it to various quantum models, including the Heisenberg model. We refer the reader to a recent review (Gelfand and Singh, 2000) for a historical perspective.

We will illustrate the method in detail in the next section, with specific model calculations. It is perhaps useful to give a general summary first. Suppose we have a lattice model with a Hamiltonian of the form

$$H = H_0 + \lambda V \tag{4.1}$$

where H_0 has a simple eigenstate spectrum which is exactly known. The ground state energy per site, in the bulk limit $N \to \infty$, can be written as a sum over cluster terms of the form

$$E_0/N = \sum_{\{g\}} c(g)\epsilon(g) \tag{4.2}$$

where $c(g)$ is the lattice constant for cluster g (this can be either the weak or strong lattice constant, as defined in Chapter 2, depending on the application), and $\epsilon(g)$ is a quantity with the dimension of energy which we term the *reduced energy* (sometimes called the *cumulant energy*). We have not defined it precisely as yet.

The key idea is to apply the same expansion (4.2) in turn to each finite cluster g. Then we have

$$E(g) = \sum_{g' \in g} c(g'/g)\epsilon(g') \tag{4.3}$$

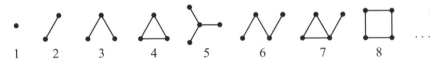

Fig. 4.1. List of low-order clusters, ordered by number of sites.

where $E(g)$ is the ground-state energy of cluster g, the sum is over all subclusters of g (including g itself) and $c(g'/g)$ is the embedding constant of g' in g. This gives a set of equations which can be solved recursively to obtain the reduced energies of clusters, which via (4.2) then gives the required bulk energy.

Let us consider a hypothetical set of clusters, as shown in Figure. 4.1. Then, by inspection, we can write

$$E_1 = \epsilon_1$$
$$E_2 = \epsilon_2 + 2\epsilon_1$$
$$E_3 = \epsilon_3 + 2\epsilon_2 + 3\epsilon_1$$
$$E_4 = \epsilon_4 + 3\epsilon_2 + 3\epsilon_1$$
$$E_5 = \epsilon_5 + 3\epsilon_3 + 3\epsilon_2 + 4\epsilon_1$$
$$E_6 = \epsilon_6 + 2\epsilon_3 + 3\epsilon_2 + 4\epsilon_1$$
$$E_7 = \epsilon_7 + \epsilon_4 + 2\epsilon_3 + 4\epsilon_2 + 4\epsilon_1$$
$$E_8 = \epsilon_8 + 4\epsilon_3 + 4\epsilon_2 + 4\epsilon_1 \tag{4.4}$$

Note that we are implicitly using a strong-embedding form, so that, for example, g_3 is not a sub-cluster of g_4. This form is necessary if the terms in V act on links between sites.

From (4.4) we can easily obtain the reduced energies by recursion, $\epsilon_1 = E_1$, $\epsilon_2 = E_2 - 2\epsilon_1$, or in general

$$\epsilon(g) = E(g) - \sum_{\substack{g' \in g \\ g' \neq g}} c(g'/g)\epsilon(g') \tag{4.5}$$

which can then be substituted into (4.2) to give the bulk energy. Other ground-state properties can be obtained by a straightforward extension of this procedure. By including an external field term in the Hamiltonian one obtains the ground-state energy as a function of the field, and differentiation yields the magnetization and susceptibility. This differentiation can be carried out at the cluster level.

It is also easy to see why only connected or linked clusters need be included in (4.2). Imagine a cluster consisting of two disconnected components g_1 and g_2. Then the energy, or any other extensive property, obeys the *cluster addition property*

$$E(g_1 \oplus g_2) = E(g_1) + E(g_2) \tag{4.6}$$

The corresponding reduced energy *vanishes*

$$\epsilon(g_1 \oplus g_2) = E(g_1 \oplus g_2) - \sum(\text{sub-cluster contributions})$$
$$= E(g_1 \oplus g_2) - E(g_1) - E(g_2) = 0 \qquad (4.7)$$

Note that we have now assembled all the ingredients required for a *proof* of the linked cluster expansion.

(i) Define the reduced energy for each cluster by equation (4.3): we have seen that this allows one to calculate the reduced energies $\epsilon(g)$ recursively in terms of the $E(g)$.

(ii) The cluster addition property (4.6) implies that the reduced energy vanishes for any disconnected cluster.

(iii) Then equation (4.2) follows from equation (4.3) in the limit as the size of cluster g goes to infinity, with

$$c(g') = \lim_{N(g) \to \infty} c(g'/g)/N(g) \qquad (4.8)$$

where $N(g)$ is the number of sites in cluster g.

The *utility* of the linked cluster expansion lies in the fact that for a local interaction, the minimum order at which a cluster contributes grows proportionally to the size of the cluster. Thus to obtain a perturbation series which is exact up to a given order, it is sufficient to sum up the reduced energies of all clusters up to the corresponding size. This gives us a systematic procedure for carrying the expansion to higher orders. This may easily be understood in terms of the connected diagram approach: the reduced energy $\epsilon(g)$ is equal to the sum of all connected diagrams which *span* the cluster g, i.e. involve every site (or bond, as the case may be) of that particular cluster. Such diagrams will only occur beyond a certain minimum order in λ, proportional to the size of the cluster. The great *advantage* of the linked cluster approach is that if we can evaluate $\epsilon(g)$ directly, it gives us the sum of many different connected diagrams in one hit.

So far we have not specified how to calculate the cluster energies. For any finite cluster the space of allowed states is usually finite, and hence H will be represented by a finite square matrix. We could compute the ground-state energy numerically, either by complete diagonalization or, for larger matrices, by a Lanczos or other procedure. However, we are interested in generating series expansions in λ. Thus we need to compute the cluster energies $E(g)$ as perturbation expansions in λ. The reduced energies will then also be series in λ, as will the final bulk properties. There exists an efficient procedure for doing high-order perturbation theory on a finite cluster, and the technical details are described in Appendix 4. This is the most demanding part of the method, both in terms of computer time and memory, and limits the length of series obtained. An alternative technique which has been applied to several models, namely the 'continuous unitary transformation' (CUT) method, is outlined in Section 11.2.3.

4.3 An example: the transverse field Ising model in one dimension

Before continuing with the general formalism let us illustrate the idea with a simple example, the transverse field Ising chain

$$H = -J \sum_i \sigma_i^z \sigma_{i+1}^z - \Gamma \sum_i \sigma_i^x, \tag{4.9}$$

introduced in Section 1.2.2. The ground-state energy of this model is known exactly (Pfeuty, 1970) and can be written as

$$-E_0/NJ = \frac{4}{\pi}(1 + \lambda)E(m) \tag{4.10}$$

where $\lambda = \Gamma/J$, $m = 4\lambda/(1 + \lambda)^2$ and $E(m)$ is the complete elliptic integral of the second kind. The function $E(m)$ is singular at $m = 1$, or $\lambda = 1$, corresponding to a quantum critical point. The magnetization or order parameter $M_z = \langle 0|\sigma_i^z|0\rangle$ can also be obtained exactly as

$$M_z = \begin{cases} [1 - \lambda^2]^{1/8}, & \lambda < 1; \\ 0, & \lambda > 1. \end{cases} \tag{4.11}$$

Expressions (4.10) and (4.11) have exactly the same functional form as seen in the classical two-dimensional Ising model, as noted in Section 1.2.2.

4.3.1 Expansion in the disordered phase

It is convenient to write the Hamiltonian in the slightly modified form

$$H = \sum_i (1 - \sigma_i^z) - \mu \sum_i \sigma_i^x \sigma_{i+1}^x$$
$$= H_0 - \mu V \tag{4.12}$$

We have interchanged the z and x indices and added a constant term, as well as set $\Gamma = 1$, and introduced $\mu = J/\Gamma = \lambda^{-1}$. The perturbation operator can be further written as

$$V = \sum_i (\sigma_i^+ \sigma_{i+1}^+ + \sigma_i^+ \sigma_{i+1}^- + \sigma_i^- \sigma_{i+1}^+ + \sigma_i^- \sigma_{i+1}^-) \tag{4.13}$$

where σ^+, σ^- are raising and lowering operators. We take as our basis set the so-called Zeeman states $|m\rangle = |m_1, m_2, \cdots, m_N\rangle$ with $m_i = \pm 1$, which are

Table 4.1. *Low-order clusters and Hamiltonian matrices in the disordered phase expansion for the transverse field Ising chain.*

| g_1 : | • | $|1\rangle = |+\rangle$ | $H = (0)$ |

g_2 : •—• $\quad \begin{aligned}|1\rangle &= |++\rangle\\|2\rangle &= |--\rangle\end{aligned}\quad H = \begin{pmatrix} 0 & -\mu \\ -\mu & 4 \end{pmatrix}$

g_3 : •—•—• $\quad \begin{aligned}|1\rangle &= |+++\rangle\\|2\rangle &= |--+\rangle\\|3\rangle &= |+--\rangle\\|4\rangle &= |-+-\rangle\end{aligned}\quad H = \begin{pmatrix} 0 & -\mu & -\mu & 0 \\ -\mu & 4 & 0 & -\mu \\ -\mu & 0 & 4 & -\mu \\ 0 & -\mu & -\mu & 4 \end{pmatrix}$

eigenstates of H_0. The unperturbed ground-state is that with all spins up, $\{m_i = 1, \text{ all } i\}$. The effect of V acting on a pair of neighbouring sites is then to flip both spins. As an aside, we remark that for computational purposes the m_i can be regarded as the bits $(0,1)$ in a single computer word and the states can be easily manipulated via logical bit operations. This is well known, and is incorporated in our computer programs.

The clusters in this simple case are trivial – they are just chains of $1, 2, \cdots$ sites. The subgraph embedding factors are also simple, with

$$c(g_m/g_n) = n - m + 1, \qquad m \leq n \tag{4.14}$$

In Table 4.1 we display the first three clusters, together with the relevant basis states, and Hamiltonian matrices in the ground-state sector for small μ.

Proceeding systematically we obtain expansions for the energies and reduced energies of clusters to some maximum order. Table 4.2 displays the results for clusters $g_1, \cdots g_7$, with $x \equiv \mu^2/4$.

There is one important point to note from these results. The reduced energy of any cluster has its leading term proportional to μ^{2p}, where p is the number of bonds. The perturbation V must flip each pair of spins at least twice, starting from and returning to the unperturbed ground-state. This cancellation of leading terms on subtraction of sub-cluster contributions acts as an important check on the correctness of the calculations.

Finally we can combine the terms in Table 4.2, to obtain the bulk ground state energy per site

$$\frac{E_0}{N} = -\tfrac{1}{4}\mu^2 - \tfrac{1}{64}\mu^4 - \tfrac{1}{256}\mu^6 - \tfrac{25}{16384}\mu^8 - \tfrac{49}{65536}\mu^{10} - \tfrac{441}{1048576}\mu^{12} + \cdots \tag{4.15}$$

The reader can verify that this is in agreement with the series obtained from the exact result (4.10).

Table 4.2. *Energy and reduced energy expansions for low-order clusters for the transverse field Ising chain.*

Cluster	Energy and reduced energy
1	$E_1 = 0, \quad \epsilon_1 = 0$
2	$E_2 = -x + \frac{1}{4}x^2 - \frac{1}{8}x^3 + \frac{5}{64}x^4 - \frac{7}{128}x^5 + \frac{21}{512}x^6 + \cdots$
	$\epsilon_2 = E_2$
3	$E_3 = -2x + \frac{1}{2}x^3 - \frac{1}{2}x^4 + \frac{1}{8}x^5 + \frac{3}{8}x^6 + \cdots$
	$\epsilon_3 = -\frac{1}{2}x^2 + \frac{3}{4}x^3 - \frac{21}{32}x^4 + \frac{15}{64}x^5 + \frac{75}{256}x^6 + \cdots$
4	$E_4 = -3x - \frac{1}{4}x^2 + \frac{1}{4}x^3 + \frac{75}{64}x^4 - \frac{225}{128}x^5 - \frac{7}{256}x^6 + \cdots$
	$\epsilon_4 = -\frac{7}{8}x^3 + \frac{9}{4}x^4 - \frac{33}{16}x^5 - \frac{377}{512}x^6 + \cdots$
5	$E_5 = -4x - \frac{1}{2}x^2 + \frac{25}{32}x^4 + \frac{49}{16}x^5 - \frac{1553}{256}x^6 + \cdots$
	$\epsilon_5 = -\frac{33}{16}x^4 + \frac{429}{64}x^5 - \frac{1443}{256}x^6 + \cdots$
6	$E_6 = -5x - \frac{3}{4}x^2 - \frac{1}{4}x^3 + \frac{25}{64}x^4 + \frac{147}{64}x^5 + \frac{2205}{256}x^6 + \cdots$
	$\epsilon_6 = -\frac{715}{128}x^5 + \frac{663}{32}x^6 + \cdots$
7	$E_7 = -6x - x^2 - \frac{1}{2}x^3 + \frac{49}{32}x^5 + \frac{441}{64}x^6 + \cdots$
	$\epsilon_7 = -\frac{4199}{256}x^6 + \cdots$

4.3.2 Expansion in the ordered phase

It is equally possible to obtain an expansion about the limit $\Gamma = 0$, i.e. in the ordered phase. In this case it is convenient to write the Hamiltonian in the modified form

$$H = \sum_i (1 - \sigma_i^z \sigma_j^z) - \lambda \sum_i \sigma_i^x \qquad (4.16)$$

where we have set $J = 1$ and again added a constant term. This time the perturbation operator

$$V = \sum_i (\sigma_i^+ + \sigma_i^-) \qquad (4.17)$$

acts by flipping the spins on individual sites. We again use the eigenstates of σ^z, and thus of H_0, as our basis. There are two degenerate ground-states of H_0, the fully magnetized states with $\{m_i = +1, \text{ all } i\}$ or $\{m_i = -1, \text{ all } i\}$. The clusters are the same as before.

As before, we display in Table 4.3 the first three clusters with their basis states and Hamiltonian matrices. We take the unperturbed ground-state as all spins up.

Again there are a number of things to note.

Table 4.3. *Low-order clusters and Hamiltonian matrices for the ordered phase expansion for the transverse field Ising chain.*

g_1 : • $|1\rangle = |+\rangle$
$|2\rangle = |-\rangle$
$$H = \begin{pmatrix} 0 & -\lambda \\ -\lambda & 4 \end{pmatrix}$$

g_2 : •—• $|1\rangle = |++\rangle$
$|2\rangle = |-+\rangle$
$|3\rangle = |+-\rangle$
$|4\rangle = |--\rangle$
$$H = \begin{pmatrix} 0 & -\lambda & -\lambda & 0 \\ -\lambda & 4 & 0 & -\lambda \\ -\lambda & 0 & 4 & -\lambda \\ 0 & -\lambda & -\lambda & 4 \end{pmatrix}$$

g_3 : •—•—• $|1\rangle = |+++\rangle$
$|2\rangle = |-++\rangle$
$|3\rangle = |++-\rangle$
$|4\rangle = |+-+\rangle$
$|5\rangle = |--+\rangle$
$|6\rangle = |+--\rangle$
$|7\rangle = |---\rangle$
$|8\rangle = |-+-\rangle$
$$H = \begin{pmatrix} 0 & -\lambda & -\lambda & -\lambda & 0 & 0 & 0 & 0 \\ -\lambda & 4 & 0 & 0 & -\lambda & 0 & 0 & -\lambda \\ -\lambda & 0 & 4 & 0 & 0 & -\lambda & 0 & -\lambda \\ -\lambda & 0 & 0 & 4 & -\lambda & -\lambda & 0 & 0 \\ 0 & -\lambda & 0 & -\lambda & 4 & 0 & -\lambda & 0 \\ 0 & 0 & -\lambda & -\lambda & 0 & 4 & -\lambda & 0 \\ 0 & 0 & 0 & 0 & -\lambda & -\lambda & 4 & -\lambda \\ 0 & -\lambda & -\lambda & 0 & 0 & 0 & -\lambda & 8 \end{pmatrix}$$

• In working out the unperturbed energies, the cluster must be considered as embedded within a chain, the rest of which is in its ground-state, i.e. each boundary spin of the cluster is coupled to an external spin in the + state.
• The number of contributing states is double that from the previous case, and represents the full Hilbert space of the cluster.
• The matrices have exactly the same form as the matrices in the previous calculation with one more site.

This last point, which the reader will have already noticed, means that no further work is needed. We can simply replace the variable $\mu = \lambda^{-1}$ in (4.15) by λ and obtain the same series. This is not just accidental but reflects an important property of the model, known as *self-duality* (Pfeuty, 1970; Fradkin and Susskind, 1978). It is possible, by means of an operator transformation, to map the two terms in H into each other, and hence to relate the small and large λ phases. More precisely, it can be shown that

$$H(\lambda) = \lambda H(\lambda^{-1}) \qquad (4.18)$$

even without knowing the exact solution (4.10). This is sufficient to locate the critical point (assuming there is only one) at $\lambda = 1$.

The general procedure for calculating series for the ground-state energy is thus straightforward.

(i) Generate a list of linked clusters, their lattice constants and embedding constants, using the methods outlined in Chapter 2.

(ii) For each cluster, calculate a perturbation series for the ground-state energy using the method outlined in Appendix 4, or using the orthogonal transformation discussed in the next section.

(iii) Derive the reduced energy for the cluster, and build up the series for the bulk ground-state energy, using Eqs (4.2) and (4.5).

4.4 Magnetization and susceptibility

We will have occasion, throughout the following chapters, to derive series for other ground-state properties such as the magnetization and susceptibility, as well as other quantities which can be expressed as ground-state expectation values of appropriate operators.

To illustrate the idea we consider the transverse field Ising chain, in the ordered phase, and obtain series for the magnetization and zero-field susceptibility. To this end we include a field term in the Hamiltonian, which becomes

$$H = \sum_{\langle ij \rangle}(1 - \sigma_i^z \sigma_{i+1}^z) - \lambda \sum_i \sigma_i^x - h \sum_i \sigma_i^z \qquad (4.19)$$

and seek to expand the ground-state energy of the bulk lattice in the form

$$E_0(h) = E_0 - Mh - \tfrac{1}{2}\chi h^2 + \cdots \qquad (4.20)$$

where the prefactors M, χ are the required magnetization and susceptibility.

The linked cluster method is used as before, where the expansion (4.20) is used at the cluster level. The formalism is explained in Appendix 4 and is incorporated in two computer programs *tim1.f* and *tim2.f*. Table 4.4 displays the bare and reduced magnetization series for low order clusters, in terms of the expansion variable $x = \lambda^2/4$ with $\lambda = \Gamma/J$.

Combining the data gives the magnetization of the bulk lattice, correct to order x^6, as

$$M = 1 - \tfrac{1}{2}x - \tfrac{7}{8}x^2 - \tfrac{35}{16}x^3 - \tfrac{805}{128}x^4 - \tfrac{4991}{256}x^5 + O(x^6) \qquad (4.21)$$

The reader can verify that this agrees with the expansion of the exact result (4.11). It is worth noting that this same result could have been obtained as a ground-state expectation value, by obtaining the ground-state wavefunctions of clusters with no field term in H.

The susceptibility series is obtained in exactly the same way. Table 4.5 gives the data for low-order clusters, from which we can obtain the bulk susceptibility to order x^6.

There is no known exact solution for the susceptibility. However, we expect a divergence at the critical point $\lambda = 1$ with exponent $\gamma = 7/4$, the susceptibility

Table 4.4. *Magnetization expansions for low-order clusters for the transverse field Ising chain.*

Cluster	Bare magnetization (M_n) and reduced magnetization (m_n)
1	$M_1 = 1 - \frac{1}{2}x + \frac{3}{8}x^2 - \frac{5}{16}x^3 + \frac{35}{128}x^4 - \frac{63}{256}x^5 + \cdots$ $m_1 = M_1$
2	$M_2 = 2 - x - \frac{1}{2}x^2 + 2x^3 - \frac{17}{8}x^4 - \frac{1}{16}x^5 + \cdots$ $m_2 = -\frac{5}{4}x^2 + \frac{21}{8}x^3 - \frac{171}{64}x^4 + \frac{55}{128}x^5 + \cdots$
3	$M_3 = 3 - \frac{3}{2}x - \frac{11}{8}x^2 - \frac{3}{16}x^3 + \frac{1131}{128}x^4 - \frac{2811}{256}x^5 + \cdots$ $m_3 = -\frac{9}{2}x^3 + \frac{855}{64}x^4 - \frac{1421}{128}x^5 + \cdots$
4	$M_4 = 4 - 2x - \frac{9}{4}x^2 - \frac{19}{8}x^3 + \frac{163}{64}x^4 + \frac{4739}{128}x^5 + \cdots$ $m_4 = -\frac{69}{4}x^4 + \frac{3771}{64}x^5 + \cdots$
5	$M_5 = 5 - \frac{5}{2}x - \frac{25}{8}x^2 - \frac{73}{16}x^3 - \frac{479}{128}x^4 + \frac{4487}{256}x^5 + \cdots$ $m_5 = -\frac{135}{2}x^5 + \cdots$

Table 4.5. *Ordered phase susceptibility expansions for low-order clusters for the transverse field Ising chain.*

Cluster	Bare susceptibility (χ_n^b) and reduced susceptibility (χ_n^r)
1	$\chi_1^b = \frac{1}{2}x - \frac{3}{4}x^2 + \frac{15}{16}x^3 - \frac{35}{32}x^4 + \frac{315}{256}x^5 + \cdots$ $\chi_1^r = \chi_1^b$
2	$\chi_2^b = x + \frac{5}{2}x^2 - 9x^3 + \frac{39}{4}x^4 + \frac{95}{16}x^5 + \cdots$ $\chi_2^r = 4x^2 - \frac{87}{8}x^3 + \frac{191}{16}x^4 + \frac{445}{128}x^5 + \cdots$
3	$\chi_3^b = \frac{3}{2}x + \frac{23}{4}x^2 + \frac{67}{8}x^3 - \frac{4433}{64}x^4 + \frac{8617}{128}x^5 + \cdots$ $\chi_3^r = \frac{437}{16}x^3 - \frac{5751}{64}x^4 + \frac{14509}{256}x^5 + \cdots$
4	$\chi_4^b = 2x + 9x^2 + \frac{103}{4}x^3 + \frac{523}{32}x^4 - \frac{227249}{512}x^5 + \cdots$ $\chi_4^r = \frac{1317}{8}x^4 - \frac{293145}{512}x^5 + \cdots$
5	$\chi_5^b = \frac{5}{2}x + \frac{49}{4}x^2 + \frac{345}{8}x^3 + \frac{6525}{64}x^4 - \frac{2613}{64}x^5 + \cdots$ $\chi_5^r = \frac{234031}{256}x^5 + \cdots$

exponent of the classical two-dimensional Ising model. It is interesting to check this. We have derived long series for the susceptibilities in both the ordered and disordered phases. The series coefficients are given in Table 4.6.

Table 4.7 shows the results of a Dlog Padé approximant analysis of these series. As is apparent, both the critical point and exponent are given to impressive accuracy.

Table 4.6. *Series coefficients for the susceptibility in both ordered and disordered phases of the transverse field Ising chain.*

ordered phase		disordered phase	
0	λ^0	1	μ^0
1/8	λ^2	2	μ^1
13/64	λ^4	11/4	μ^2
139/512	λ^6	7/2	μ^3
5479/16384	λ^8	799/192	μ^4
206345/524288	λ^{10}	77/16	μ^5
15095657/33554432	λ^{12}	37469/6912	μ^6
270545973/536870912	λ^{14}	62387/10368	μ^7
38215162539/68719476736	λ^{16}	8741521/1327104	μ^8
8004747663347/13194139533312	λ^{18}	7113365/995328	μ^9
0.655904370113	λ^{20}	1101794737/143327232	μ^{10}
0.703916017137	λ^{22}	294520877/35831808	μ^{11}
0.750858763618	λ^{24}	901634212603/103195607040	μ^{12}
0.796842023210	λ^{26}	23857725259/2579890176	μ^{13}
0.841956551768	λ^{28}	9.746312296109	μ^{14}
0.886278723184	λ^{30}	10.23885210944	μ^{15}
0.929873608956	λ^{32}	10.72169676414	μ^{16}
0.972797246695	λ^{34}	11.19909157534	μ^{17}
1.015098348101	λ^{36}	11.66829719168	μ^{18}
1.056819607122	λ^{38}	12.13263076659	μ^{19}
1.097998718648	λ^{40}	12.58990783042	μ^{20}
1.138669182549	λ^{42}	13.04281179629	μ^{21}
		13.48952826353	μ^{22}

The convergence of the estimates to the exact results, with increasing number of terms, is also worth noting.

4.5 One-particle excitations

The low-energy excited states of quantum many body systems can often be identified with 'quasiparticles', stable or long-lived excitations which carry a momentum **k** and have a well-defined energy $\epsilon(\mathbf{k})$. Examples include the 'magnon' states in magnetized systems, and triplet excitations in dimerized systems. These states control the low-temperature thermodynamic behaviour of the system and the dynamics at large distances, and are directly observable via scattering experiments. It is thus of considerable importance to calculate the energy spectrum and other properties of these elementary excitations for quantum lattice models. This is traditionally

Table 4.7. *Dlog-Padé approximants to the susceptibility in both ordered and disordered phases. The position of the singularity (λ_c^2 and μ_c) and the critical index (in brackets) are given. An asterisk denotes a defective approximant. To save space, only even N results are given.*

	$[N-1, N]$	$[N, N]$	$[N+1, N]$
		Ordered phase	
N = 2	0.999039(−1.7352)	0.998796(−1.7331)	0.999500(−1.7406)
N = 4	0.999875(−1.7463)	0.999896(−1.7467)	0.999931(−1.7475)
N = 6	0.999975(−1.7487)	0.999979(−1.7488)	0.999984(−1.7490)
N = 8	0.999993(−1.7494)	0.999994(−1.7495)	0.999995(−1.7495)
N = 10	0.999997(−1.7497)		
		Disordered phase	
N = 2	0.997002(−1.7395)	1.004876(−1.7814)	0.997972(−1.7323)
N = 4	1.000074(−1.7505)	0.999965(−1.7494)	0.999976(−1.7495)
N = 6	0.999975(−1.7495)*	0.999856(−1.7513)*	0.999989(−1.7497)
N = 8	0.999999(−1.7499)	0.999997(−1.7499)	0.999998(−1.7499)
N = 10	0.999999(−1.7499)	0.999999(−1.7499)	0.999998(−1.7499)*

done via operator transformations which express the Hamiltonian as a quadratic part plus correction terms. The quadratic part can then be diagonalized, yielding independent quasiparticles and their energy spectrum, as in spin-wave theory, for example.

We will show, in this section, how quasiparticle excitations can be computed perturbatively to high order via series expansion methods. An important by-product is that we can compute the energy gap between the ground-state and the lowest excited state, and hence locate any quantum critical points, where the gap vanishes. Our discussion will be general, but will be illustrated via the transverse field Ising chain.

Historically, there was a gap of many years before a true linked cluster method was developed for excited states. A technique involving 'overlapping clusters' was proposed by Nickel (1980, unpublished), and was used in a number of applications (e.g. Hamer and Irving, 1984; He *et al.*, 1990), but it was not until 1996 that Gelfand developed a genuine linked cluster formalism.

Let us start from the unperturbed system H_0, in which the eigenstates can be written as products of states of n distinct units (sites, dimers, plaquettes, $\cdots$) which we shall refer to generically as 'sites'. The lowest excited state of the unperturbed system, which is n-fold degenerate, contains one site in its lowest excited state while all the others are in their ground-state. When the perturbation is switched on, this excitation is able to propagate to neighbouring sites and thereby throughout the entire system. The degeneracy will be lifted, leading to a band of one-particle states.

The calculation of the one-particle excitation spectrum proceeds in the following two steps.

4.5.1 Block diagonalization

The first step is to block diagonalize the Hamiltonian for a given cluster, to form an 'effective Hamiltonian'. There is no unique transformation to achieve this, and different transformations may give different effective Hamiltonians, but the final series expansions for physical quantities should be unique, and independent of the transformation used, as long as the cluster expansion works correctly. Gelfand (1996) used a similarity transformation, not necessarily unitary, which works correctly for excited states which have different quantum numbers to the ground-state. For excited states in the ground-state sector, however, this approach fails, and one needs to use a unitary transformation to obtain correct results. The similarity transformation approach is also less convenient for the calculation of expectation values, since it does not yield the wavefunction directly, and we therefore do not discuss it any further.

Consider a particular cluster g, with Hamiltonian $H = H_0 + \lambda V$, as usual. We construct a complete basis from eigenstates of H_0: the n degenerate one-particle states, referred to above, and all other states generated from them by repeated applications of V. Within this basis the matrix for H_0 is diagonal, i.e.

$$(H_0)_{ij} = e_i \delta_{ij}, \quad i, j = 1, \cdots, N \tag{4.22}$$

where N is the total number of basis states. The full matrix H is, of course, not diagonal. The idea now is to construct a unitary transformation

$$\widetilde{H} = O^{-1} H O \tag{4.23}$$

giving the following block diagonalized form for $\widetilde{H}$

$$\widetilde{H} = \begin{pmatrix} h_{\text{eff}} & 0 & 0 \\ 0 & \blacksquare & 0 \\ 0 & 0 & \blacksquare \end{pmatrix} \tag{4.24}$$

where each diagonal block corresponds to a different set of degenerate eigenstates of H_0, and all off-diagonal blocks are zero. The upper left $n \times n$ matrix, denoted h_{eff}, is an effective Hamiltonian for the one-particle states of cluster g, and will yield the one-particle excitation energies, as we show below.

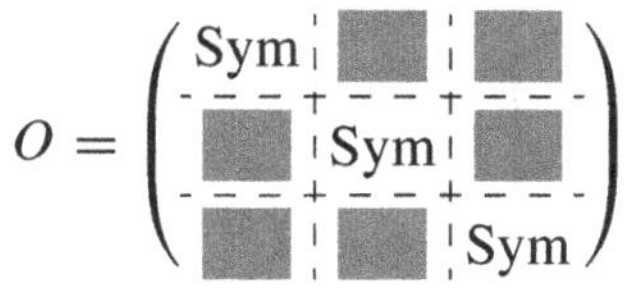

$$O = \begin{pmatrix} \text{Sym} & \blacksquare & \blacksquare \\ \blacksquare & \text{Sym} & \blacksquare \\ \blacksquare & \blacksquare & \text{Sym} \end{pmatrix}$$

Fig. 4.2. The block structure for matrices O in the two-block method, where Sym denotes a symmetric sub-matrix.

There are several points to note. In all of the applications considered in this book the matrices are real, and thus the matrix O in (4.23) can be chosen to be a real, orthogonal matrix, satisfying $O^T O = I$, with I the identity matrix. For many purposes it suffices to consider a two-block decomposition in (4.24), the upper left block corresponding to the one-particle sector and the other block containing all other states.

The orthogonal transformation in (4.23) can be constructed order-by-order in perturbation theory. We leave the technical details to Appendix 5; however, it is worth noting that the construction of the orthogonal matrix is not unique. Zheng *et al.* (2001b) developed two different approaches. The first of these, which we refer to as the 'two-block' orthogonal transformation (TBOT) is simple to implement. The matrix O is assumed to have the structure shown in Figure 4.2 with the diagonal blocks chosen to be symmetric. In practice, one only needs to compute and store the left $N \times n$ submatrix of O, which makes the approach computationally efficient. However, the TBOT method is found to fail in some cases, in particular when the one-particle states have the same quantum numbers as the ground-state. The signal for this is when the minimum order at which a cluster contributes does not grow as the size of the cluster increases. The reason for the failure is found to be that at high orders a one-particle excitation can annihilate from one cluster and reappear on another disconnected cluster, violating the assumptions for the cluster expansion to hold. After block diagonalization, this should be impossible for the 'renormalized' one-particle states but it seems that some subtle violation occurs in the two-block method; the reason for this is not yet fully understood.

The second approach, termed the 'multiblock' orthogonal transformation (MBOT), is computationally more demanding but appears to work correctly in all cases we have considered. In this approach the orthogonality of O is ensured by writing $O = e^S$, where S is a real antisymmetric matrix ($S^T = -S$), and is assumed to have a structure where the diagonal blocks of S are set to zero (Figure 4.3). This suffices to determine the diagonal blocks of O, order-by-order in perturbation theory. The technical details are given in Appendix 5.

$$S = \begin{pmatrix} 0 & -s_1^T & -s_2^T \\ s_1 & 0 & -s_3^T \\ s_2 & s_3 & 0 \end{pmatrix}$$

Fig. 4.3. The block structure of S matrices in the multiblock orthogonal transformation method.

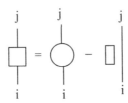

Fig. 4.4. Decomposition of a one-particle matrix element into irreducible components. The round box denotes the full matrix element, the square boxes the irreducible matrix elements, and the single line denotes a delta function.

4.5.2 Linked cluster expansion

The second step is to formulate a linked cluster expansion for the elements of the effective Hamiltonian matrix, which we write in the form

$$h_{\mathrm{eff}} = \begin{pmatrix} h_{11} & h_{12} & \cdots & h_{1n} \\ h_{21} & h_{22} & \cdots & h_{2n} \\ \vdots & \vdots & \ddots & \vdots \\ h_{n1} & h_{n2} & \cdots & h_{nn} \end{pmatrix} \tag{4.25}$$

Each element h_{ij} represents all processes in which the excitation travels from site i to site j in the cluster. Gelfand (1996) realized (see also Gelfand and Singh, 2000) that while h_{ij} itself does not satisfy the key requirement of cluster additivity, the 'irreducible' one-particle matrix Δ, defined by

$$\Delta_{ij} = h_{ij} - E_0 \delta_{ij} \tag{4.26}$$

where E_0 is the ground-state energy of the cluster, does satisfy cluster additivity. Figure 4.4 illustrates this decomposition. Consider a cluster consisting of two disconnected components g_1, and g_2, where sites i and j are included in cluster g_1. Assuming that a one-particle state cannot annihilate from one sub-cluster and appear on the other, the matrix elements will be given by

$$(h_{\mathrm{eff}})_{ij}(g_1 \oplus g_2) = (h_{\mathrm{eff}})_{ij}(g_1) + E_0(g_2)\delta_{ij} \tag{4.27}$$

and therefore

$$\Delta_{ij}(g_1 \oplus g_2) = \Delta_{ij}(g_1) + \Delta_{ij}(g_2) \tag{4.28}$$

(cluster additivity), where Δ_{ij} vanishes for any cluster which does not include sites i and j. Hence we can form a linked cluster expansion for Δ_{ij}, with non-zero contributions coming only from linked clusters which are 'rooted at' i and j (i.e., include the sites i, j).

The computational procedure is then as follows.

(i) Generate a list of clusters, with their lattice constants, up to some maximum number of sites, as discussed in Chapter 2. Note, however, that different geometrical embeddings of each cluster must be distinguished, unlike the ground-state calculation, where clusters were only distinguished on the basis of their topology. For instance, ⌐•••—• must be distinguished from •—•—•, because the distance between pairs of sites (i,j) depends on the embedding. This leads to a substantial increase in the list of clusters, and limits the length of the series which can be calculated.

(ii) For each cluster compute the elements of the effective Hamiltonian matrix as series in λ, to the order chosen.

(iii) For each cluster g, compute the auxiliary quantities

$$\Delta^g(\mathbf{0}) = \sum_{i=0}^{n} h_{ii} - nE_0 \tag{4.29a}$$

$$\Delta^g(\mathbf{r}) = \sum_{\substack{i,j \\ \mathbf{r}=\mathbf{r}_i-\mathbf{r}_j}} h_{ij} \tag{4.29b}$$

where the sum over i can be interpreted as a sum over all positions of the cluster g relative to a fixed initial position i. We refer to these as *transition amplitudes* for the cluster. We assume that the ground-state energy E_0 has been computed previously.

(iv) Subtract off sub-graph contributions to obtain the reduced transition amplitudes

$$d^g(\mathbf{r}) = \Delta^g(\mathbf{r}) - \sum_{g' \in g} c(g'/g) d^{g'}(\mathbf{r}) \tag{4.30}$$

[analogous to Eq. (4.3)].

(v) Combine the reduced transition amplitude series for all clusters to obtain the corresponding quantities $\Delta(\mathbf{r})$ for the bulk lattice

$$\Delta(\mathbf{r}) = \sum_{g} c(g) d^g(\mathbf{r}) \tag{4.31}$$

where $c(g)$ is the lattice constant. Then

$$\langle \mathbf{r} + \mathbf{r}' | h_{\text{eff}} | \mathbf{r}' \rangle = \Delta(\mathbf{r}) \tag{4.32}$$

for all $\mathbf{r}'$, by translation invariance. The one-particle excited states will be eigenstates of momentum

$$|\mathbf{k}\rangle = \frac{1}{\sqrt{N}} \sum_{\mathbf{r}} e^{i\mathbf{k}\cdot\mathbf{r}} |\mathbf{r}\rangle \tag{4.33}$$

where N is the number of sites, and hence the excitation energy is easily found

$$\begin{aligned}\epsilon(\mathbf{k}) &= \langle\mathbf{k}|h_{\text{eff}}|\mathbf{k}\rangle \\ &= \sum_{\mathbf{r}} e^{i\mathbf{k}\cdot\mathbf{r}} \Delta(\mathbf{r})\end{aligned} \tag{4.34}$$

4.5.3 Example: the transverse field Ising chain

To make the preceding discussion transparent let us consider the transverse field Ising chain. We choose the disordered phase form

$$H = \sum_i (1 - \sigma_i^z) - \mu \sum_i (\sigma_i^+ \sigma_{i+1}^+ + \sigma_i^+ \sigma_{i+1}^- + \sigma_i^- \sigma_{i+1}^+ + \sigma_i^- \sigma_{i+1}^-) \tag{4.35}$$

and consider, in turn, the three clusters of Table 4.1.

We consider the TBOT method. The results for each cluster are (note that we make use of the inversion symmetry $\Delta(\mathbf{r}) = \Delta(-\mathbf{r})$ to lump both together into $\Delta(\mathbf{r})$)

g_1 : • The ground-state has spin $\sigma^z = +1$. The excited state is $|1\rangle = |-\rangle$ and the 1×1 Hamiltonian matrix is $H = 2$. The only transition amplitude is $d^{g_1}(0) = 2$.

g_2 : $\overset{1}{\bullet}\!\!-\!\!\overset{2}{\bullet}$ The two unperturbed one-particle states are

$$|1\rangle = |-+\rangle \quad \text{and} \quad |2\rangle = |+-\rangle \tag{4.36}$$

No additional states are generated by V. Hence the 2×2 H matrix, which is also the effective Hamiltonian, is

$$H \equiv h_{\text{eff}} = \begin{pmatrix} 2 & -\mu \\ -\mu & 2 \end{pmatrix} \tag{4.37}$$

Subtracting off the ground-state energy (Table 4.2) and the contribution of cluster 1 gives

$$\begin{aligned}d^{g_2}(0) &= h_{11} + h_{22} - 2E_0^{g_2} - 2d^{g_1}(0) \\ &= \tfrac{1}{2}\mu^2 + \tfrac{1}{32}\mu^4 + \tfrac{1}{256}\mu^6 + \cdots \end{aligned} \tag{4.38a}$$
$$d^{g_2}(1) = h_{12} + h_{21} = -2\mu \tag{4.38b}$$

g_3 : The basis has four states, the three-unperturbed one-particle states plus another. The states and Hamiltonian matrix are

$$
\begin{array}{ll}
|1\rangle = |-++\rangle \\
|2\rangle = |+-+\rangle \\
|3\rangle = |++-\rangle \\
|4\rangle = |---\rangle
\end{array}
\qquad
H = \begin{pmatrix}
2 & -\mu & 0 & -\mu \\
-\mu & 2 & -\mu & 0 \\
0 & -\mu & 2 & -\mu \\
-\mu & 0 & -\mu & 6
\end{pmatrix}
$$

Solving the TBOT equations gives the 3×3 effective Hamiltonian matrix

$$
h_{\text{eff}} = \begin{pmatrix}
h_{11} & h_{12} & h_{13} \\
h_{21} & h_{22} & h_{23} \\
h_{31} & h_{32} & h_{33}
\end{pmatrix}
\tag{4.39}
$$

with

$$
\begin{aligned}
h_{11} &= h_{33} = 2 - \tfrac{1}{4}\mu^2 + 0\mu^4 + \tfrac{5}{1024}\mu^6 + \cdots \\
h_{22} &= 2 + 0\mu^2 + 0\mu^4 - \tfrac{1}{512}\mu^6 + \cdots \\
h_{12} &= h_{21} = h_{23} = h_{32} = -\mu + \tfrac{1}{16}\mu^3 - \tfrac{3}{512}\mu^5 + \cdots \\
h_{13} &= h_{31} = -\tfrac{1}{4}\mu^2 + 0\mu^4 + \tfrac{5}{1024}\mu^6 + \cdots
\end{aligned}
\tag{4.40}
$$

The non-zero matrix elements for O are

$$
\begin{aligned}
O_{11} &= O_{33} = 1 - \tfrac{1}{32}\mu^2 + \tfrac{3}{1024}\mu^4 + \tfrac{17}{16384}\mu^6 + \cdots \\
O_{13} &= O_{31} = -\tfrac{1}{32}\mu^2 + \tfrac{3}{1024}\mu^4 + \tfrac{17}{16384}\mu^6 + \cdots \\
O_{12} &= O_{21} = O_{23} = O_{32} = \tfrac{1}{64}\mu^3 - \tfrac{7}{2048}\mu^5 + \cdots \\
O_{22} &= 1 + 0\mu^2 - \tfrac{1}{128}\mu^4 + \tfrac{11}{4096}\mu^6 + \cdots \\
O_{41} &= O_{43} = \tfrac{1}{4}\mu - \tfrac{1}{64}\mu^3 - \tfrac{9}{2048}\mu^5 + \cdots \\
O_{42} &= -\tfrac{1}{8}\mu^2 + \tfrac{3}{128}\mu^4 - \tfrac{3}{4096}\mu^6 + \cdots
\end{aligned}
\tag{4.41}
$$

From these we obtain the reduced transition amplitudes

$$
\begin{aligned}
d^{g_3}(0) &= \tfrac{1}{16}\mu^4 - \tfrac{3}{128}\mu^6 + \cdots \\
d^{g_3}(1) &= \tfrac{1}{4}\mu^3 - \tfrac{3}{128}\mu^5 + \cdots \\
d^{g_3}(2) &= -\tfrac{1}{2}\mu^2 + 0\mu^4 + \tfrac{5}{512}\mu^6 + \cdots
\end{aligned}
\tag{4.42}
$$

We note that the effective Hamiltonian matrix (4.40) is symmetric. In the original Gelfand method this is not the case, although the final physical results are the same. Note also that the lowest order contribution from a cluster of n sites is of order μ^{n-1}. This is an indication that the method is working correctly.

To proceed much further it is necessary to computerize the procedure. This is discussed in Appendix 5, and a sample program *tim3.f* is provided at www.cambridge.org/9780521842426. In Table 4.8 we give the transition amplitudes for the bulk chain, to order μ^8. This requires clusters of up to nine sites.

Table 4.8. *Transition amplitude series for one-particle excitations in the transverse field Ising chain.*

$$\Delta(0) = 2 + \tfrac{1}{2}\mu^2 + \tfrac{1}{32}\mu^4 + \tfrac{1}{128}\mu^6 + \tfrac{25}{8192}\mu^8 + \cdots$$

$$\Delta(1) = -2\mu + \tfrac{1}{4}\mu^3 + \tfrac{1}{32}\mu^5 + \tfrac{5}{512}\mu^7 + \cdots$$

$$\Delta(2) = -\tfrac{1}{2}\mu^2 + \tfrac{1}{8}\mu^4 + \tfrac{5}{256}\mu^6 + \tfrac{7}{1024}\mu^8 + \cdots$$

$$\Delta(3) = -\tfrac{1}{4}\mu^3 + \tfrac{5}{64}\mu^5 + \tfrac{7}{512}\mu^7 + \cdots$$

$$\Delta(4) = -\tfrac{5}{32}\mu^4 + \tfrac{7}{128}\mu^6 + \tfrac{21}{2048}\mu^8 + \cdots$$

$$\Delta(5) = -\tfrac{7}{64}\mu^5 + \tfrac{21}{512}\mu^7 + \cdots$$

$$\Delta(6) = -\tfrac{21}{256}\mu^6 + \tfrac{33}{1024}\mu^8 + \cdots$$

$$\Delta(7) = -\tfrac{33}{512}\mu^7 + \cdots$$

$$\Delta(8) = -\tfrac{429}{8192}\mu^8 + \cdots$$

From these transition amplitudes we finally obtain the one-particle excitation energies

$$\epsilon(k) = \sum_n \Delta(n) \cos nk$$
$$= 2 - 2\mu \cos k + \mu^2 \left(1 - \cos^2 k\right) + \mu^3 \left(\cos k - \cos^3 k\right)$$
$$+ \tfrac{1}{4}\mu^4 \left(1 - 6\cos^2 k - 5\cos^4 k\right) + \cdots \tag{4.43}$$

The reader may confirm that this agrees, to the order given, with the exact result (Pfeuty, 1970)

$$\epsilon(k) = 2(1 + \mu^2 - 2\mu \cos k)^{1/2} \tag{4.44}$$

For this particular example the more complex MBOT approach gives identical results, For the benefit of the reader who wishes to explore this further, we provide the leading terms for cluster g_3, for the non-zero matrix elements of S

$$S_{41} = S_{43} = -S_{14} = -S_{34} = \tfrac{1}{4}\mu - \tfrac{1}{96}\mu^3 - \tfrac{17}{3840}\mu^5 + \cdots$$
$$S_{42} = S_{24} = -\tfrac{1}{8}\mu^2 + \tfrac{1}{48}\mu^4 - \tfrac{1}{2560}\mu^6 + \cdots \tag{4.45}$$

4.6 The transverse field Ising model in two and three dimensions

To conclude this chapter we consider the transverse field Ising model in higher spatial dimensions. These cases cannot be solved exactly. Series expansions at $T = 0$ can be computed for ground-state bulk properties and for excitations, and these provide the most accurate means of locating the quantum critical point, computing critical exponents and determining excitation spectra. We consider, specifically, the square (SQ) and simple cubic (SC) lattices.

4.6.1 Expansions in the disordered phase

We write the Hamiltonian in the form

$$H = \sum_i (1 - \sigma_i^z) - \mu \sum_{\langle ij \rangle} \sigma_i^x \sigma_{i+1}^x - h \sum_i \sigma_i^x \tag{4.46}$$

where $\mu = J/\Gamma$ and where we have introduced a field h which couples to the order parameter $\langle \sigma_i^x \rangle$. Our expansion will be convergent for small μ. This is the disordered, or large Γ, region of the model, where $\langle \sigma_i^x \rangle = 0$ when $h = 0$. In the literature it has also been referred to as the 'strong coupling' or 'high-temperature' phase. The quantum critical point μ_c can be most effectively located from the susceptibility series

$$\chi = -\frac{\partial^2 E_0}{\partial h^2}\bigg|_{h=0} \tag{4.47}$$

where E_0 is the ground-state energy. This is expected to diverge at the critical point as

$$\chi \sim C(\mu_c - \mu)^{-\gamma}; \quad \mu \to \mu_c - \tag{4.48}$$

where C is a non-universal amplitude, and γ is a univeral exponent. Following the idea of 'quantum-classical correspondence' discussed earlier, we expect that $\gamma_{2D} \simeq 1.239$ and $\gamma_{3D} \simeq 1$, the classical exponents in three and four dimensions for the $N = 1$ universality class. In the latter case logarithmic correction terms are expected.

Series for the ground-state energy and susceptibility are known to 16th order for the SQ lattice (He *et al.*, 1990), and to 14th order for the SC lattice (Zheng *et al.*, 1994). The coefficients of the susceptibility series are reproduced in Table 4.9, in the form $\chi = 1 + \sum_n c_n \mu^n$

For a detailed analysis of these series we refer to the original papers. The ratio plot in Figure 4.5 shows that the series are regular, with a clear even–odd oscillation, characteristic of bipartite lattices. A simple visual extrapolation yields estimates $\mu_c^{-1} = 3.05, 5.15$ for the SQ, SC lattices respectively, corresponding to the quantum critical points. The estimated critical exponents are in line with the expected values.

The papers cited above also computed series for the minimum energy gap Δ, via a formalism which required disconnected as well as connected graphs. However, this approach has been rendered largely obsolete with the development of linked cluster expansions for excitations.

Table 4.9. *Coefficients of the disordered phase susceptibility for the transverse field Ising model on the square (SQ) and simple cubic (SC) lattices.*

Order	SQ lattice	SC lattice
0	1.000000000	1.000000000
1	4.000000000	6.000000000
2	1.350000000×10^1	3.225000000×10^1
3	4.500000000×10^1	1.725000000×10^2
4	1.448437500×10^2	9.080468750×10^2
5	4.644444444×10^2	4.775895833×10^3
6	1.469358507×10^3	2.496107682×10^4
7	4.639482350×10^3	1.304100139×10^5
8	1.454411924×10^4	6.791495856×10^5
9	4.553747966×10^4	3.536163534×10^6
10	1.419477302×10^5	1.837669188×10^7
11	4.421005347×10^5	9.548768040×10^7
12	1.372818041×10^6	4.955398252×10^8
13	4.260277897×10^6	2.571415145×10^9
14	1.319245910×10^7	$1.333149283 \times 10^{10}$
15	4.083288839×10^7	
16	1.261792858×10^8	

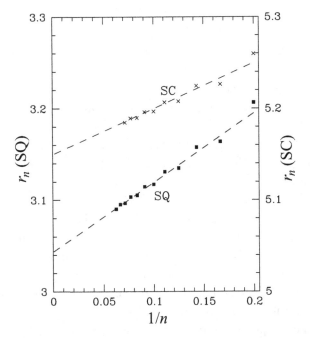

Fig. 4.5. Ratio plot for the disordered phase susceptibility for the square (SQ) and simple cubic (SC) lattices. The dotted lines are simple visual extrapolations.

Table 4.10. *Coefficients of the magnetization series for the transverse field Ising model on the square (SQ) and simple cubic (SC) lattices. The expansion variable is $x = \lambda^2$.*

Order	SQ lattice	SC lattice
0	$5.000000000 \times 10^{-1}$	$5.000000000 \times 10^{-1}$
1	$-2.500000000 \times 10^{-1}$	$-1.111111111 \times 10^{-1}$
2	$-1.736111111 \times 10^{-1}$	$-2.518518519 \times 10^{-2}$
3	$-1.467013889 \times 10^{-1}$	$-7.334621091 \times 10^{-3}$
4	$-2.022858796 \times 10^{-1}$	$-3.133155006 \times 10^{-3}$
5	$-2.250832954 \times 10^{-1}$	$-1.235233225 \times 10^{-3}$
6	$-3.481680145 \times 10^{-1}$	$-6.177842807 \times 10^{-4}$
7	$-4.168506552 \times 10^{-1}$	$-2.790562317 \times 10^{-4}$
8	$-7.194408082 \times 10^{-1}$	$-1.452183554 \times 10^{-4}$
9	$-8.958967253 \times 10^{-1}$	$-7.023368200 \times 10^{-5}$
10	-1.538030379	$-3.773175617 \times 10^{-5}$
11	-2.151144545	$-1.907194514 \times 10^{-5}$
12	-3.467380088	$-1.039438214 \times 10^{-5}$
13	-5.171535767	

4.6.2 Expansions in the ordered phase

The ordered phase (small Γ/J) Hamiltonian is most conveniently written as

$$H = \sum_{\langle ij \rangle} (1 - \sigma_i^z \sigma_j^z) - \lambda \sum_i \sigma_i^x - h \sum_i \sigma_i^z \qquad (4.49)$$

where $\lambda = \mu^{-1} = \Gamma/J$ and again a field term has been included. Note that we have interchanged x and z axes between (4.46) and (4.49). Expansions can now be obtained in powers of λ: in the literature such expansions have been termed 'weak coupling' or 'low temperature'.

Expansions for the ground-state energy E_0, the magnetization $M = \langle \sigma^z \rangle$, and the susceptibility for the SQ lattice have been obtained to order x^{13} by Oitmaa *et al.* (1991) (where $x = \lambda^2$). The same quantities for the SC lattice have been computed to order x^{12} by Zheng *et al.* (1994). In Table 4.10 we give the coefficients of the magnetization series.

Again we refer the reader to the original papers for a detailed series analysis.

4.6.3 One-particle excitations

Following the methods described in Section 4.5 we have computed the one-particle excitation energies for both the square and simple cubic lattices. On these lattices the model is not self-dual and hence separate series are needed in the ordered and

Table 4.11. *Leading terms in the series for one-particle excitations in the transverse field Ising model on the square (SQ) and simple cubic (SC) lattices. (The notation is $c_{nx} = \cos(nk_x)$, etc.).*

SQ – disordered phase:
$$\epsilon(\mathbf{k})/\Gamma = 2 - 2\mu(c_{1x} + c_{1y}) + \mu^2[1 - 2c_{1x}c_{1y} - \tfrac{1}{2}(c_{2x} + c_{2y})]$$
$$+ \tfrac{1}{4}\mu^3(c_{1x} + c_{1y} - 6(c_{2x}c_{1y} + c_{1x}c_{2y}) - (c_{3x} + c_{3y}))$$
$$+ \tfrac{1}{32}\mu^4[70 - 16c_{1x}c_{1y} - 60c_{2x}c_{2y} - 24(c_{2x} + c_{2y})$$
$$- 40(c_{3x}c_{1y} + c_{1x}c_{3y}) - 5(c_{4x} + c_{4y})] + \cdots$$

SQ – ordered phase:
$$\epsilon(\mathbf{k})/J = 8 - \tfrac{1}{4}\lambda^2(1 + c_{1x} + c_{1y}) + \tfrac{1}{768}\lambda^4[19 + 12(c_{1x} + c_{1y})]$$
$$- \tfrac{1}{884736}\lambda^6[4745 + 4176c_{1x}c_{1y} + 4710(c_{1x} + c_{1y}) + 504(c_{2x} + c_{2y})$$
$$+ 276(c_{2x}c_{1y} + c_{1x}c_{2y}) + 46(c_{3x} + c_{3y})] + \cdots$$

SC – disordered phase:
$$\epsilon(\mathbf{k})/\Gamma = 2 - 2\mu(c_{1x} + c_{1y} + c_{1z}) + \tfrac{1}{2}\mu^2[3 - 4(c_{1x}c_{1y} + c_{1x}c_{1z} + c_{1y}c_{1z})$$
$$- (c_{2x} + c_{2y} + c_{2z})] + \tfrac{1}{4}\mu^3[-24c_{1x}c_{1y}c_{1z} + (c_{1x} + c_{1y} + c_{1z})$$
$$- 6(c_{2x}c_{1y} + c_{1x}c_{2y} + c_{2x}c_{1z} + c_{1x}c_{2z} + c_{2y}c_{1z} + c_{1y}c_{2z})$$
$$- (c_{3x} + c_{3y} + c_{3z})] + \cdots$$

SC – ordered phase:
$$\epsilon(\mathbf{k})/J = 12 - \tfrac{1}{12}\lambda^2(1 + c_{1x} + c_{1y} + c_{1z})$$
$$+ \tfrac{1}{69120}\lambda^4[151 + 60(c_{1x} + c_{1y} + c_{1z}) - 5(c_{2x} + c_{2y} + c_{2z})$$
$$- 20(c_{1x}c_{1y} + c_{1x}c_{1z} + c_{1y}c_{1z})] + \cdots$$

disordered phases. The Gelfand method or two-block method can be used in the disordered phase, but fails in the ordered phase because the one-particle excitations occur in the same sector as the ground-state. Here we have used the multi-block method.

In the disordered phase we have obtained series in powers of $\mu = J/\Gamma$ to order μ^{14} and μ^9 for the SQ and SC lattices respectively, involving 4 654 284 and 208 060 clusters, up to 15 and 10 sites, respectively, while in the ordered phase the series have been computed in powers of $\lambda = \Gamma/J$ to order λ^{18} and λ^{16} for the two lattices, involving 6473 and 29 977 clusters, up to 10 and 9 sites, respectively. We provide, in Table 4.11, the leading terms of these series.

Figure 4.6 shows the dispersion curves for the square lattice, along symmetry lines in the Brillouin zone, for three parameter ratios $\mu = J/\Gamma = 0.2$, 0.3286 (the critical point), 0.4. At the critical point, both expansions about ordered and disordered phases are identical. We note that the spectrum has an energy gap to the

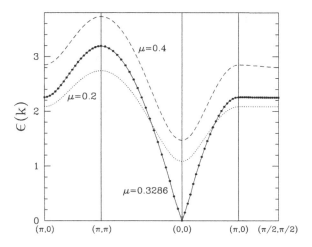

Fig. 4.6. One-particle excitation energies throughout the Brillouin zone for the square lattice.

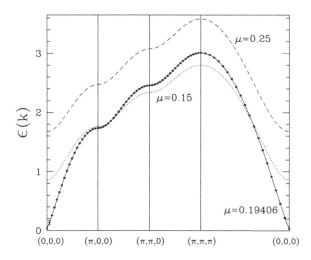

Fig. 4.7. One-particle excitation energies throughout the Brillouin zone for the simple cubic lattice.

ground-state throughout the zone, except at the critical point ($\mu = 0.3286$) where the gap closes at $\mathbf{k} = 0$. Asymptotically we expect

$$\Delta \sim |\mu - \mu_c|^{\nu} \quad \mu \to \mu_c \pm \tag{4.50}$$

where ν is the usual critical exponent for the three-dimensional classical Ising model ($\nu \simeq 0.63$). Analysis of our series confirms this expectation.

The dispersion curve for one-particle excitations for the simple cubic lattice is shown in Figure 4.7, again for three values of μ: 0.15, $0.19406(\mu_c)$, and 0.25. Again the gap is non-zero, except at the critical point where the $\mathbf{k} = 0$ gap closes. The exponent v is found to be consistent with the mean-field value $v = 1/2$ for the 4D Ising model, modulo logarithmic corrections (Zheng *et al.*, 1994).

The results of this subsection are published here for the first time. They illustrate the type of detailed information about excited states that can be obtained using series methods. We will show many further examples in later chapters.

5

Quantum antiferromagnets at $T = 0$

5.1 Introduction: simple antiferromagnets

In the previous chapter we showed how to derive perturbation expansions for quantum spin models at $T = 0$, for both ground state bulk properties and for excitations. We used, as an example, the transverse field Ising model. This is arguably the simplest model to show a quantum phase transition. Moreover the model has a natural perturbation parameter, Γ/J or J/Γ, and the series are relatively straightforward to derive and interpret.

In the present chapter we turn to quantum antiferromagnets. These are systems in which the exchange interaction favours anti-alignment of neighbouring spins. The physics will turn out to be much richer, and we will use series methods to explore not only simple antiferromagnets on bipartite lattices but also the effects of frustration, competing interactions, destruction of antiferromagnetism due to singlet formation on dimers or plaquettes, and other related topics. With some minor modifications the techniques are the same as in Chapter 4. In this chapter, however, we will provide fewer details and concentrate more on results.

5.1.1 The Ising expansion

Let us start by considering a spin-$\frac{1}{2}$ system on a bipartite lattice, with nearest-neighbour antiferromagnetic exchange. A bipartite lattice is one which consists of two interpenetrating sub-lattices A, B such that the nearest neighbours of A spins all lie on sub-lattice B and vice versa. Examples are the linear chain, the square and honeycomb lattices in two dimensions, the simple cubic and body-centred cubic lattices in three dimensions.

The Hamiltonian is taken to be

$$H = J \sum_{\langle ij \rangle} \mathbf{S}_i \cdot \mathbf{S}_j \tag{5.1}$$

with $J > 0$ and the summation over nearest-neighbour pairs. For classical vector spins the lowest energy state will have all A spins aligned in some arbitrary direction and all B spins aligned in precisely the opposite direction. This is termed the Néel state after one of the pioneers of the subject of antiferromagnetism. For quantum spins things are not so simple, as the Néel state is not an eigenstate of H and, hence, cannot be the true ground state. The true ground state is only known exactly in one dimension (see e.g. Takahashi, 1999) and it has no long-range magnetic order, but has long-range correlations. In higher dimension the ground state is ordered but the order parameter (sub-lattice magnetization) is reduced by quantum fluctuations. The relative effect of quantum fluctuations decreases with increasing dimension, with increasing number of neighbours, and also with increasing spin quantum number S. In one dimension quantum fluctuations prevent long-range order even at $T = 0$, in two dimensions order is present at $T = 0$ but is destroyed at any finite temperature (the Mermin–Wagner theorem), whereas in three dimensions antiferromagnetic order persists to a finite critical temperature, the Néel temperature.

We would like to use series expansion methods to calculate properties of the model described by (5.1). The first question is how to write this in the form $H_0 + \lambda V$, where H_0 is solvable exactly. There are a number of ways one might choose, but for magnetically ordered phases the so-called *Ising expansion* seems most natural. We write the Hamiltonian in the form (setting $J = 1$)

$$
\begin{aligned}
H &= \sum_{\langle ij \rangle} S_i^z S_j^z + \lambda \sum_{\langle ij \rangle} (S_i^x S_j^x + S_i^y S_j^y) \\
&= \sum_{\langle ij \rangle} S_i^z S_j^z + \frac{\lambda}{2} \sum_{\langle ij \rangle} (S_i^+ S_j^- + S_i^- S_j^+)
\end{aligned}
\tag{5.2}
$$

where S_i^+, S_i^- are the usual spin-raising and lowering operators and we have included an exchange anisotropy parameter $\lambda \leq 1$. The first term is equivalent to a classical Ising model and has a trivial up-down ground state. Hence the name 'Ising expansion'. The other term, which becomes the perturbation V, incorporates the quantum fluctuations. Before proceeding to the calculations, there is one further technical point. It is convenient to perform a spin rotation $(S^x, S^y, S^z) \rightarrow (S^x, -S^y, -S^z)$ on sub-lattice B, leading to the modified form

$$
H = -\sum_{\langle ij \rangle} S_i^z S_j^z + \frac{\lambda}{2} \sum_{\langle ij \rangle} (S_i^+ S_j^+ + S_i^- S_j^-)
\tag{5.3}
$$

With this form both sub-lattices are identical and we avoid the need to keep track of different sub-lattices in the clusters.

Series will be obtained in powers of λ, and can be evaluated directly for a system with easy-axis anisotropy. Usually, however, we will be interested in the isotropic limit $\lambda = 1$. As this is a singular point, special care may be needed in the analysis (c.f. test function $f_2(\lambda)$ in Section 1.5).

5.1.2 The energy and magnetization

Using the linked cluster perturbation formalism one can derive expansions in powers of λ for various bulk properties at $T = 0$ for any lattice, in the form

$$F(\lambda) = \sum_{\{g\}} c_g f_g(\lambda) \qquad (5.4)$$

where the sum is over all connected clusters which are embeddable in the lattice, c_g is the strong ('low-temperature') lattice constant, and $f_g(\lambda)$ are the reduced quantities for each cluster. After accumulating the data we obtain for the ground-state energy per site an expansion of the form

$$E_0 = \sum_{r=0}^{\infty} e_r \lambda^r \qquad (5.5)$$

with the coefficients known to some maximum order r_{max}. By including a field term in the Hamiltonian (5.3), keeping terms up to h^2 and differentiating we can also obtain series for the staggered magnetization (order parameter)

$$M_s = \sum_{r=0}^{\infty} \mu_r \lambda^r \qquad (5.6)$$

as well as for the parallel and perpendicular susceptibilities.

Although attempts to derive such series go back to the 1960s, the first relatively long series (10 terms) for the square lattice were derived by Singh (1989). We have extended the series by several terms (Zheng *et al.*, 1991; Oitmaa *et al.*, 1994). In Table 5.1 we gives the series coefficients for the square (SQ) and simple cubic (SC) lattices (the highest order coefficients have not been published previously). The number of distinct clusters to this order is 185690 (to 16 sites) and 180252 (to 14 sites) for the SQ and SC lattices respectively. The reader is referred to the literature for further results: in particular for susceptibilities, for other lattices, and for higher S.

In analysing the series it pays to recall that $\lambda = 1$ is a singular point. Spin-wave theory predicts singularities of the form $(1 - \lambda^2)^{n/2}$ where n is an odd integer. To overcome this we first transform to a new variable y (Huse, 1988) with

$$1 - y = (1 - \lambda^2)^{1/2} \qquad (5.7)$$

before using Padé and first-order differential approximants. Figure 5.1 shows our estimated ground-state energy and magnetization for the square lattice as a function

Table 5.1. *Ising expansion coefficients for ground-state energy and staggered magnetization for the simple $S = \frac{1}{2}$ Heisenberg antiferromagnet on the square (SQ) and simple cubic (SC) lattices.*

Order	E_0	M_s
	SQ lattice	
0	$-5.000000000000000 \times 10^{-1}$	$5.000000000000000 \times 10^{-1}$
2	$-1.666666666666667 \times 10^{-1}$	$-1.111111111111111 \times 10^{-1}$
4	$9.259259259259259 \times 10^{-4}$	$-1.777777777777778 \times 10^{-2}$
6	$-1.581569664902998 \times 10^{-3}$	$-9.471293496822597 \times 10^{-3}$
8	$-8.252128462349078 \times 10^{-4}$	$-7.442915293814974 \times 10^{-3}$
10	$-3.118506492685293 \times 10^{-4}$	$-4.376910244841174 \times 10^{-3}$
12	$-2.419420797199354 \times 10^{-4}$	$-3.605706354342920 \times 10^{-3}$
14	$-1.511226784769453 \times 10^{-4}$	$-2.801005151457947 \times 10^{-3}$
16	$-1.057629398659745 \times 10^{-4}$	$-2.258085008344180 \times 10^{-3}$
	SC lattice	
0	$-7.500000000000000 \times 10^{-1}$	$5.000000000000000 \times 10^{-1}$
2	$-1.500000000000000 \times 10^{-1}$	$-6.000000000000000 \times 10^{-2}$
4	$5.000000000000000 \times 10^{-4}$	$-4.011111111111111 \times 10^{-3}$
6	$-1.565320944487611 \times 10^{-3}$	$-3.724431496527780 \times 10^{-3}$
8	$-4.704476286332721 \times 10^{-4}$	$-1.921150493840936 \times 10^{-3}$
10	$-2.617515568853395 \times 10^{-4}$	$-1.282583243920382 \times 10^{-3}$
12	$-1.500007148263468 \times 10^{-4}$	$-9.023859053625596 \times 10^{-4}$
14	$-9.512504585233385 \times 10^{-5}$	$-6.716166729715401 \times 10^{-4}$

of the anisotropy parameter λ. The series results are given by the solid line. For comparison we also show the results from first- and second-order spin-wave theory. As is clear from the figure, first-order spin-wave theory is not very accurate, whereas the second-order spin-wave theory agrees very well with the series results. The third-order spin-wave theory curves are indistiguishable from series on the scale of the figures, and are not shown. Similar results for the SC lattice are shown in Figure 5.2. We note the effect of increasing quantum fluctuations in both lowering the ground-state energy and in reducing the magnetization from the classical Néel value. At the isotropic point ($\lambda = 1$) the magnetization reduction is some 38% for the SQ lattice, but only 15% for the SC lattice. Quantum fluctuations have a smaller effect in high dimension and for higher coordination number.

Another important quantity which characterizes the antiferromagnetic state, and which is one of the three fundamental parameters in an effective low-energy field theory, is the spin-stiffness, ρ_s. It is defined through the expression

$$E(\theta) = E(0) + \frac{1}{2}\rho_s\theta^2 + 0(\theta^4) \tag{5.8}$$

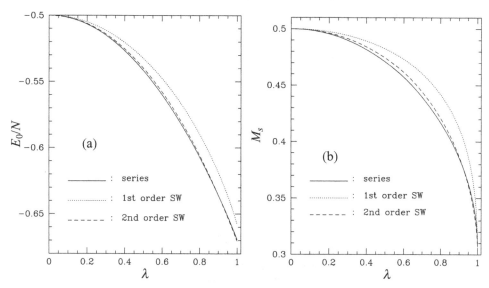

Fig. 5.1. Ground-state energy (a) and staggered magnetization (b) versus anisotropy for the square lattice $S = \frac{1}{2}$ Heisenberg antiferromagnet.

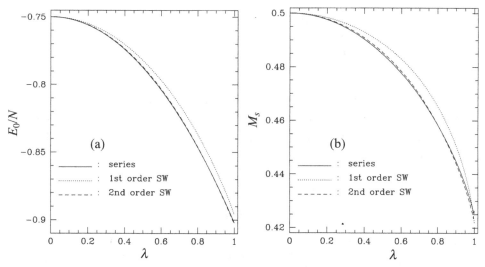

Fig. 5.2. Ground-state energy (a) and staggered magnetization (b) versus anisotropy for the simple cubic lattice $S = \frac{1}{2}$ Heisenberg antiferromagnet.

for the increase in energy per site when the order parameter is given a twist by an angle θ per unit length in some direction. A series expansion for the spin-stiffness can be obtained by re-expressing the Hamiltonian in terms of a twisting local quantization axis. We refer to the literature (Singh and Huse, 1989; Hamer *et al.*, 1994b) for further details and for the series coefficients.

Table 5.2. *Comparison of results for physical quantities obtained from series and other methods: third and fourth order spin-wave theory (SW), exact diagonalizations (ED) and quantum Monte Carlo (QMC) calculations.*

Method	E_0	M_s	$\chi_\perp$	ρ_s
		SQ lattice		
Series[a]	−0.6693(1)	0.307(1)	0.0659(10)	0.182(5)
SW[b]	−0.6695	0.307	0.063	0.175
QMC[c]	−0.669437(5)	0.3070(3)	0.0625(9)	0.175(2)
QMC[d]	−0.669442(26)	0.3075(2)		
ED[e]	−0.66944(6)	0.3055(5)	0.075(2)	0.180(2)
		SC lattice		
Series[f]	−0.9021(2)	0.424(2)	0.0653(5)	
SW[b]	−0.902498	0.4227	0.06438	0.2343

[a] Zheng *et al.* (1991); Hamer *et al.* (1994b)
[b] Fourth-order spin-wave theory for E_0 (Zheng and Hamer, 1993), third-order for others (Hamer *et al.*, 1992)
[c] Sandvik (1997)
[d] Buonaura and Sorella (1998)
[e] Lin *et al.* (2001)
[f] Oitmaa *et al.* (1994)

To conclude this subsection we present in Table 5.2 collected data for various bulk quantities for the $S = \frac{1}{2}$ Heisenberg antiferromagnet, comparing results obtained via series expansions, spin-wave theory, exact diagonalizations and quantum Monte Carlo calculations.

5.1.3 One-magnon excitations

Continuing our discussion of simple antiferromagnets in two and three dimensions, we now look at the energies of one-particle excited states. We write the Hamiltonian as $H = H_0 + \lambda V$ where, as in the previous section, H_0 consists of Ising terms and V contains the transverse terms. If we take the unperturbed ground state as having all spins down then, for $\lambda = 0$, the degenerate one-particle excitations consist of single up spins on any of the N sites of the cluster. Turning on V allows these spin-flipped states to propagate throughout the lattice, forming a coherent excitation branch throughout the Brillouin zone. At the isotropic point this become an $S = 1$ magnon.

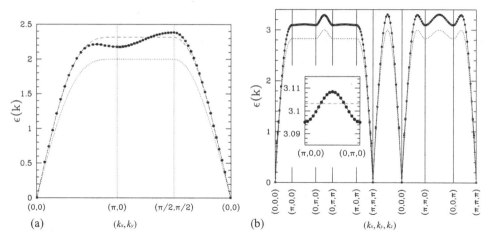

Fig. 5.3. The one-magnon excitation spectrum $\epsilon(\mathbf{k})$ along high-symmetry cuts through the Brillouin zone for the Heisenberg antiferromagnet on square (a) and simple cubic (b) lattices. Also shown are the results of first-order (dotted line), second-order (dashed line), third-order (solid line, for square lattice only) spin-wave theory.

We use the two-block approach, as discussed in the previous chapter, based on connected clusters. As in the examples discussed there, we need to keep track of all different spatial configurations. The results shown below are obtained from 14th-order series for the square lattice and tenth-order series for the simple cubic lattice, and involve 4 654 284 and 1 487 597 clusters, up to 15 and 11 sites, respectively.

Figure 5.3 shows the one-magnon excitation spectrum $\epsilon(\mathbf{k})$ along high-symmetry cuts through the Brillouin zone for the Heisenberg antiferromagnet on the square and simple cubic lattices, together with the results of first-, second- and third-order spin-wave theory. We can see that for the square lattice the spectrum is gapless at $(0, 0)$ or the equivalent point (π, π), which, of course, reflects the Goldstone nature of the magnon excitations. First- and second-order spin-wave theory give a flat dispersion curve along the zone edge $(\pi, 0)$ to $(\pi/2, \pi/2)$, whereas series results indicate that the curve is not flat, with the excitation energy at $(\pi/2, \pi/2)$ about 9% higher. It is interesting to note that experimental results for $Cu(DCOO)_2 \cdot 4D_2O$ (CFTD) and $Sr_2Cu_3O_4Cl_2$ also show a higher energy at $(\pi/2, \pi/2)$. For references to the experimental work, and for a fit of the theoretical results to the data the reader is referred to Zheng *et al.* (2005).

For the simple cubic lattice the series results are almost indistinguishable from second-order spin-wave theory, although some difference is apparent in certain directions on magnification [inset to Figure 5.3(b)].

5.2 Dimerized systems and quantum phase transitions

In this section we broaden our discussion of quantum antiferromagnets at $T = 0$, discuss the important physical phenomenon of dimerization, and introduce a new expansion method, namely 'dimer expansions'.

As an example, let us consider the spin-$\frac{1}{2}$ antiferromagnetic chain, with Hamiltonian

$$H = J \sum_i \mathbf{S}_i \cdot \mathbf{S}_{i+1} \tag{5.9}$$

As mentioned in Chapter 1, this model does not have true long-range order, even at $T = 0$, but does have long-range correlations, decaying via a power law. It is said to have *quasi-long-range* order, and the excitation spectrum is gapless. Some of the $T = 0$ properties can be calculated exactly (see e.g. Takahashi, 1999). However real one-dimensional antiferromagnetic systems, such as $CuGeO_3$ and NaV_2O_5, undergo a 'spin-Peierls' transition, and lower their total energy via a lattice distortion to a dimerized state, as shown in Figure 5.4(a). This costs lattice energy but gains magnetic energy. In the fully dimerized state each alternating pair of spins forms a spin $S = 0$ dimer singlet, magnetic correlations become short ranged (one lattice spacing) and the excitation spectrum develops a gap. In one dimension an arbitrarily weak spin-phonon interaction will favour the spin-Peierls phase.

Alternatively one could consider a chain with explicit dimerization

$$H = J_0 \sum_i [1 + (-1)^i \delta] \mathbf{S}_i \cdot \mathbf{S}_{i+1} \tag{5.10}$$

with alternating interactions $J_0(1 + \delta)$, $J_0(1 - \delta)$. In either case we can develop perturbation expansions by taking

$$H = H_0 + \lambda V \tag{5.11}$$

with $H_0 = J \sum_i \mathbf{S}_{2i} \cdot \mathbf{S}_{2i+1}$, $V = J \sum_i \mathbf{S}_{2i-1} \cdot \mathbf{S}_{2i}$, $J = J_0(1 + \delta)$, and $\lambda = (1 - \delta)/(1 + \delta)$.

The unperturbed system ($\lambda = 0$) now consists of independent dimers on half of the bonds, and the unperturbed ground state is a product of dimer singlets. The uniform chain is recovered at $\lambda = 1$. Such 'dimer expansions' were first used by Singh *et al.* (1988), and have proved valuable in subsequent studies of low-dimensional quantum antiferromagnets, particularly for magnetically disordered phases. We will illustrated the method in the next subsection.

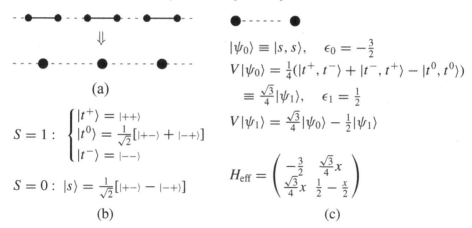

Fig. 5.4. (a) Mapping from alternating chain to coupled dimers; (b) the quantum states of an isolated dimer; (c) the effect of V on a pair of dimers, and the effective Hamiltonian.

5.2.1 Dimer expansions and the alternating chain

As a simple example, let us consider the alternating (dimerized) spin-$\frac{1}{2}$ chain discussed above. We take the Hamiltonian to be (setting $J = 1$)

$$H = \sum_i \mathbf{S}_{2i} \cdot \mathbf{S}_{2i+1} + \lambda \sum_i \mathbf{S}_{2i-1} \cdot \mathbf{S}_{2i} \tag{5.12}$$

and seek an expansion in powers of λ. Each dimer is now treated as a single site or 'object', with four quantum states, and the perturbation V couples adjacent dimers. This is illustrated in Figure 5.4. Series are now computed in the usual way: we consider a sequence of clusters of $1, 2, 3 \cdots$ dimers, evaluate the ground-state energy, or other quantity, as a power series in λ, and then subtract the sub-graph contributions to give reduced energies. These are then combined, weighted with lattice constants, to obtain the bulk energy per site. In Figure 5.4(c) we illustrate this for the cluster of two dimers. In this case we end up with a 2×2 matrix, and the ground-state energy is obtained exactly as

$$E_0 = -\tfrac{1}{2} - \tfrac{1}{4}\lambda - \left(1 - \tfrac{1}{2}\lambda + \tfrac{1}{4}\lambda^2\right)^{1/2} \tag{5.13}$$

In Table 5.3 we give the series for the bare and reduced energies for clusters containing up to four dimers. As can be seen the leading order in the reduced energies is $\lambda^{2(p-1)}$, where p is the number of sites. Thus the data in Table 5.3 suffice to give the energy series correct to λ^7. The series for the bulk ground state

Table 5.3. *Energy and reduced energy for low-order clusters in the dimer expansion for the alternating Heisenberg chain. The variable $y = \lambda/8$ is used to avoid large denominators.*

Cluster	Energy and reduced energy
•	$E_1 = -3/4, \quad \epsilon_1 = -3/4$
•—•	$E_2 = -\frac{3}{2} - 6y^2 - 12y^3 - 6y^4 + 60y^5 + 228y^6 + 168y^7 + \cdots$
	$\epsilon_2 = -6y^2 - 12y^3 - 6y^4 + 60y^5 + 228y^6 + 168y^7 + \cdots$
•—•—•	$E_3 = -\frac{9}{4} - 12y^2 - 24y^3 - 32y^4 - \frac{176}{3}y^5 - \frac{1712}{9}y^6 + \frac{11008}{27}y^7 + \cdots$
	$\epsilon_3 = -20y^4 - \frac{536}{3}y^5 - \frac{5816}{9}y^6 + \frac{1936}{27}y^7 + \cdots$
•—•—•—•	$E_4 = -3 - 18y^2 - 36y^3 - 58y^4 - \frac{532}{3}y^5 - \frac{7268}{9}y^6 - \frac{70549}{27}y^7 + \cdots$
	$\epsilon_4 = -\frac{1792}{9}y^6 - \frac{9781}{3}y^7 + \cdots$

energy per site, correct to λ^7, is then

$$E_0/N = -\frac{3}{8} - \frac{3}{64}\lambda^2 - \frac{3}{256}\lambda^3 - \frac{13}{4096}\lambda^4 - \frac{89}{49152}\lambda^5$$
$$-\frac{463}{393216}\lambda^6 - \frac{81557}{113246208}\lambda^7 + \cdots \tag{5.14}$$

This series is currently known to order λ^{27}!

From this series we could, of course, use Padé approximants to evaluate the energy of the uniform chain ($\lambda = 1$) to high accuracy. Rather than present such results, we show, in Figure 5.5, the sequence of partial sums, up to terms λ^{27}. As can be seen the convergence is uniform, but relatively slow. This is not too surprising since the uniform chain has long-range correlations and we are starting from a totally uncorrelated dimer state. For comparison we show similar results from an Ising expansion, which we have computed through order λ^{30}. Convergence is more rapid in this case.

Dimer expansions can be also be useful for elementary excitations. A simple physical picture can be given. In the unperturbed limit of decoupled dimers, the ground state has each dimer in its singlet state $|0\rangle$, with energy $-3J/4$. The lowest excited states are those in which a single dimer, at any position, is excited into one of its three triplet states $|t^+\rangle, |t^0\rangle, |t^-\rangle$. These excitations, which have been termed 'triplons', cannot propagate and hence there will be a flat band (highly degenerate level) with an energy $+J$ above the ground state. When the interaction between dimers is 'turned on', such a triplon can propagate through the lattice, and will become a quasiparticle with dispersion, generating a band of quasiparticle states. The minimum energy gap will remain nonzero for any $\lambda < 1$, but will close at $\lambda = 1$.

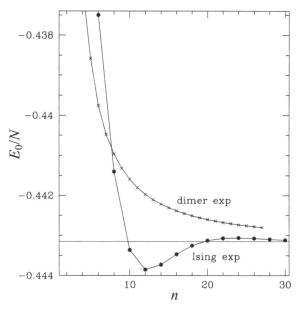

Fig. 5.5. Convergence of partial sums of the first n terms from dimer and Ising expansions for the $S = \frac{1}{2}$ Heisenberg chain ground state energy. The horizontal line is the exact result $\frac{1}{4} - \ln 2 = -0.443147$.

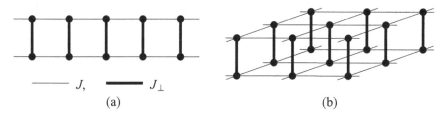

Fig. 5.6. (a) A two-leg spin ladder and (b) a spin bilayer. The exchange within chains or planes (J) is taken to be antiferromagnetic, while the interchain or interplane exchange ($J_\perp$) can have either sign.

We will not show specific results for this example, but will use dimer expansions for excitations in later subsections.

5.2.2 Coupled chains and coupled planes

We continue with two examples which show interesting new physics, the cases of two coupled spin chains or planes (Figure 5.6). There has been a great deal of theoretical work on these systems, using a variety of methods. There are also experimental realizations: $SrCu_2O_3$, $Cu_2(C_5H_{12}N_2)_2Cl_4$ and $KCuCl_3$ are well

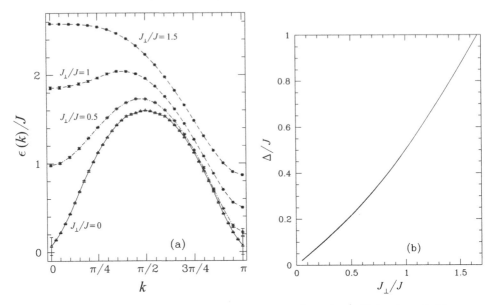

Fig. 5.7. (a) Elementary spin-triplet spectra $\epsilon(k)$ for spin-$\frac{1}{2}$ Heisenberg ladder. (b) The minimum triplet energy gap Δ/J versus $J_\perp/J$. The results are estimated from several integrated differential approximants to the dimer series up to order $(J/J_\perp)^{15}$.

approximated as spin-$\frac{1}{2}$ ladders while the cuprate superconductor $YBa_2Cu_3O_{6-x}$ contains pairs of weakly coupled CuO_2 planes.

The Hamiltonian can be taken as

$$H = J \sum_{\langle ij \rangle} \mathbf{S}_i \cdot \mathbf{S}_j + J_\perp \sum_{[kl]} \mathbf{S}_k \cdot \mathbf{S}_l \qquad (5.15)$$

where the first sum is over the interactions within chains or planes and the second is between chains or between planes. We take $J, J_\perp > 0$, i.e. both antiferromagnetic.

Interest in ladder systems began when it was argued that the excitation spectrum should be gapped, unlike the individual chains which are gapless (see e.g. Dagotto and Rice, 1996). This is not surprising in the limit $J_\perp \gg J$, when the spins on rungs will form $S = 0$ dimers, but it appears to be true for any finite $J_\perp$. The elementary excitations will be triplons, excitations on a single rung, which can propagate along the ladder via the J interaction. This system has been studied using both Ising and dimer expansions (Oitmaa *et al.*, 1996). We show, in Figure 5.7, the triplon dispersion curves and the spin gap versus $J_\perp/J$, obtained from these expansions.

Interestingly it appears to be the case that multi-leg ladders are gapless with the number of legs n odd and gapped with n even, with the gap decreasing with

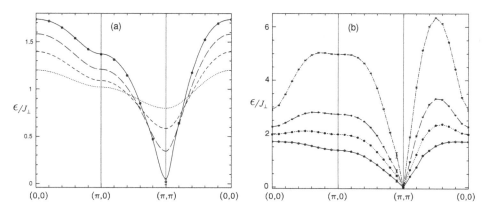

Fig. 5.8. Elementary spin-triplet spectra $\epsilon(k_x, k_y)$ for high-symmetry lines in the Brillouin zone for the spin-$\frac{1}{2}$ Heisenberg bilayer: (a) $\alpha = 10, 5, 10/3, 2.537$ shown in the figure from the top to the bottom at (π, π), respectively; and (b) $\alpha < \alpha_c$ for $\alpha = 0.5, 1, 1.5, 2.5$ (shown in the figure from the top to the bottom at $(\pi, 0)$, respectively). From Zheng (1997).

increasing n and vanishing in the limit $n \to \infty$, where an anisotropic square lattice is reached. This system is itself interesting, concerning the onset of long-range order. We refer the interested reader to the original literature for further information.

Turning to the case of two coupled spin-$\frac{1}{2}$ planes, we start with the following observation. In the limit $J_\perp \gg J$ we would again expect $S = 0$ dimers on the rungs, a gapped spectrum, and short-range order. On the other hand for $J_\perp = 0$ each plane has long-range order and a gapless spectrum. We would therefore expect a quantum phase transition at some value $\alpha_c = (J_\perp/J)_c$, between these two phases. This is borne out by many calculations, of various kinds. The most extensive series study is by Zheng (1997), who finds $(J_\perp/J)_c = 2.537(5)$. We show, in Figure 5.8, dispersion curves for elementary spin-triplet excitations in the Brillouin zone obtained by Ising expansions (for $\alpha < \alpha_c$) and dimer expansions (for $\alpha > \alpha_c$). We note the existence of a gap in the spectrum in the dimer phase ($\alpha > \alpha_c$), which closes at α_c and remains zero in the Néel phase at (π, π) for all $\alpha < \alpha_c$.

We are aware of two experimental realizations of spin-$\frac{1}{2}$ bilayer systems. Neutron scattering data for the high T_c material $YBa_2Cu_3O_{6.2}$ (Reznik *et al.*, 1996) suggests a ratio $J_\perp/J \simeq 0.08$, which is well within the gapless phase. The material $BaCuSi_2O_6$ on the other hand, lies in the gapped phase with $J_\perp/J \simeq 7.7$ (Jaime *et al.*, 2004).

When more layers are added there is again an 'odd–even' effect. Systems with an odd number of layers remain in the Néel phase for any $J_\perp/J$, whereas those with an even number show a quantum phase transition. Zheng (1999) has studied such multi-layer systems using series methods.

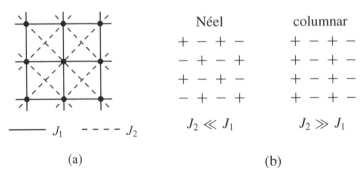

Fig. 5.9. (a) The J_1–J_2 square lattice antiferromagnet; (b) the ordered phases for small and large J_2. These are respectively two- and four-fold degenerate.

5.3 The J_1–J_2 square lattice antiferromagnet

In the preceding two sections we have shown how Ising and dimer expansions can be used to study zero-temperature properties of quantum antiferromagnets. In this section we will use these approaches to investigate a richer system, a spin-$\frac{1}{2}$ square lattice with both nearest- and next-nearest-neighbour antiferromagnetic exchange interactions. This is illustrated in Figure 5.9(a) and has the Hamiltonian

$$H = J_1 \sum_{\langle ij \rangle} \mathbf{S}_i \cdot \mathbf{S}_j + J_2 \sum_{[kl]} \mathbf{S} \cdot \mathbf{S}_l \tag{5.16}$$

For $J_2 = 0$ the system has Néel order, with non-zero staggered magnetization M_s in some spontaneously chosen direction. As the frustrating J_2 interaction is increased, it will destabilize this order, which will vanish at a critical point $(J_2/J_1)_{c_1} \simeq 0.4$. For large J_2, on the other hand, the system will order in a collinear phase in a 'columnar' pattern, as shown in Figure 5.9(b). The (appropriately defined) staggered magnetization will be non-zero. This state will be destabilized by decreasing J_2 and will vanish at a second transition point $(J_2/J_1)_{c_2} \simeq 0.6$. The so-called 'intermediate phase' (or phases?) is still not well understood, despite much recent work.

For the corresponding classical model there is a single transition point, at $(J_2/J_1)_c = 0.5$. The splitting of this point, and the existence of an intermediate 'spin-liquid' phase is a striking effect of strong quantum fluctuations in two dimensions.

The first extensive studies of this model, in the late 1980s, were motivated by the destruction of Néel order in the cuprate high-T_c superconductors upon doping, with the J_2 interaction mimicking the frustration effect of doping. Subsequent work, until recently, has been entirely theoretical. Recently, however, two materials have been discovered, Li_2VOSiO_4 and Li_2VOGeO_4, which appear to be good examples of quasi two-dimensional J_1-J_2 square lattice systems (Melzi *et al.*, 2000; Rosner *et al.*, 2002). Current evidence suggests that $J_2/J_1 \sim 5 - 10$ for these materials, putting them well into the columnar region of the phase diagram.

Early work, mostly based on exact diagonalizations, gave clear evidence of a gapped intermediate phase, in the range $J_2/J_1 \simeq 0.4 - 0.6$, but was unable to locate the critical points with any precision and was unable to shed much light on the nature of the intermediate phase. More recent work, using quantum Monte Carlo approaches and series expansions, has partially but not completely resolved this. We will describe the series work in the following subsections.

5.3.1 The Néel and columnar phases

The ordered phases of the system are most conveniently studied by Ising expansions. In the Néel region, the Hamiltonian is written as

$$H = H_0 + \lambda(V_1 + V_2) \tag{5.17}$$

where

$$H_0 = -\sum_{\langle ij \rangle} S_i^z S_j^z + g \sum_{[kl]} S_k^z S_l^z \tag{5.18a}$$

$$V_1 = \frac{1}{2} \sum_{\langle ij \rangle} (S_i^+ S_j^+ + S_i^- S_j^-) \tag{5.18b}$$

$$V_2 = \frac{g}{2} \sum_{[kl]} (S_k^+ S_l^- + S_k^- S_l^+) \tag{5.18c}$$

In obtaining this form we have carried out a spin rotation on sub-lattice B, set $J_1 = 1$ and introduced a parameter $g = J_2/J_1$. The expansion parameter λ is, as usual, the exchange anisotropy. The series must be extrapolated to $\lambda = 1$. For the large J_2 region, where we expect columnar order, a slightly different decomposition of the Hamiltonian is needed.

We have carried out such expansions for various ground state properties, to order λ^{10} (Oitmaa and Zheng, 1996), using the methods described previously. Figure 5.10 shows the order parameter (staggered magnetization) at $\lambda = 1$, as a function of the ratio $g = J_2/J_1$. The following observations can be made. The series clearly show the existence of long-range order for $g \lesssim 0.4$ and $g \gtrsim 0.6$, and the absence of magnetic order in the intermediate region. However, the series are rather irregular and difficult to analyse and our error bars (confidence limits) are quite large near the transition points. Series for second and third neighbour correlators indicate that these vanish as $g \to 0.4$ from below and $g \to 0.6$ from above, confirming the above picture.

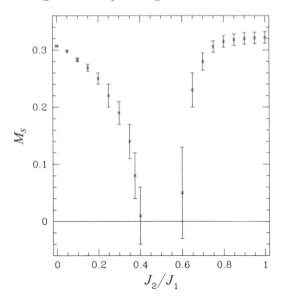

Fig. 5.10. The staggered/columnar magnetization M_s for the J_1–J_2 model, versus J_2/J_1. From Oitmaa and Zheng (1996).

5.3.2 The intermediate phase

While there is general aqreement on the existence of a magnetically disordered ('spin liquid') intermediate phase in the approximate range $0.4 < g < 0.6$, the nature of the phase remains elusive. Quantum Monte Carlo calculations have favoured, variously, a dimerized phase, a 'resonating plaquette' phase, or some combination of the two. In this sub-section we describe recent series work.

The first results come from series for the ground-state energy, from expansions about separated dimers or plaquettes [Figure 5.11(a)]. These results, extrapolated to $\lambda = 1$, are compared to energies obtained from Ising expansions in Figure 5.11(b). The energies of dimer and plaquette phases are very similar, and it is difficult from these results to favour one or the other. There appears to be a smooth joining at $g \simeq 0.4$, suggesting a second-order transition and a crossing at $g \simeq 0.6$, suggesting that the intermediate/columnar transition is first order.

Next we investigate the triplet excitation spectrum, which is obtained via a dimer expansion to order 8. In Figure 5.12 we show dispersion curves for $g = 0.4, 0.5, 0.6$ along symmetry lines in the reduced Brillouin zone of the dimerized lattice. We note that the triplet gap for $g = 0.4$ becomes small at $\mathbf{k} = (0, \pi)$ (which corresponds to the Néel ordering wavevector (π, π) of the undimerized lattice), whereas at $g \simeq 0.6$ the minimum gap occurs at $\mathbf{k} = (0, 0)$ (which corresponds to $(\pi, 0)$ of the original lattice, the wavevector for columnar order). This is further emphasized in Figure 5.13(a), where we show the variation of these two gaps with g. The

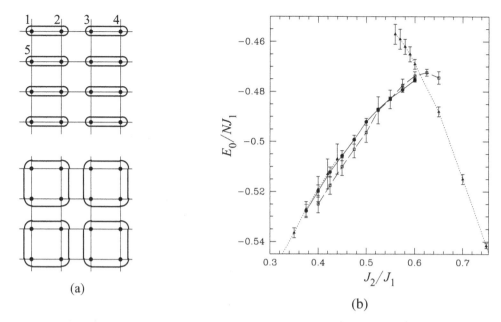

(a)

(b)

Fig. 5.11. (a) Starting states for dimer and plaquette expansions for J_1–J_2 model; (b) Comparison of ground-state energies obtained from the Ising expansion (dotted lines), the columnar dimer expansion (dashed lines), and the plaquette expansion (solid lines). From Singh *et al.* (1999).

conclusion is that the triplet gap is non-zero throughout the intermediate phase, as expected for a phase with no long-range order. The 'columnar gap' appears to vanish at $g \simeq 0.62$, whereas the Néel gap vanishes at $g \simeq 0.35$, rather that at $g \simeq 0.4$. This suggests a more complex scenario for the Néel/intermediate transition.

To shed further light on this possibility we have derived series for two different dimer order parameters.

$$D_x = \langle \mathbf{S}_2 \cdot \mathbf{S}_3 \rangle - \langle \mathbf{S}_1 \cdot \mathbf{S}_2 \rangle \qquad (5.19)$$

$$D_y = \langle \mathbf{S}_1 \cdot \mathbf{S}_5 \rangle - \langle \mathbf{S}_1 \cdot \mathbf{S}_2 \rangle \qquad (5.20)$$

[see Figure 5.11(a) for the notation]. The variation of these quantities with g is shown in Figure 5.13(b). It appears that D_y is nearly zero for $g \lesssim 0.5$, and then becomes non-zero, while D_x is non-zero throughout but vanishes around 0.4. This suggests the development of a 'plaquette like' phase at $g \simeq 0.5$, with alternating correlations in both directions.

Finally we have computed series expansions for several generalized susceptibilities, with respect to fields which break different spatial symmetries (Sushkov

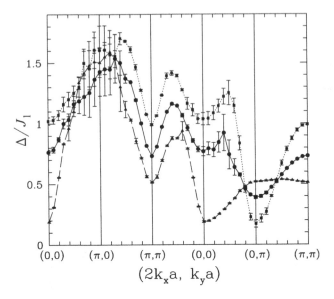

Fig. 5.12. Triplet excitation spectrum in the quantum disordered phase for $J_2/J_1 = 0.4$ (dotted line), 0.5 (solid line) and 0.6 (dashed lined). Note that we work in the Brillouin zone of the dimerized lattice, where $\mathbf{k}_{AF} = (0, \pi)$ and $\mathbf{k}_{COL} = (0, 0)$. From Kotov *et al.* (2000).

et al., 2001). The 'dimer susceptibility' χ_D is the response to a field term

$$F \sum_{ij} (-1)^i \left(S_{i,j}^x S_{i+1,j}^x + S_{i,j}^y S_{i+1,j}^y \right) \tag{5.21}$$

in the Hamiltonian, while the 'plaquette susceptibility' χ_p gives the response to a field term

$$F \sum_{ij} (-1)^j \mathbf{S}_{i,j} \cdot \mathbf{S}_{i,j+1} \tag{5.22}$$

Figure 5.14(a) shows a plot of the inverse dimer susceptibility, obtained from Ising expansions in the Néel phase. Although the error bars are large, χ_D appears to diverge somewhere in the region $g = 0.35 - 0.4$, corresponding to the onset of dimerization. Figure 5.14(b) shows the inverse plaquette susceptibility. There is evidence of a divergence at $g \simeq 0.5$, consistent with the behaviour of D_y above.

5.3.3 Summary

Series studies of the J_1–J_2 frustrated antiferromagnet have provided evidence for a complex and subtle phase structure. While the existence of ordered Néel and columnar phases and a disordered intermediate phase are clear, there are signs that

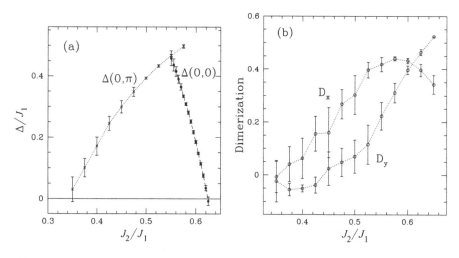

Fig. 5.13. The minimum gap (a) and the dimerization (b) versus J_2/J_1 obtained from dimer expansions.

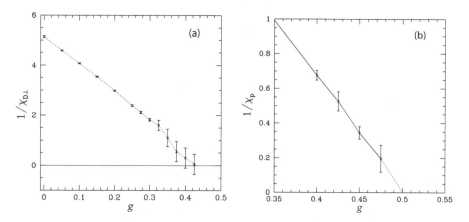

Fig. 5.14. (a) The inverse dimer susceptibility, calculated in the Néel phase from Ising expansions; (b) The inverse of plaquette susceptibility, calculated via dimer expansion about a simple columnar dimerized state. From Sushkov *et al.* (2002).

the intermediate phase itself may consist of two parts. There are also indications that the Néel phase may develop partial dimer order before the magnetization vanishes. Figure 5.15 shows a schematic possible phase diagram with a sequence of four quantum phase transitions. Whether this proves to be true remains to be seen.

Another interesting and controversial issue concerns the columnar phase at finite temperature. This state breaks the 90° rotational symmetry of the square lattice and this broken symmetry can, in principle, survive at finite temperatures, the Mermin–Wagner theorem notwithstanding. Chandra *et al.* (1990) have argued in favour of

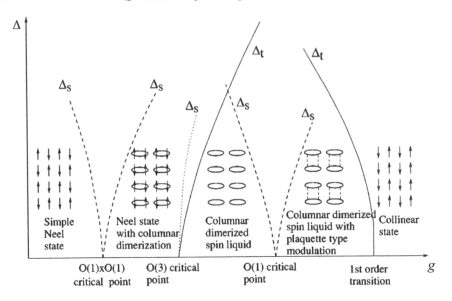

Fig. 5.15. Schematic phase diagram and the excitation spectra of the J_1–J_2 model. Solid lines show the triplet gap, dashed lines show the gaps of the relevant singlets, the dotted line shows the gap of the irrelevant singlet. From Sushkov *et al.* (2001).

a finite temperature Ising transition to a high-temperature disordered phase with restored spatial symmetry. We will return to this issue in Chapter 7.

5.4 Other systems

As final examples of novel magnetic systems, which have received considerable study in recent years and where series expansion methods have played a significant role, we will discuss the materials CaV_4O_9 (CAVO) and $SrCu_2(BO_3)_2$ (SCBO).

5.4.1 The CAVO system

In 1995 Taniguchi and co-workers found that the quasi-two-dimensional material CaV_4O_9 has an energy gap for spin excitations. This was subsequently confirmed by a number of other groups, using a variety of experimental probes. Since the material contains layers of $S = \frac{1}{2}$ V^{4+} ions, the existence of a spin gap indicates the formation of local $S = 0$ complexes (dimers or plaquettes). Much theoretical work has been done to identify the nature of this non-magnetic ground state.

The structure of the V^{4+} layers can be represented as a planar network, shown in Figure 5.16. This is a square lattice from which 1/5 of the sites have been removed.

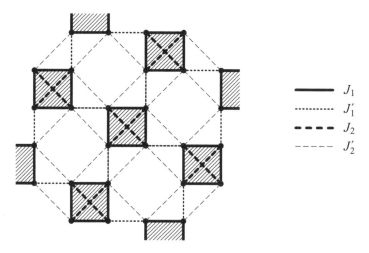

Fig. 5.16. The network of $S = \frac{1}{2}$ V^{4+} ions in CAVO. The shaded squares show elementary plaquettes, the thin dashed squares are 'metaplaquettes'. Four exchange interactions are used in model studies.

Model studies have involved up to four different exchange constants, as shown in Figure 5.16. Independent attempts to calculate the exchange constants from *ab initio* density functional theory are not in complete agreement. However it appears that all four exchange constants J_1, J_1', J_2, J_2' (shown in Figure 5.16) are antiferromagnetic and of comparable magnitude, roughly 9, 9, 4, 15 meV.

The first model calculations considered only the nearest-neighbour interactions J_1, J_1' and identified a phase diagram with three phases: a plaquette phase, on the shaded plaquettes of Figure 5.16, for small J_1', a dimer phase for larger J_1' and a Néel phase in a window around $J_1' \simeq J$. However the failure to fit susceptibility data and the *ab initio* calculations indicate the need to include frustrating second-neighbour terms, and later calculations have used all four interactions.

We have derived a number of $T = 0$ series expansions, starting from Ising, dimer and plaquette states (Zheng *et al.*, 1997, 1998). The phase diagram is defined within the three-parameter space (x, y_1, y_2) where

$$x = J_1/J_2', \quad y_1 = J_1'/J_1, \quad y_2 = J_2/J_1 \qquad (5.23)$$

Surfaces within this space then separate different phases. Some representative results are shown below. Figure 5.17(a) shows approximate phase boundaries between Néel and 'metaplaquette' phases. The later is a phase based on the larger plaquettes shown in Figure 5.16, which is the expected phase when the interaction J_2' is dominant, as appears to be the case with CAVO. Figure 5.17(b) shows a selection of triplet excitation spectra for the choice $x = 0.4$, $y_1 = 1$ and various y_2.

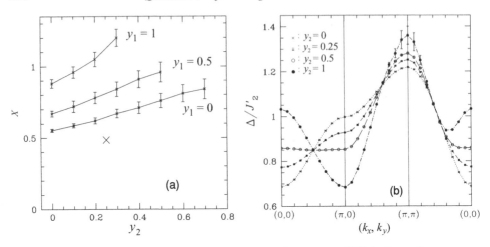

Fig. 5.17. (a) Phase boundary separating metaplaquette and Néel phases for various y_2. The cross marks the approximate position of CaV$_4$O$_9$ in the parameter space; (b) Triplet excitation spectra obtained from a metaplaquette expansion for $x = 0.4$, $y_1 = 1.0$, and various y_2. From Zheng *et al.* (1998).

The results for small y_2 are similar to the experimental dispersion curves of CAVO obtained by neutron scattering.

5.4.2 The SCBO system

A second novel material, which is also quasi-two-dimensional, and was found recently to have a spin gap, is SrCu$_2$(BO$_3$)$_2$ (Kageyama *et al.*, 1999). The magnetic component is Cu^{2+}, with spin $S = \frac{1}{2}$, and so the absence of magnetic order in the ground state is again attributable to local singlet formation, in this case via dimers. Figure 5.18(a) shows, schematically, the structure of SCBO and the dominant exchange interactions J, J'. In the real material it appears that $J \simeq 7$meV and $J'/J \simeq 0.67$. However, the adequacy of this two parameter model, and the precise values of the parameters, remain somewhat uncertain. Nevertheless, if we use this model, it is easy to see that it is equivalent to the so-called Shastry–Sutherland model (Shastry and Sutherland, 1981a). This model, proposed for theoretical reasons, is a square lattice with alternating frustrating diagonal interactions on every second plaquette [Fig.5.18(b)].

The Hamiltonian of the model is then taken as

$$H = J' \sum_{\langle ij \rangle} \mathbf{S}_i \cdot \mathbf{S}_j + J \sum_{[kl]} \mathbf{S}_k \cdot \mathbf{S}_l \tag{5.24}$$

where the first sum is over the bonds of the square lattice and the second is over the

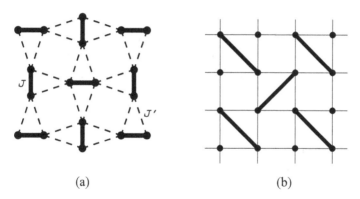

(a)   (b)

Fig. 5.18. (a) Structure of the SCBO lattice. Cu^{2+} sites only are shown. The thick
and thin lines represent the interactions J, J'; (b) The Shastry–Sutherland lattice.

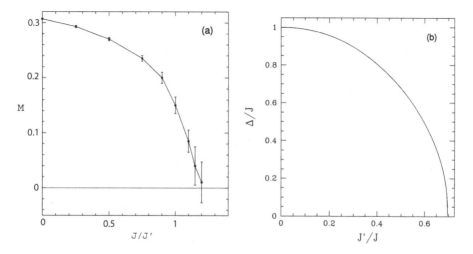

Fig. 5.19. (a) Staggered magnetization obtained from Ising expansions in the Néel
phase; (b) minimum triplet gap versus J'/J, obtained from dimer expansions. From
Zheng et al. (2002b).

diagonal interactions, which, in SCBO, are actually the shorter bonds and hence
expected to be the stronger. Shastry and Sutherland showed that the simple product
state formed by singlet dimers on the J-bonds is an exact eigenstate of H for
all J, J' and that it is the ground state at least for $J/J' > 2$. On the other hand
for $J = 0$ the ground state will have Néel order. Thus we anticipate at least one
quantum phase transition as the diagonal interaction is increased.

Following earlier work which estimated the transition point to be at $J/J' \simeq$
1.43 ± 0.02, and to be first order, we have studied this problem more closely using
longer series of various kinds (Zheng et al., 2002b). From this, and other work, it

appears that there may be two separate transitions: a transition (probably second order) from the Néel phase to an intermediate phase at $J/J' \simeq 1.2$, followed by a first-order transition to the dimer phase at $J/J' \simeq 1.5$. Evidence for this scenario is shown in Figure. 5.19(a), which shows the staggered magnetization in the Néel phase vanishing at $J/J' \simeq 1.2$, and in Figure. 5.19(b) which shows the minimum triplet gap Δ in the dimer phase vanishing at $J/J' \simeq 1.44$.

The nature of the intermediate phase remains elusive. Various suggestions have been made: helical order, columnar dimer order, plaquette order. We have computed the ground-state energies for these states from series expansions. The results are inconclusive, as none of these fall obviously below the Néel energy. The columnar dimer phase appears to be the strongest candidate.

How do these questions pertain to the real material $SrCu_2(BO_3)_2$? Early estimates, obtained by fitting high-temperature data, gave $(J/J')_{SCBO} \simeq 1.47$, very close to the limit of stability of the dimer phase. It was suggested that some of the unusual properties of the material were a consequence of this closeness. A more recent estimate, including also interactions between planes, gives 1.66, which puts the material clearly in the dimer phase. This question, as well as the precise nature of the phase diagram of the Shastry–Sutherland model, remains unresolved.

5.5 Open questions

There has been a great deal of progress in recent years, both experimentally and theoretically, in exploring the properties of strongly correlated systems in low dimensions. Experimentalists have shown great ingenuity in synthesizing new materials made up of quasi-one dimensional chains or ladders, or quasi-two dimensional layer systems such as the superconducting cuprates. Theorists have been exploring beyond the simple Heisenberg model to look at systems involving competing interactions and frustration, giving rise to new patterns of spontaneous symmetry breaking and the formation of new phases such as spin liquids or valence bond solid systems breaking translational symmetry. We have discussed some examples in this chapter.

Many open questions remain. The nature of the intermediate phase in the $J_1–J_2$ model on the square lattice is still controversial, as is the nature of the phase transitions at the boundary of this phase. Similar questions arise for other systems, such as the SCBO model, or the Shastry–Sutherland model on the square lattice (Shastry and Sutherland, 1981a).

Recently, theoretical predictions have been made of new universality classes of quantum phase transitions in two-dimensional systems which lie beyond the

Landau–Ginzburg–Wilson paradigm, where fractionalized excitations such as *spinons* may become deconfined at the transition (Laughlin, 1998; Senthil *et al.*, 2004). The challenge now is to find specific model systems where these predictions may be realized, and to verify numerically whether or not they occur. These are examples of the many interesting questions that remain to be explored in the years ahead.

6

Correlators, dynamical structure factors and multi-particle excitations

6.1 Introduction

In the previous chapters we showed how series expansion methods can be used to study ground state properties and elementary excitations in simple (and not so simple) antiferromagnets. In the present chapter we extend this to a consideration of more complex properties. We begin by discussing various spin-spin correlation functions, or 'correlators'. These play an important role in characterizing the nature of the ground state and, as we shall show, can be used to locate quantum phase transition points. The correlators can be combined to form the dynamical two-spin structure factors $S(\mathbf{k}, \omega)$, which are measured in inelastic neutron scattering experiments. We will discuss how these can be computed via series methods, and show some results for the integrated (static) forms.

While one-particle excited states are usually the dominant excitations in quantum systems, there are situations in which two-particle excitations, including bound states, and more general multi-particle excitations also play an important role. There are a number of experimental probes which are beginning to show features associated with multi-particle continuum and bound states, including two-magnon Raman spectroscopy, photo-emission and inelastic neutron scattering. For example, Tennant *et al.* (2003) have measured two-magnon states in copper nitrate, a quasi one-dimensional antiferromagnet (see Section 6.4.4). Thus one can hope to build up a detailed picture of the dynamics of these quasiparticle excitations, and to construct an effective Hamiltonian to describe them. In the last part of the chapter we outline how one may calculate series expansions for the spectrum of multi-particle excitations in a quantum lattice model at $T = 0$, and hence how to calculate their contributions to the dynamical structure factors which are measured in scattering experiments.

6.2 Two-spin correlators for the Heisenberg antiferromagnet

We consider the simple antiferromagnet, with exchange anisotropy, discussed in Section 5.1. The Hamiltonian is

$$H = \sum_{\langle ij \rangle} S_i^z S_j^z + \lambda \sum_{\langle ij \rangle} (S_i^x S_j^x + S_i^y S_j^y) \tag{6.1}$$

The basic quantities are the correlators

$$C(\mathbf{r}) = \langle \mathbf{S_i} \cdot \mathbf{S_{i+r}} \rangle_0 \tag{6.2}$$

where the subscript 0 denotes a ground state expectation value. By translational invariance this quantity is independent of the site $\mathbf{i}$. It is useful to separate (6.2) into 'longitudinal' and 'transverse' parts

$$C(\mathbf{r}) = C_l(\mathbf{r}) + C_t(\mathbf{r}) \tag{6.3}$$

where

$$C_l(\mathbf{r}) = \langle S_i^z S_{i+r}^z \rangle_0 \tag{6.4a}$$

$$C_t(\mathbf{r}) = \langle S_i^x S_{i+r}^x + S_i^y S_{i+r}^y \rangle_0$$

$$= \frac{1}{2} \langle S_i^+ S_{i+r}^- + S_i^- S_{i+r}^+ \rangle_0 \tag{6.4b}$$

The linked cluster approach can be used to obtain Ising expansions for these quantities, in powers of λ, by including a field term in the Hamiltonian

$$H = H_0 + \lambda V + h \sum_i S_i^\alpha S_{i+r}^\alpha \tag{6.5}$$

and obtaining the ground-state energy as

$$\mathcal{E}_0(\lambda, h) = E_0(\lambda) + N C_\alpha(r) h + \cdots \tag{6.6}$$

Alternatively, if we are using the orthogonal transformation method discussed in Section 4.6.3, we compute the expectation value directly using the wave function of the ground state. This latter approach is more convenient if one needs to compute many correlators for different $\mathbf{r}$.

Our group (Zheng and Oitmaa, 2001) have used this approach to compute short-range correlators (up to four lattice spacings) for the square and simple cubic lattices, to order λ^{14} and λ^{12} respectively. We refer the reader to that paper for detailed results.

Figure 6.1 shows nearest- and next-nearest-neighbour correlators for the square lattice, as functions of λ. We note that at the isotropic point $\lambda = 1$, where the Hamiltonian has full spin rotation symmetry, the correlators $\langle S^z S^z \rangle$ and $\langle S^x S^x \rangle$

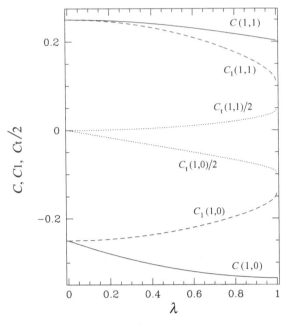

Fig. 6.1. Nearest-neighbour and next-nearest-neighbour longitudinal and transverse correlators for the anisotropic $S = \frac{1}{2}$ antiferromagnet on the square lattice. With no symmetry breaking $C_t(\mathbf{r}) = 2C_1(\mathbf{r})$.

remain unequal. This is a signature of spontaneous symmetry breaking, and a preferred z direction for the sublattice magnetization. On the other hand, a similar calculation for the J_1–J_2 model (Section 5.3) shows that the correlators become equal at $g \simeq 0.4$, where the broken symmetry is restored (Sushkov *et al.*, 2001).

6.3 Dynamical and static structure factors

One of the central issues in the theory of strongly correlated systems is the question whether well-defined quasiparticles exist, or not. This question is best addressed by studying the structure factor $S(\mathbf{k}, \omega)$, which gives the scattering probability of a probe particle at energy transfer ω and momentum transfer $\mathbf{k}$. It provides a powerful tool for studying the excitation spectrum in strongly correlated systems.

6.3.1 Formalism

In neutron scattering, the neutron couples magnetically to the atomic spin, and the scattering cross-section is proportional to the dynamical structure factor $S^{\alpha\beta}(\mathbf{k}, \omega)$, defined by (Marshall and Lovesey, 1971)

$$S^{\alpha\beta}(\mathbf{k}, \omega) = \frac{1}{2\pi N} \sum_{i,j} \int_{-\infty}^{\infty} e^{i(\omega t + \mathbf{k} \cdot (\mathbf{r}_i - \mathbf{r}_j))} \langle S_j^{\alpha}(t) S_i^{\beta}(0) \rangle dt \tag{6.7}$$

where i, j label lattice sites, $\alpha, \beta = x, y, z$ are Cartesian components, N is the number of scattering centres, and the angular bracket denotes the thermal expectation value, or at $T = 0$ the ground state expectation value. The time-dependent operator in (6.7) is, as usual,

$$S_j^{\alpha}(t) = e^{iHt} S_j^{\alpha}(0) e^{-iHt} \tag{6.8}$$

with $\hbar = 1$.

Integrating (6.7) over energy gives the 'integrated' or 'static' structure factor

$$S^{\alpha\beta}(\mathbf{k}) = \int_{-\infty}^{\infty} d\omega S^{\alpha\beta}(\mathbf{k}, \omega)$$
$$= \frac{1}{N} \sum_{i,j} e^{i\mathbf{k} \cdot (\mathbf{r}_i - \mathbf{r}_j)} \langle S_j^{\alpha} S_i^{\beta} \rangle \tag{6.9}$$

which is just the Fourier transform of the two-spin correlator, discussed in the previous section. This quantity would be measured in a neutron scattering experiment where all energies were included.

Often one is less interested in the total structure factors (6.7), (6.9) than in the contributions to them from particular classes of excitations, one-particle or multi-particle. We will show how these can be calculated via linked cluster expansions. To this end we introduce a more general notation for the dynamical structure factor at $T = 0$

$$S(\mathbf{k}, \omega) = \frac{1}{2\pi N} \sum_{i,j} \int_{-\infty}^{\infty} dt e^{i(\omega t + \mathbf{k} \cdot (\mathbf{r}_i - \mathbf{r}_j))} \langle \psi_0 | \mathcal{O}_j^{\dagger}(t) \mathcal{O}_i(0) | \psi_0 \rangle \tag{6.10}$$

where $|\psi_0\rangle$ is the ground-state wavefunction and $\mathcal{O}$ is a general operator, which depends on the scattering probe.

By introducing a complete set of eigenstates $|\psi_\Lambda\rangle$ of H between the operators in (6.10), using the time dependence as in (6.8), and integrating over time, it follows that

$$S(\mathbf{k}, \omega) = \sum_{\Lambda} S_{\Lambda}(\mathbf{k}, \omega) \tag{6.11}$$

i.e., a sum over *exclusive* structure factors (sometime also called spectral weights) $S_\Lambda(\mathbf{k}, \omega)$, given by

$$S_\Lambda(\mathbf{k}, \omega) = \frac{1}{N}\delta(\omega - E_\Lambda + E_0)\left|\sum_i \langle\psi_\Lambda|\mathcal{O}_i|\psi_0\rangle e^{i\mathbf{k}\cdot\mathbf{r}_i}\right|^2 \tag{6.12}$$

6.3.2 Calculation of structure factors

A series approach to the calculation of structure factors was first developed by Singh and Gelfand (1995), who computed the one-magnon spectral weight for the square lattice antiferromagnet. Instead of describing their method we will follow a more general formalism (Zheng *et al.*, 2003), which has the virtue that it can be used for multi-particle as well as for one-particle excitations. Later in this chapter we will give an example, for the case of the alternating Heisenberg chain, of the calculation of two-particle excitations and corresponding structure factors. For the moment, however, we limit the discussion to one-particle states.

Let us now consider the matrix element $\langle\psi_\Lambda|\mathcal{O}_i|\psi_0\rangle$ in (6.12). A one-particle eigenstate $|\psi_\Lambda\rangle$ of H can be labelled by a momentum $\mathbf{k}$, and can be written in the form

$$|\psi_\Lambda(\mathbf{k})\rangle = \frac{1}{\sqrt{N_c}}\sum_m e^{i\mathbf{k}\cdot\mathbf{r}_m}|\psi_\Lambda(m)\rangle \tag{6.13}$$

where $|\psi_\Lambda(m)\rangle$ is the state which evolves from the unperturbed one-particle excited state of H_0, at position m. N_c is the number of excitation sites, and can differ from the number of lattice sites N: for instance in a dimerized system each excitation covers two lattice sites. We can then write

$$\langle\psi_\Lambda(\mathbf{k})|\mathcal{O}_i|\psi_0\rangle = \frac{1}{\sqrt{N_c}}\sum_m e^{-i\mathbf{k}\cdot\mathbf{r}_m}\langle\psi_\Lambda(m)|\mathcal{O}_i|\psi_0\rangle \tag{6.14}$$

where we have introduced the 'exclusive matrix element'

$$\Omega_\Lambda(\delta) \equiv \langle\psi_\Lambda(m)|\mathcal{O}_i|\psi_0\rangle \tag{6.15}$$

which depends, because of translational invariance, only on the relative distance $\delta = \mathbf{r}_i - \mathbf{r}_m$. The exclusive structure factor (6.12) is given by

$$S_\Lambda(\mathbf{k}) = \frac{N_c}{N}\left|\sum_\delta \Omega_\Lambda(\delta)e^{i\mathbf{k}\cdot\delta}\right|^2 \tag{6.16}$$

This result forms the starting point of our linked cluster expansion.

It is easy to see that the exclusive matrix elements (6.15) allow a linked cluster expansion. If a cluster g is made up of two disconnected components g_1 and g_2, then

$$\Omega_\Lambda^g(i; m) = \Omega_\Lambda^{g_1}(i; m) + \Omega_\Lambda^{g_2}(i; m) \tag{6.17}$$

Through particle conservation, these matrix elements vanish unless i and m are in the same cluster, and hence one or other of the terms $\Omega_\Lambda^{g_1}$, $\Omega_\Lambda^{g_2}$ will be zero. It follows that the exclusive matrix elements for the bulk lattice can be obtained via a linked cluster expansion

$$\Omega_\Lambda = \sum_g c(g) \Omega_\Lambda^g \tag{6.18}$$

An efficient linked cluster algorithm for the calculation of the structure factors can now be formulated.

(i) Generate a list of connected clusters appropriate to the problem at hand, together with lattice constants and sub-graph data.

(ii) For each cluster g, construct matrices for the Hamiltonian H and operators $\mathcal{O}_i$, in the basis of eigenstate of H_0.

(iii) 'Block diagonalize' the matrix H by an *orthogonal* transformation, as in Section 4.5.1, and compute the exclusive matrix elements Ω_Λ for cluster g. Note that one cannot use Gelfand's similarity transformation here, since we need to use the orthogonal wavefunctions for the ground and excited states.

(iv) Subtract sub-cluster contributions to get the reduced matrix elements.

(v) Combine these via (6.18) to obtain the bulk quantities.

The one-particle structure factor can be obtained immediately and directly from (6.16). For the two-particle case further work is required, and we return to this in Section 6.5.

6.3.3 Example: the transverse field Ising chain

As a simple example to illustrate the calculation of the one-particle structure factor, we consider the transverse field Ising chain in the disordered phase. The Hamiltonian is, as in (4.12),

$$H = \sum_i (1 - \sigma_i^z) - \mu \sum_i (\sigma_i^+ \sigma_{i+1}^+ + \sigma_i^+ \sigma_{i+1}^- + \sigma_i^- \sigma_{i+1}^+ + \sigma_i^- \sigma_{i+1}^-). \tag{6.19}$$

We choose the operator

$$\mathcal{O}_i = \sigma_i^- \tag{6.20}$$

Our structure factor will then decribe spin fluctuation processes in the plane perpendicular to the applied field.

Let us consider explicitly the third cluster of Table 4.1, viz. g_3, $\overset{1}{\bullet}\!\!-\!\!\overset{2}{\bullet}\!\!-\!\!\overset{3}{\bullet}$. There are four states in the ground-state sector, and the Hamiltonian matrix is given in

Table 4.1. Using the TBOT method to block-diagonalize this (see Section 4.5.1) gives the ground-state wavefunction as

$$
\begin{aligned}
|\psi_0\rangle = {}& |+++\rangle + \tfrac{1}{4}(|--+\rangle + |+--\rangle)\mu \\
& + \tfrac{1}{16}(2|-+-\rangle - |+++\rangle)\mu^2 - \tfrac{1}{64}(|--+\rangle + |+--\rangle)\mu^3 \\
& - \tfrac{1}{512}(|+++\rangle + 12|-+-\rangle)\mu^4 - \tfrac{9}{2048}(|--+\rangle + |+--\rangle)\mu^5 \\
& - \tfrac{1}{8192}(39|+++\rangle + 6|-+-\rangle)\mu^6 + \cdots
\end{aligned}
\tag{6.21}
$$

The one-particle sector for this cluster also has four states. The states and the Hamiltonian matrix are given in Section 4.5.3. Again using the TBOT method to block-diagonalize this, we obtain three wavefunctions $|\psi_{1p}(m)\rangle$ which evolve from the unperturbed one-particle excited states at positions m ($m = 1, 2, 3$). In particular, the wavefunction $|\psi_{1p}(1)\rangle$ is

$$
\begin{aligned}
|\psi_{1p}(1)\rangle = {}& |-++\rangle + \tfrac{1}{4}|---\rangle\mu - \tfrac{1}{32}(|-++\rangle + |++-\rangle)\mu^2 \\
& + \tfrac{1}{64}(|+-+\rangle - |---\rangle)\mu^3 + \tfrac{3}{1024}(|-++\rangle + |++-\rangle)\mu^4 \\
& - \tfrac{1}{2048}(7|+-+\rangle + 9|---\rangle)\mu^5 \\
& + \tfrac{17}{163840}(|-++\rangle + |++-\rangle)\mu^6 + \cdots
\end{aligned}
\tag{6.22}
$$

From these wavefunctions one can easily obtain the matrix elements

$$
\Omega_{1p}(m, i) \equiv \langle\psi_{1p}(m)|\sigma_i^-|\psi_0\rangle
\tag{6.23}
$$

as

$$
\begin{aligned}
\Omega_{1p}(1, 1) = \Omega_{1p}(3, 3) &= 1 - \tfrac{1}{32}\mu^2 - \tfrac{5}{1024}\mu^4 + \tfrac{61}{16384}\mu^6 + \cdots \\
\Omega_{1p}(2, 2) &= 1 - \tfrac{1}{16}\mu^2 - \tfrac{13}{512}\mu^4 + \tfrac{113}{8192}\mu^6 + \cdots \\
\Omega_{1p}(1, 2) = \Omega_{1p}(3, 2) &= \tfrac{3}{64}\mu^3 - \tfrac{25}{2048}\mu^5 + \cdots \\
\Omega_{1p}(2, 1) = \Omega_{1p}(2, 3) &= -\tfrac{1}{64}\mu^3 + \tfrac{7}{2048}\mu^5 + \cdots \\
\Omega_{1p}(1, 3) = \Omega_{1p}(3, 1) &= \tfrac{1}{32}\mu^2 - \tfrac{3}{1024}\mu^4 - \tfrac{17}{16384}\mu^6 + \cdots
\end{aligned}
\tag{6.24}
$$

or, in the translationally invariant form $\Omega_{1p}(\delta)$

$$
\begin{aligned}
\Omega_{1p}(0) = \Omega_{1p}(1, 1) &+ \Omega_{1p}(2, 2) + \Omega_{1p}(3, 3) \\
&= 3 - \tfrac{1}{8}\mu^2 - \tfrac{9}{256}\mu^4 + \tfrac{87}{4096}\mu^6 + \cdots \\
\Omega_{1p}(1) = \Omega_{1p}(-1) &= \Omega_{1p}(1, 2) + \Omega_{1p}(2, 3) \\
&= \tfrac{1}{32}\mu^3 - \tfrac{9}{1024}\mu^5 + \cdots \\
\Omega_{1p}(2) = \Omega_{1p}(-2) &= \Omega_{1p}(1, 3) \\
&= \tfrac{1}{32}\mu^2 - \tfrac{3}{1024}\mu^4 - \tfrac{17}{16384}\mu^6 + \cdots
\end{aligned}
\tag{6.25}
$$

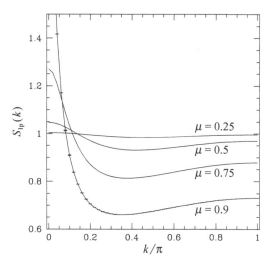

Fig. 6.2. The one-particle structure factor S_{1p} for the transverse field Ising chain at $\mu = 0.25, 0.5, 0.75$, and 0.9.

From this result, and similar calculations for other clusters, followed by the usual subtraction of sub-graph contributions, we obtain the one-particle reduced matrix element for the bulk lattice , to order μ^6, as

$$\Omega_{1p} \equiv \sum_{\delta} \Omega_{1p}(\delta)e^{i\delta k}$$

$$= 1 - \tfrac{1}{16}\mu^2 - \tfrac{47}{1024}\mu^4 - \tfrac{513}{16384}\mu^6 + \cdots + (\tfrac{1}{16}\mu^3 + \tfrac{1}{64}\mu^5 + \cdots)\cos(k)$$

$$+ (\tfrac{1}{16}\mu^2 + \tfrac{9}{256}\mu^4 + \tfrac{315}{32768}\mu^6 + \cdots)\cos(2k)$$

$$+ (\tfrac{1}{16}\mu^3 + \tfrac{13}{512}\mu^5 + \cdots)\cos(3k) + (\tfrac{59}{1024}\mu^4 + \tfrac{329}{16384}\mu^6 + \cdots)\cos(4k)$$

$$+ (\tfrac{27}{512}\mu^5 + \cdots)\cos(5k) + (\tfrac{1589}{32768}\mu^6 + \cdots)\cos(6k) + \cdots \tag{6.26}$$

This is obtained from clusters up to and including six sites.

The one-particle structure factor is then simply $|\Omega_{1p}|^2$. In Figure 6.2 we show a plot of $S(k)$ for several values of μ. As is evident, $S(k)$ is flat over most of the Brillouin zone but begins to develop a peak at $k = 0$ as μ increases towards the critical point $\mu = 1$.

6.3.4 The $S = \tfrac{1}{2}$ square lattice antiferromagnet

In the previous sub-section we showed, in some detail, how the linked cluster expansion method can be used to compute the one-particle structure factors for a simple model. The same approach can be used for more complex models, such as the Heisenberg antiferromagnet. The $S = \tfrac{1}{2}$ antiferromagnet on the square lattice is relevant to a number of real materials and this motivates us to present here some

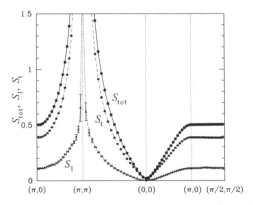

Fig. 6.3. The various integrated structure factors S_{tot} (unpolarized), S_{t} (transverse) and S_{l} (longitudinal) along high-symmetry cuts through the Brillouin zone for the Heisenberg antiferromagnet on the square lattice.

recent results on the structure factors for this model. The one-magnon excitation spectrum of the model, and comparisons with real materials, were discussed in Section 5.1.3.

From the general definition (6.7) one is led to consider distinct longitudinal and transverse structure factors defined by

$$S_{\text{l}}(\mathbf{k}, \omega) = S^{zz}(\mathbf{k}, \omega)$$
$$S_{\text{t}}(\mathbf{k}, \omega) = S^{xx}(\mathbf{k}, \omega) + S^{yy}(\mathbf{k}, \omega). \tag{6.27}$$

For an isotropic system, in the absence of long-range magnetic order, $S_{\text{t}}(\mathbf{k}, \omega) = 2S_{\text{l}}(\mathbf{k}, \omega)$. This will not be the case if magnetic order is present; in particular for the square lattice at $T = 0$. Scattering experiments with polarized neutrons can distinguish between S_{l} and S_{t}, but unpolarized neutrons will measure the total structure factor $S_{\text{tot}} = S_{\text{l}} + S_{\text{t}}$.

The dominant contribution to the transverse dynamical structure factor will arise from one-magnon excitations, and $S_{\text{t}}(\mathbf{k}, \omega)$ will have the form

$$S_{\text{t}}(\mathbf{k}, \omega) = A_1(\mathbf{k})\delta(\omega - \omega(\mathbf{k})) + S_{\text{inc}}(\mathbf{k}, \omega) \tag{6.28}$$

where $A_1(\mathbf{k})$ is the one-magnon spectral weight or exclusive structure factor and $S_{\text{inc}}(\mathbf{k}, \omega)$ is a smooth incoherent background term, arising from multi-magnon processes.

We have recently obtained 14-term series (Ising expansions) for the various static structure factors of the square lattice (Zheng *et al.* (2005)). The reader is referred to the original paper for many details. We present here only the main results. Figure 6.3 shows the total, transverse, and longitudinal structure factors along high symmetry lines in the Brillouin zone.

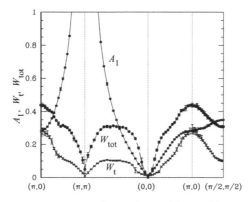

Fig. 6.4. The one-magnon structure factor A_1, and the multi-magnon relative spectral weights W_t and W_{tot} for the spin-$\frac{1}{2}$ Heisenberg antiferromagnet on the square lattice.

Note that all structure factors diverge at the Néel wavevector $\mathbf{k} = (\pi, \pi)$. If the Néel state were the exact ground state the static longitudinal structure factor would vanish, except for a δ-function peak at (π, π). The actual shape reflects the additional contribution from quantum fluctuations. Note also that the transverse structure factor, whose dominant contribution comes from one-magnon states, substantially exceeds the longitudinal structure factor, and gives the dominant contribution to the total unpolarized neutron scattering intensity.

Our results can be analysed to determine the one-magnon spectral weight $A_1(\mathbf{k})$ and the relative multi-magnon spectral weights

$$W_t(\mathbf{k}) = 1 - A_1(\mathbf{k})/S_t(\mathbf{k}),$$
$$W_{tot}(\mathbf{k}) = 1 - A_1(\mathbf{k})/S_{tot}(\mathbf{k}) \tag{6.29}$$

These are shown in Figure 6.4.

We note that the maximum multiple-magnon contribution to the structure factors occurs at the $(\pi, 0)$ point and is approximately 44%/29% for unpolarized/polarized neutrons. This is a significant contribution and needs to be allowed for in analysis of experimental data.

The one-magnon transverse structure factor has recently been measured (Christensen *et al.*, 2004) for the material $Cu(DCOO)_2 \cdot 4D_2O$ (CFTD). Figure 6.5 shows a comparison between theory and experiment, with the parameter value $J = 6.13$ meV. As is evident, the overall agreement is excellent.

The results shown in this subsection will give the reader an indication of the current state of theoretical calculations, via series methods, and of comparison with real materials. Further details, together with analogous results for the simple cubic lattice, are given in our paper (Zheng *et al.*, 2005).

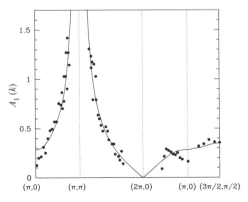

Fig. 6.5. Comparison of the one-magnon transverse structure factor $A_1(\mathbf{k})$ for CFTD (solid points) and our series results.

6.4 Two-particle and multi-particle excitations

As mentioned in the introduction to this chapter, experiments are now beginning to probe in detail two-particle and multi-particle states of condensed matter systems. It is therefore important to developed systematic methods for calculating the energies and structure factors associated with such states, for specific lattice models. Historically there were two independent approaches developed at more or less the same time. The Cologne group (Knetter and Uhrig, 2000) developed a 'continuous unitary transformation' (CUT) approach, for both one and two-particle excitations. We describe this method briefly in Section 11.2.3. A collaboration between Bonn, Davis and UNSW developed a more general linked cluster approach, which we will discuss below. We consider specifically the two-particle case, following closely the treatment of Zheng *et al.* (2001b). The method could in principle be generalized to three or more particles.

6.4.1 Linked cluster expansion

We start again with the Hamiltonian written as

$$H = H_0 + \lambda V \tag{6.30}$$

acting on a cluster of N sites. The unperturbed states of two 'particles' are labelled by their locations $\mathbf{i}, \mathbf{j}$. For simplicity, here we assume that they are identical, and cannot reside on the same site. There are then $N(N-1)/2$ distinct unperturbed states, and we may assume $\mathbf{i} > \mathbf{j}$. This will be the case for two-magnon or two-triplet states, for example. Other cases will require slight modifications of the formalism – see Zheng *et al.* (2001b).

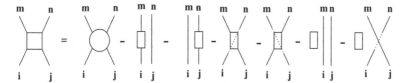

Fig. 6.6. Decomposition of the scattering matrix element for two identical particles into irreducible components. Notation as in Figure 4.4.

As in the one-particle case, one can block diagonalize the Hamiltonian to get the effective Hamiltonian h_{eff} for two-particle states. Let

$$E_2(\mathbf{i}, \mathbf{j}; \mathbf{m}, \mathbf{n}) = \langle \mathbf{i}, \mathbf{j} | h_{\text{eff}} | \mathbf{m}, \mathbf{n} \rangle \tag{6.31}$$

be the matrix element between initial two-particle state $|\mathbf{i}, \mathbf{j}\rangle$ and final state $|\mathbf{m}, \mathbf{n}\rangle$. As in the one-particle case, $E_2(\mathbf{i}, \mathbf{j}; \mathbf{m}, \mathbf{n})$ is not an extensive quantity. To obtain a quantity obeying the cluster addition property, one must subtract the ground-state energy and one-particle contributions, to form the 'irreducible two-particle matrix element', as illustrated in Figure 6.6.

$$\Delta_2(\mathbf{i}, \mathbf{j}; \mathbf{m}, \mathbf{n}) = E_2(\mathbf{i}, \mathbf{j}; \mathbf{m}, \mathbf{n}) - \Delta_1(\mathbf{i}, \mathbf{m})\delta_{\mathbf{j}, \mathbf{n}} - \Delta_1(\mathbf{i}, \mathbf{n})\delta_{\mathbf{j}, \mathbf{m}} - \Delta_1(\mathbf{j}, \mathbf{m})\delta_{\mathbf{i}, \mathbf{n}}$$
$$- \Delta_1(\mathbf{j}, \mathbf{n})\delta_{\mathbf{i}, \mathbf{m}} - E_0(\delta_{\mathbf{i}, \mathbf{m}}\delta_{\mathbf{j}, \mathbf{n}} + \delta_{\mathbf{i}, \mathbf{n}}\delta_{\mathbf{j}, \mathbf{m}}) \tag{6.32}$$

where E_0 is the ground state energy, and Δ_1 is the irreducible one-particle matrix element. This quantity can easily be shown to obey the cluster addition property, as long as the block diagonalization ensures that two particles cannot 'annihilate' from one cluster and 'reappear' on another disconnected one. Thus in principle, one can develop series expansions for the matrix elements $\Delta_2(\mathbf{i}, \mathbf{j}; \mathbf{m}, \mathbf{n})$ in terms of connected clusters alone, where the only non-zero contributions come from clusters which contain, or are 'rooted to', all four sites $\mathbf{i}, \mathbf{j}; \mathbf{m}, \mathbf{n}$.

Because of translational invariance, Δ_2 is a function of three variables, i.e. the relative coordinates:

$$\Delta_2(\mathbf{i}, \mathbf{j}; \mathbf{m}, \mathbf{n}) \equiv \Delta_2(\mathbf{r}, \delta_1, \delta_2) \tag{6.33}$$

where $\delta_1 = \mathbf{i} - \mathbf{j}$, $\delta_2 = \mathbf{m} - \mathbf{n}$ are the distances between the two particles in the initial and final states, and $\mathbf{r} = \frac{1}{2}(\mathbf{m} + \mathbf{n} - \mathbf{i} - \mathbf{j})$ is the relative change in the two-particle centre of mass. We refer to the Δ_2 as 'two-particle transition amplitudes'.

In the one-particle case, the dispersion relation can be written down immediately by Fourier transforming the transition amplitudes in position space, as in Eq. (4.34). For the two-particle states, however, it is necessary to solve the two-body Schrödinger equation

$$h_{\text{eff}} |\Psi\rangle = E |\Psi\rangle \tag{6.34}$$

There are two methods which we have used to treat this problem. The most useful one is the 'finite lattice' approach in coordinate space, which we discuss below. The other is to formulate the two-body Schrödinger equation as an integral equation in momentum space. In the interests of brevity, we relegate the discussion of this second method to Appendix 7.

Note that, unlike the case for the irreducible one-particle matrix element, the series for Δ_2 may depend on the transformation used to block diagonalize the Hamiltonian, but the resulting series for the dispersion of the two-particle bound/antibound states, obtained from solving the Schrödinger equation, should be independent of the transformation used, up to the order computed.

Singular behaviour may arise in the solution of the Schrödinger equation at thresholds where new bound states emerge from the continuum. Our method should be able to deal with new bound states arising as the total two-particle momentum $\mathbf{k}$ varies. Obtaining a numerical solution rather than a series solution to the Schrödinger equation makes it possible, also, to explore new bound states emerging as the coupling λ increases, as long as the naive sum to the series converges.

6.4.2 Finite lattice approach to the two-particle Schrödinger equation

In this approach, the Schrödinger equation in the two-particle subspace is solved for a finite but large system with periodic boundary conditions. The Schrödinger equation then becomes a finite-dimensional matrix equation. The cluster expansion provides each matrix element of the effective Hamiltonian as a power series in the expansion parameter. The centre of mass momentum is a conserved quantity, so the only variable left in the equation is the separation of the two particles.

Consider two identical particles on a finite lattice of N sites with periodic boundary conditions. The two-particle eigenstates can be labelled by the total momentum $\mathbf{k}$ and separation δ, and can be expressed as

$$|\mathbf{k}, \delta\rangle = \frac{1}{\sqrt{N}} \sum_{\mathbf{j}=1}^{N} e^{i\mathbf{k}\cdot(2\mathbf{j}+\delta)/2}|\mathbf{j} + \delta, \mathbf{j}\rangle \tag{6.35}$$

in terms of position states, with the particles at $\mathbf{j}$ and $\mathbf{j} + \delta$. The metrix elements of the two-particle effective Hamiltonian can then be written as

$$\langle \mathbf{k}, \delta_1|h_{\text{eff}}|\mathbf{k}, \delta_2\rangle = \frac{1}{N} \sum_{\mathbf{n}=1}^{N}\sum_{\mathbf{j}=1}^{N} e^{i\mathbf{k}\cdot(2\mathbf{n}+\delta_2-2\mathbf{j}-\delta_1)/2} E_2(\mathbf{j} + \delta_1, \mathbf{j}; \mathbf{n} + \delta_2, \mathbf{n}) \tag{6.36}$$

Using (6.32), and using translational invariance to eliminate the summation over $\mathbf{n}$ in (6.36), gives finally

$$\langle \mathbf{k}, \delta_1 | h_{\text{eff}} | \mathbf{k}, \delta_2 \rangle = e^{i\mathbf{k}\cdot(\delta_2 - \delta_1)/2} \Delta_1(\delta_2 - \delta_1) + e^{i\mathbf{k}\cdot(\delta_1 - \delta_2)/2} \Delta_1(\delta_1 - \delta_2)$$
$$+ e^{i\mathbf{k}\cdot(\delta_1 + \delta_2)/2} \Delta_1(\delta_1 + \delta_2) + e^{-i\mathbf{k}\cdot(\delta_1 + \delta_2)/2} \Delta_1(-\delta_1 - \delta_2) + E_0 \delta_{\delta_1, \delta_2}$$
$$+ \sum_{j=1}^{N} e^{i\mathbf{k}\cdot[1 - j + (\delta_2 - \delta_1)/2]} \Delta_2(1 - j + \delta_2/2 - \delta_1/2, \delta_1, \delta_2) \qquad (6.37)$$

The diagonal term for the ground-state energy E_0 can simply be dropped, if we want to get the energy gap. Finally, we need to diagonalize the matrix h_{eff} in (6.37) for each given value of $\mathbf{k}$ to obtain the energy eigenvalues. This may be done by numerical diagonalization, or by a degenerate perturbation theory. An example is given below.

The two-particle spectrum will, in general, consist of a continuum plus discrete states, below or above the continuum. The continuum is delimited by the maximum and minimum energy of two one-particle excitations whose total momentum is $\mathbf{k}$. Discrete states below the bottom edge of the continuum are 'bound states', and the *binding energy* is defined relative to the lower edge of the continuum at the particular $\mathbf{k}$ value. Discrete states above the upper edge of the continuum are conventionally termed 'antibound states', and one can define the *antibinding energy* relative to the upper edge of the continuum.

The approach described above gives us explicit two-particle wave functions in real-space, from which we can compute a 'coherence length', defined by

$$L = \frac{\sum_\delta |\delta| f(\delta)^2}{\sum_\delta f(\delta)^2} \qquad (6.38)$$

where $f(\delta)$ is the amplitude (i.e. two-particle wave functions) for separation δ. We can also compute two-particle structure factors, and other properties, to which we return later.

6.4.3 Example: the transverse field Ising chain

Let us take the transverse field Ising chain as an example, to show how to develop a cluster expansion for the two-particle transition amplitude $\Delta_2(\mathbf{r}, \delta_1, \delta_2)$. We use the same Hamiltonian, Eq. (6.19), as in the calculation of one-particle states. The unperturbed two-particle states are those with two down spins. Since this system can be mapped exactly onto a system of free spinless fermions there are no two-particle bound states. The possible two-particle excitation energies are, exactly,

$$E_2(k_1, k_2) = 2(1 + \mu^2 - 2\mu \cos k_1)^{1/2} + 2(1 + \mu^2 - 2\mu \cos k_2)^{1/2} \qquad (6.39)$$

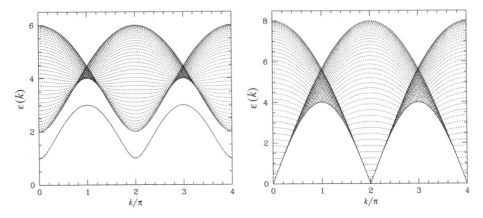

Fig. 6.7. One-particle dispersion relation and two-particle continuum for the one-dimensional transverse Ising model at $\mu = 0.5$ (left) and $\mu = 1$ (right). At the critical point $\mu = 1$ the one-particle dispersion merges with the lower edge of the two-particle continuum, and the system becomes gapless.

The form of the exact two-particle continuum is shown in Figure 6.7.

Note that this two-particle state is in the same sector as the ground state, so the similarity transformation fails, but both the multi-block method and two-block orthogonal transformation work. In fact they give identical results, since the diagonal blocks of O in the multi-block method turn out to be symmetric in this particular example.

Let us consider the clusters g_2 and g_3 of Table 4.1 (g_1 is not involved in the calculation).

g_2 : ●——● The basis has two states: one unperturbed two-particle state and the unperturbed ground state

$$|1\rangle = |--\rangle \quad \text{and} \quad |2\rangle = |++\rangle \tag{6.40}$$

The unperturbed two-particle state $|1\rangle$ can also be labelled by the location of the excitations, i.e. $|2, 1\rangle$. The 2×2 H matrix is

$$H = \begin{pmatrix} 4 & -\mu \\ -\mu & 0 \end{pmatrix} \tag{6.41}$$

The effective Hamiltonian for the two-particle state is

$$h_{\text{eff}} \equiv E_2(2, 1; 2, 1) = 4 + \tfrac{1}{4}\mu^2 - \tfrac{1}{64}\mu^4 + \tfrac{1}{512}\mu^6 + \cdots \tag{6.42}$$

This turns out to be just $4 - E_0$, where E_0 is the ground-state series for this cluster. The irreducible two-particle matrix element

$$\Delta_2(2, 1; 2, 1) = E_2(2, 1; 2, 1) - E_0 - \Delta_1(1, 1) - \Delta_1(2, 2) \qquad (6.43)$$

is easily shown to vanish for this cluster.

g_3 : $\overset{1}{\bullet}\!\!-\!\!\overset{2}{\bullet}\!\!-\!\!\overset{3}{\bullet}$ The unperturbed basis has four states, the three two-particle states plus another. The states and Hamiltonian matrix are

$$
\begin{aligned}
|1\rangle &= |--+\rangle \equiv |2, 1\rangle \\
|2\rangle &= |-+-\rangle \equiv |3, 1\rangle \\
|3\rangle &= |+--\rangle \equiv |3, 2\rangle \\
|4\rangle &= |+++\rangle
\end{aligned}
\qquad
H =
\begin{pmatrix}
4 & -\mu & 0 & -\mu \\
-\mu & 4 & -\mu & 0 \\
0 & -\mu & 4 & -\mu \\
-\mu & 0 & -\mu & 0
\end{pmatrix}
$$

Solving the matrix equation with the two-block method, results in the 3×3 effective Hamiltonian matrix

$$
h_{\text{eff}} =
\begin{pmatrix}
h_{11} & h_{12} & h_{13} \\
h_{21} & h_{22} & h_{23} \\
h_{31} & h_{32} & h_{33}
\end{pmatrix}
\qquad (6.44)
$$

with

$$
\begin{aligned}
h_{11} &= h_{33} = h_{13} = h_{31} = 4 + \tfrac{1}{4}\mu^2 + 0\mu^4 - \tfrac{5}{1024}\mu^6 + \cdots \\
h_{22} &= 4 + 0 \times \mu^2 + 0 \times \mu^4 + \tfrac{1}{512}\mu^6 + \cdots \\
h_{12} &= h_{21} = h_{23} = h_{32} = -\mu + \tfrac{1}{16}\mu^3 - \tfrac{3}{512}\mu^5 + \cdots
\end{aligned}
\qquad (6.45)
$$

From these we can obtain the non-zero irreducible two-particle matrix elements $\Delta_2(i, j; m, n)$ as

$$\Delta_2(2, 1; 3, 2) = \Delta_2(3, 2; 2, 1) = \tfrac{1}{2}\mu^2 + 0\mu^4 - \tfrac{5}{512}\mu^6 + \cdots \qquad (6.46)$$

and so the non-zero $\Delta_2(r, \delta_1, \delta_2)$ are

$$\Delta_2(1, 1, 1) = \Delta_2(-1, 1, 1) = \tfrac{1}{2}\mu^2 + 0\mu^4 - \tfrac{5}{512}\mu^6 + \cdots \qquad (6.47)$$

Note that no subtraction of sub-cluster contributions is necessary here since there is no contribution from the smaller clusters g_1, g_2.

Proceeding in this way, and including contributions from clusters of up to six sites, gives the irreducible two-particle matrix elements $\Delta_2(r, \delta_1, \delta_2)$ for the bulk chain, to order μ^5, as

$$\Delta_2(1, 1, 1) = \Delta_2(-1, 1, 1) = \tfrac{1}{2}\mu^2 - \tfrac{1}{8}\mu^4 + \cdots$$

$$\Delta_2(\tfrac{3}{2}, 1, 2) = \Delta_2(-\tfrac{3}{2}, 1, 2) = \Delta_2(\tfrac{3}{2}, 2, 1) = \Delta_2(-\tfrac{3}{2}, 2, 1)$$
$$= \tfrac{1}{4}\mu^3 - \tfrac{5}{64}\mu^5 + \cdots$$

$$\Delta_2(2, 1, 3) = \Delta_2(-2, 1, 3) = \Delta_2(2, 3, 1) = \Delta_2(2, 1, 3)$$
$$= \Delta_2(2, 2, 2) = \Delta_2(-2, 2, 2) = \tfrac{5}{32}\mu^4 + \cdots$$

$$\Delta_2(\tfrac{5}{2}, 1, 4) = \Delta_2(-\tfrac{5}{2}, 1, 4) = \Delta_2(\tfrac{5}{2}, 4, 1) = \Delta_2(-\tfrac{5}{2}, 4, 1)$$
$$= \Delta_2(\tfrac{5}{2}, 2, 3) = \Delta_2(-\tfrac{5}{2}, 2, 3) = \Delta_2(\tfrac{5}{2}, 3, 2)$$
$$= \Delta_2(-\tfrac{5}{2}, 3, 2) = \tfrac{7}{64}\mu^5 + \cdots \tag{6.48}$$

To solve the two-particle Schrödinger equation, let us consider a small finite lattice of $N = 8$ sites with periodic boundary conditions, where the possible two-particle separations are $d = 1, 2, 3, 4$. Inserting the Δ_1 and Δ_2 matrix elements obtained before into Eq. (6.37), one obtains a 4×4 matrix for the two-particle effective Hamiltonian

$$(h_{\text{eff}})_{ij} = h_{ij} \tag{6.49}$$

with

$$h_{11} = 4 + \mu^2 + \tfrac{1}{16}\mu^4 + \tfrac{1}{2}\mu^2 \cos(k) - \tfrac{1}{8}\mu^4 \cos(k) + \cdots$$

$$h_{12} = h_{21} = -2\mu \cos(k/2) + \tfrac{1}{4}\mu^3 \cos(k/2) + \tfrac{1}{4}\mu^3 \cos(3k/2) + \cdots$$

$$h_{13} = h_{31} = -\tfrac{1}{2}\mu^2 \cos(k) + \tfrac{1}{8}\mu^4 \cos(k) + \tfrac{5}{32}\mu^4 \cos(2k) + \cdots$$

$$h_{14} = h_{41} = -\tfrac{1}{4}\mu^3 \cos(3k/2) + \cdots$$

$$h_{22} = 4 + \mu^2 + \tfrac{1}{16}\mu^4 + \tfrac{5}{32}\mu^4 \cos(2k) + \cdots$$

$$h_{23} = h_{32} = -2\mu \cos(k/2) + \tfrac{1}{4}\mu^3 \cos(k/2) + \cdots$$

$$h_{24} = h_{42} = -\tfrac{1}{2}\mu^2 \cos(k) + \tfrac{1}{8}\mu^4 \cos(k) + \cdots$$

$$h_{33} = h_{44} = 4 + \mu^2 + \tfrac{1}{16}\mu^4 + \cdots$$

$$h_{34} = h_{43} = -2\mu \cos(k/2) + \tfrac{1}{4}\mu^3 \cos(k/2) + \cdots \tag{6.50}$$

where k is the total two-particle momentum.

One can numerically diagonalize this symmetric matrix for given k, μ (truncating the expansions at the largest power of μ), to get approximate values of two-particle energies. It is then necessary to repeat this for larger N and longer series, and scale the results as $N \to \infty$. The results of such a calculation are shown in Figure 6.8, where we show binding and antibinding energies (relative to the continuum) for $k = 0, \pi$. As is apparent, the binding/antibinding energies are non-zero for finite N,

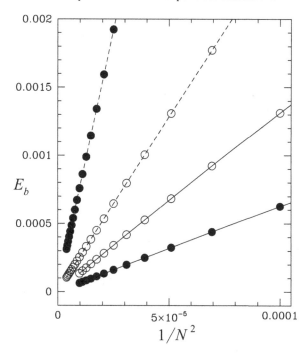

Fig. 6.8. The binding energy (full points) and antibinding energy (open points) E_b versus $1/N^2$ (N is the size of lattice) for the transverse Ising model with coupling $\lambda = 0.5$ and momentum $k = 0$ (dashed lines), π (solid lines). From Zheng *et al.* (2001a).

but scale to zero as $1/N^2$. This is consistent with the absence of bound states in this model, as noted earlier.

The eigenvalues of the 4×4 matrix (6.49) can also be obtained in series form. This requires degenerate state perturbation theory to lift the degeneracy, followed by a normal perturbation expansion. The resulting eigenvalues for $k = 0$ are

$$\Lambda_1 = 4 - (1 + \sqrt{5})\mu + 0.621885\mu^2 + 0.437431\mu^3 + 0.269742\mu^4 + \cdots$$
$$\Lambda_2 = 4 - (\sqrt{5} - 1)\mu + 1.628115\mu^2 + 0.437431\mu^3 - 0.129117\mu^4 + \cdots$$
$$\Lambda_3 = 4 + (\sqrt{5} - 1)\mu + 1.628115\mu^2 - 0.437431\mu^3 - 0.129117\mu^4 + \cdots$$
$$\Lambda_4 = 4 + (1 + \sqrt{5})\mu + 0.621885\mu^2 - 0.437431\mu^3 + 0.269742\mu^4 + \cdots$$

This calculation would then need to be repeated for a sequence of N values, when the series coefficients would change. One would find the series for largest/smallest eigenvalues approaching those for the lower and upper edges of the continuum, as $N \to \infty$.

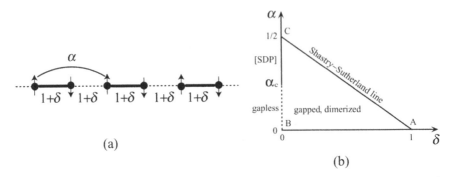

Fig. 6.9. The J_1–J_2–δ chain (a) and its phase diagram (b).

6.4.4 The J_1–J_2–δ chain

In the previous subsection, we have illustrated in detail how to compute two-particle spectra for the simplest model, the transverse field Ising chain. Unfortunately there are no bound states in this model, and the two-particle dynamics is uninteresting. A much more interesting model, which exhibits bound states, is the J_1–J_2–δ chain, which includes both bond alternation and frustration. The Hamiltonian for this model is [see Figure 6.9(a)]

$$H = \sum_i [(1 + (-1)^i \delta) \mathbf{S}_i \cdot \mathbf{S}_{i+1} + \alpha \mathbf{S}_i \cdot \mathbf{S}_{i+2}], \qquad (6.52)$$

where the $\mathbf{S}_i$ are spin-$\frac{1}{2}$ operators at sites i, α parametrizes a next-nearest neighbour coupling and δ is the alternating dimerization. We rewrite the Hamiltonian as

$$H/(1 + \delta) = \sum_i [\mathbf{S}_{2i} \cdot \mathbf{S}_{2i+1} + \lambda (\mathbf{S}_{2i} \cdot \mathbf{S}_{2i-1} + y \mathbf{S}_i \cdot \mathbf{S}_{i+2})], \qquad (6.53)$$

The parameter space (λ, y) is equivalent to the parameter space (δ, α) with $\lambda \equiv (1 - \delta)/(1 + \delta)$ and $y \equiv \alpha/(1 - \delta)$. The former parametrization makes explicit that, for $\lambda = 0$, the model consists of decoupled dimers: we take this to be our unperturbed Hamiltonian H_0. The rest of the Hamiltonian is treated as a perturbation, and various physical quantities are expanded in powers of λ. This is, then, a 'dimer expansion', as discussed in Section 5.2.1. We denote the lattice spacing to be a, the distance between dimers is then $d = 2a$.

The phase diagram for this model is shown in Figure 6.9(b). When there is no bond alternation (i.e. at $\delta = 0$) there are two regimes: for $\alpha < \alpha_c$ (where $\alpha_c \simeq 0.2411$), the system is in the same universality class as the uniform chain (point B), and the elementary excitations are gapless spin-$\frac{1}{2}$ spinons. As the frustration increases to $\alpha = \alpha_c$, there is a 'spin-Peierls' transition to a spontaneously dimerized phase (SDP), where the ground state is two-fold degenerate, and the

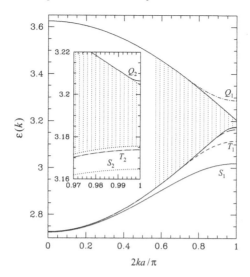

Fig. 6.10. The two-particle excitation spectrum of the J_1–J_2–δ chain with $\alpha = 0$ and $\delta = 0.6$. Note the bound and antibound states lying outside the two-particle continuum (shaded). From Zheng *et al.* (2001a).

lowest excitations are massive spin-$\frac{1}{2}$ solitons. Point C at $\alpha = \frac{1}{2}$ is the so-called Majumdar–Ghosh point, where the ground state is exactly known as a product of singlet dimers.

Adding a bond alternation ($\delta > 0$) to the exchange constants of the model leads to confinement of the spin-$\frac{1}{2}$ excitations. The system is dimerized, the elementary excitations are triplet excitations of a single dimer ('triplons') and the spinons or solitons are bound into pairs. There is a special line in the parameter space, $\alpha = (1 - \delta)/2$, where the dimerized ground state is known exactly (Shastry and Sutherland, 1981b).

We consider dimer expansions along two special lines in the phase diagram: (i) $\alpha = 0$, corresponding to the nearest-neighbour alternating chain without frustration, and (ii) $\alpha = (1 - \delta)/2$, i.e. along the Shastry–Sutherland line. The case $\lambda = 1$ ($\delta = 0$) then corresponds to the uniform Heisenberg chain and the Majumdar–Ghosh model, respectively.

The dimer expansion for the one-particle (triplet) excitation spectrum has been obtained to order λ^{13} for general y. For $y = 1/2$ (i.e. along the Shastry–Sutherland line), the series can be extended much further, and is known to order λ^{23} (Singh and Zheng, 1999). We do not discuss this further, since our focus here is on two-particle states.

Using the formalism described above, we have investigated the occurrence of two-particle bound states in this model. Figure 6.10 shows our computed two-particle excitation spectrum for the case $\alpha = 0$, $\delta = 0.6$; i.e. no frustration and

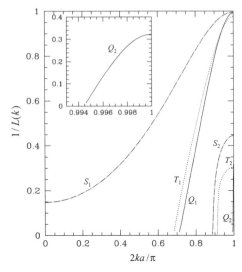

Fig. 6.11. The inverse of the coherence length $1/L$ versus momentum k for singlet (S_1 and S_2), triplet (T_1 and T_2) and quintet (Q_1 and Q_2) bound/antibound states of the $J_1-J_2-\delta$ chain with $\alpha = 0$ and $\delta = 0.6$. The inset enlarges the region near $ka = \pi/2$. From Zheng *et al.* (2001a).

relatively large degree of dimerization. We note from the figure the clear presence of bound states. There appear to be two singlet-bound states (S_1 and S_2) as well as two triplet-bound states (T_1 and T_2) and quintet ($S = 2$) antibound states (Q_1 and Q_2), at these parameter values. The singlet state S_1 is present at all momenta, while all other bound/antibound states are present only in the neighbourhood of $ka = \pi/2$, and merge with the continuum at smaller momenta. Historically, the occurrence of bound states in this model was first noted by Uhrig and Schulz (1996), using an approximate analytic method. The details, elucidated in Figure 6.10, were obtained via linked cluster series (Trebst *et al.*, 2000; Zheng *et al.*, 2001a). Bound states are also found at other points in the phase diagram, and it appears that the existence of bound states, such as these, is generic in frustrated and/or dimerized models.

Before leaving this topic we wish to show two further important features. Figure 6.11 shows the inverse of the coherence length (Eq. 6.38) for these bound states (at $\alpha = 0$, $\delta = 0.6$). It appears that at $ka = \pi/2$ the states S_1, T_1 and Q_1 are fully *localized*, with wavefunctions extending only across a pair of dimers, whereas the states S_2, T_2 and Q_2 are more extended. As the momentum approaches the threshold value at which a bound state merges with the continuum, the coherence length diverges. In Figure 6.12 we show the scaled binding energies of the various bound states at $ka = \pi/2$, as function of δ. It is evident that as $\delta \to 0$ all bound states disappear (i.e. merge into the continuum), with S_1 vanishing linearly and the others more rapidly. This is consistent with a recent approximate calculation

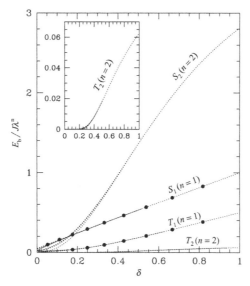

Fig. 6.12. The scaled binding energy $E_b/J\lambda^n$ at $ka = \pi/2$ versus dimerization δ for singlet (S_1 and S_2) and triplet (T_1 and T_2) bound states of the $J_1-J_2-\delta$ chain with $\alpha = 0$. The solid points are numerical exact diagonalization results (Barnes *et al.*, 1999). The inset enlarges the region for T_2. Several different integrated differential approximants to the series are shown. From Zheng *et al.* (2001a).

using the field theoretic bosonization approach (Affleck, 1998), which predicts one $S = 1$ 'bound state' (a bound state of two spinons) and one $S = 0$ bound state, for small δ.

For further details regarding bound states in this model, including results along the Shastry–Sutherland line, we refer the reader to the literature (Zheng *et al.*, 2001a).

6.5 Two-particle structure factors

In the preceding section we showed how two-particle bound states can be calculated using linked cluster series methods, and illustrated this via an example, the $J_1-J_2-\delta$ chain. Experimentally, such bound states can be observed through their contribution to the appropriate dynamical structure factors. It is useful and instructive therefore, to show how to compute two-particle structure factors, based on the general formalism of Section 6.3.

From the general results (6.12) we write the two-particle exclusive structure factors as

$$S_\Lambda(\mathbf{k}, \omega) = \frac{1}{N}\delta(\omega - E_\Lambda + E_0)\left|\sum_i \langle\psi_\Lambda|\mathcal{O}_i|\psi_0\rangle e^{i\mathbf{k}\cdot\mathbf{r}_i}\right|^2 \qquad (6.54)$$

where $|\psi_\Lambda\rangle$ is now a two-particle state. This can be written in momentum space as

$$|\psi_\Lambda(\mathbf{k})\rangle = \frac{1}{\sqrt{N_c}} \sum_{m,n} f_\Lambda(\mathbf{r}_m - \mathbf{r}_n) e^{i\mathbf{k}\cdot(\mathbf{r}_m+\mathbf{r}_n)/2}|\psi_\Lambda(m,n)\rangle \qquad (6.55)$$

where $f_\Lambda(\mathbf{r}_m - \mathbf{r}_n)$ is the 'two-particle wavefunction', depending only on the relative distance $\mathbf{r}_m - \mathbf{r}_n$, and $|\psi_\Lambda(m,n)\rangle$ is a two-particle state in real space, with particles initially excited at positions m, n. One can, by analogy with (6.15), then introduce two-particle exclusive matrix elements

$$\Omega_\Lambda^{2p}(i;m,n) \equiv \langle\psi_\Lambda(m,n)|\mathcal{O}_i|\psi_0\rangle \qquad (6.56)$$

Translation invariance implies that $\Omega_\Lambda^{2p}(i;m,n)$ only depends on the relative distance between m, n and i, i.e.

$$\Omega_\Lambda^{2p}(i;m,n) \equiv \Omega_\Lambda^{2p}(\mathbf{r},\delta) \qquad (6.57)$$

where $\mathbf{r} = (2\mathbf{r}_i - \mathbf{r}_m - \mathbf{r}_n)/2$ and $\delta = \mathbf{r}_m - \mathbf{r}_n$.

The two-particle exclusive structure factor can then be written as

$$S_\Lambda^{2p}(\mathbf{k},\omega) = \frac{N_c}{N}\delta(\omega - E_\Lambda + E_0)\left|\sum_{\mathbf{r},\delta} \Omega_\Lambda^{2p}(\mathbf{r},\delta) f_\Lambda(\delta) e^{i\mathbf{k}\cdot\mathbf{r}}\right|^2 \qquad (6.58)$$

It remains to solve the effective two-particle Schrödinger equation to obtain the wavefunctions $f_\Lambda(\delta)$. This can be done using the finite lattice approach described in Section 6.4.2, and then one can compute the two-particle structure factor from (6.58). Note that in the finite lattice approach, where the energies are discrete, the structure function for the two-particle continuum is represented by a grid of delta functions. To smooth the results, we replace the δ-function $S\delta(\omega - \omega_0)$ at energy ω_0 by the continuous Gaussian function

$$S\eta/\pi[\eta^2 + (\omega - \omega_0)^2] \qquad (6.59)$$

where the width parameter η can be tuned to give a smooth function for large lattices.

We illustrate this formalism with an example.

6.5.1 The alternating Heisenberg chain

Let us again consider the alternating Heisenberg chain, with Hamiltonian

$$H = \sum_i \mathbf{S}_{2i} \cdot \mathbf{S}_{2i+1} + \lambda \sum_i \mathbf{S}_{2i-1} \cdot \mathbf{S}_{2i} \qquad (6.60)$$

which has, as we saw in the previous section, singlet- and triplet-bound states of elementary triplet excitations below the two-triplon continuum. We use a dimer expansion to compute series in powers of λ. For this model there is no breaking of spin rotational symmetry and hence

$$S^{\alpha\beta}(\mathbf{k}, \omega) = \tfrac{1}{2}S^{-+}(\mathbf{k}, \omega)\delta_{\alpha\beta} \tag{6.61}$$

which corresponds, in the formalism of Section 6.3, to the choice $\mathcal{O}_i \equiv S_i^-$.

Series for the one and two-particle spectral weight have been derived to order λ^{13} by Hamer *et al.* (2003). The leading terms are

$$S_{1p}(kd) = \sin^2\left(\frac{kd}{4}\right)\left[2 + \lambda\,\cos(kd) + \frac{\lambda^2}{8}\left(-1 + 8\cos\left(\frac{kd}{2}\right) + 2\cos(kd)\right.\right.$$
$$+3\cos(2kd)\right) + \frac{\lambda^3}{96}\left(36\cos\left(\frac{kd}{2}\right) + 17\cos(kd) + 8\cos\left(\frac{3kd}{2}\right)\right.$$
$$+32\cos(2kd) + 15\cos(3kd)\right) + O(\lambda^4)\left.\right] \tag{6.62}$$

and

$$S_{2p}(kd) = \sin^2\left(\frac{kd}{4}\right)\left[\lambda^2\sin^4\left(\frac{kd}{4}\right) + \frac{\lambda^3}{12}\left(3 + 2\cos\left(\frac{3\,kd}{2}\right)\right)\sin^2\left(\frac{kd}{4}\right)\right.$$
$$+O(\lambda^4)\left.\right] \tag{6.63}$$

where d is the spacing between dimers. Note that the integrated two-particle spectral weight has a leading term proportional to λ^2, and hence for small λ (large dimerization) the total structure factor is dominated by the one-particle contribution.

We present here a few characteristic results. Firstly, in Figure 6.13, we show the structure factor for the two-particle continuum, as a function of k and ω for $\lambda = 0.27$, which is the coupling believed to occur in the material $Cu(NO_3)_2 \cdot 2.5D_2O$, discussed by Tennant *et al.* (2003). The heavy solid line traces the dispersion relation for the triplet-bound state T_1, and an interesting feature to note is the sharp spike in the structure factor at the threshold where the bound state merges with the continuum. Figure 6.14 shows the explicit contribution to the two-particle structure factor from the triplet-bound states T_1 and T_2. We note, in particular, that the contribution of T_2 has two peaks, and is one to two orders of magnitude below that of T_1. Hence it seems unlikely that it will be detected by experiments, at least in the near future.

6.6 Summary and further work

In this chapter we have shown how linked cluster series methods can be used to compute correlation functions at $T = 0$, and their spatial and temporal Fourier transforms – the dynamical structure factors. These latter quantities are measured

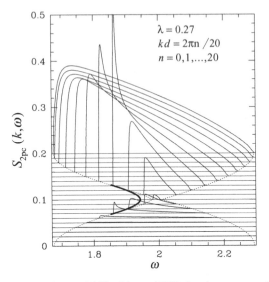

Fig. 6.13. The structure factor (shifted by $n/100$) for the two-particle continuum versus energy ω at $\lambda = 0.27$ and $kd = 2\pi n/20$, $n = 0, 1, 2, \cdots, 20$. Also shown as a bold solid line is the dispersion relation for the triplet-bound state T_1. From Zheng *et al.* (2003).

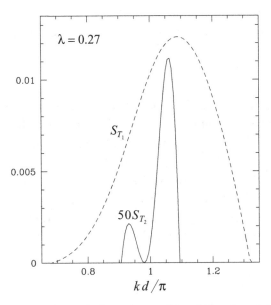

Fig. 6.14. The structure factor for bound states T_1 and T_2 at $\lambda = 0.27$. From Zheng *et al.* (2003).

directly in scattering experiments. A new generation of neutron scattering and other facilities is starting to produce a large amount of detailed high-resolution data, and this area will continue to provide a very fruitful exchange between theory and experiment.

The dynamical structure factor contains contributions from all excited states of the system. It is now becoming possible to isolate the contributions from 'one-particle' excitations, which usually give the dominant contribution at low temperatures, 'two-particle' excitations, and so on. We have shown, in this chapter, how these separate contributions can be calculated. Of particular interest here is the possibility of bound states.

The series work discussed here is quite recent and much further work is to be expected. In particular, calculations of two-particle excitations for other models, and in higher dimensions, will predictably yield much insight and many interesting results.

7

Quantum spin models at finite temperature

7.1 Introduction

In the previous three chapters we have shown how one can study the properties of quantum spin models at $T = 0$, via long perturbation expansions. Of course, it is also of great interest to study the finite-temperature thermodynamic properties of such models. One reason is to compare with experimental data on real materials, such as those considered is Section 5.4, and hence to determine both the applicability of the model and values of the unknown parameters. We will discuss this aspect further in Section 7.7 below. A second reason is that, at least in three spatial dimensions, these models have finite-temperature phase transitions and it is of interest to compute transition temperatures and critical exponents. Comparison with related classical models, as discussed in Chapters 2 and 3, is then possible.

We will develop, in some detail, two different series expansion methods for quantum models at finite temperature. These were introduced briefly in Chapter 1. The first approach, which is a true high-temperature expansion, yields series of the form

$$-\beta f(K) = \sum_{r=0}^{\infty} a_r K^r \qquad (7.1a)$$

$$\chi(K) = \sum_{r=0}^{\infty} c_r K^r \qquad (7.1b)$$

where $f(K)$, $\chi(K)$ are the zero-field free energy and susceptibility, $K = J/k_B T$ is the usual high-temperature expansion variable and a_r, c_r are numerical coefficients which are computed *exactly* to some maximum order. The second approach, which we term thermodynamic perturbation theory, is an expansion in terms of a parameter in the Hamiltonian, in which the coefficients are complete functions of temperature. The resulting series can be re-expressed as a true high-temperature expansion, but can also be extrapolated to quite low temperatures, sometimes to $T = 0$.

We take as our archetypal model the Heisenberg ferromagnet in a field

$$H = -J \sum_{\langle ij \rangle} \mathbf{S}_i \cdot \mathbf{S}_j - g\mu_B B \sum_i S_i^z \tag{7.2}$$

where the $\mathbf{S}_i = (S_i^x, S_i^y, S_i^z)$ are quantum spin-S operators, the exchange is assumed isotropic and restricted to nearest neighbours, B is an external magnetic field and g, μ_B are the usual g-factor and Bohr magneton.

In the 1960s Rushbrooke and colleagues, at the University of Newcastle, England, developed high-temperature expansions for this model, for general spin. At the same time the calculations were extended to included second-neighbour interactions and series were also obtained for the antiferromagnetic or 'staggered' susceptibility. Despite the shortness of the series (at most seven to eight terms) the first estimates of the critical temperatures and exponents were made. Also in the 1960s Baker, together with the Newcastle group, derived expansions to order 10 for the $S = \frac{1}{2}$ case, for which simplifying features exist. This body of work is well described in a review (Rushbrooke *et al.*, 1974). The estimated critical exponent γ was found to be ~ 1.43, similar to that of the classical O(3) model, discussed in Chapter 3. By general principles, they should be identical and, as we shall show, the most recent longer series lead to closer agreement. In recent years we have extended these series somewhat, to 14 terms for $S = \frac{1}{2}$ and 12 terms for $S = 1$. These results will be described in Section 7.3.

Longer series can be computed for one dimension. Bühler *et al.* (2000) have reported 16-term expansions for the frustrated $S = \frac{1}{2}$ chain with second-neighbour interactions, and 24-term expansions for the unfrustrated chain. The latter case is, in fact, exactly solvable via the thermodynamic Bethe ansatz method (see e.g. Takahashi, 1999). By means of a recently discovered integral equation, Shiroishi and Takahashi (2002) have computed the high-temperature susceptibility and specific heat to 100 terms!

7.2 Derivation of high-temperature series

In this section we will explain in detail two different methods to derive high-temperature series for models such as (7.2). The first of these, known as the 'moment' method, is akin to the primitive method of deriving high-temperature series for the Ising model (Section 2.1). The second, more powerful, method is a linked cluster method, akin to the approaches of Chapters 4 and 5 at zero temperature. We will consider, for purposes of illustration, the $S = \frac{1}{2}$ ferromagnetic chain.

7.2.1 The moment method

We write the Hamiltonian in the form

$$-\beta H = KP + \beta h M \tag{7.3}$$

where $K = \beta J$, $h = g\mu_B B$ and

$$P = \sum_i \mathbf{S}_i \cdot \mathbf{S}_{i+1} = \sum_{i,\alpha} S_i^\alpha S_{i+1}^\alpha \quad \alpha = x, y, z \tag{7.4a}$$

$$M = \sum_i S_i^z \tag{7.4b}$$

As the two terms in (7.3) commute, we can write the partition function as

$$\mathcal{Z} = \text{Tr}\{e^{-\beta H}\} = \text{Tr}\{e^{KP} e^{\beta h M}\}$$

$$= \mathcal{Z}_0 + \frac{1}{2}(\beta h)^2 \mathcal{Z}_2 + \cdots \tag{7.5}$$

where

$$\mathcal{Z}_0 = \text{Tr}\{e^{KP}\}, \quad \mathcal{Z}_2 = \text{Tr}\{e^{KP} M^2\} \tag{7.6}$$

Note that we have expanded in powers of magnetic field, to order h^2. This is sufficient for the zero-field susceptibility.

Let us first consider $\mathcal{Z}_0$, i.e. the zero-field case. Expanding the exponential operator gives

$$\mathcal{Z}_0 = \text{Tr}\left\{1 + \sum_{r=1}^\infty \frac{K^r}{r!} P^r\right\}$$

$$= 2^N \left\{1 + \sum_{r=1}^\infty \frac{K^r}{r!} \langle P^r \rangle\right\} \tag{7.7}$$

where the angular bracket denotes

$$\langle A \rangle = 2^{-N} \text{Tr}\{A\} \tag{7.8}$$

As usual we can use a graphical description. Denoting the terms in P by bonds joining nearest neighbour sites leads to graphs representing all possible terms in P^r. Noting, at the outset, that the trace of a single spin operator S^α vanishes allows us to exclude graphs with any degree-one vertex. The list of graphs will then start with those given in Table 7.1.

It is instructive to evaluate the contributions of these first few graphs by hand. The spin traces are for $S = \frac{1}{2}$ only. Corresponding results for general S are given in Rushbrooke *et al.* (1974).

Table 7.1. *Lowest-order graphs for zero-field free energy.*

$$\begin{array}{ccccc} 1 & 2 & 3 & 4 & 5 \end{array}$$

(1)
$$X_1 = \tfrac{1}{4}\mathrm{Tr}_{12}\big[(\mathbf{S}_1 \cdot \mathbf{S}_2)^2\big]$$
$$= \tfrac{1}{4}\mathrm{Tr}_{12}\Big[\sum_{\alpha\beta} S_1^\alpha S_2^\alpha S_1^\beta S_2^\beta\Big]; \quad \alpha,\beta = x,y,z$$
$$= \sum_{\alpha\beta}(\alpha\beta)^2 = \tfrac{3}{16} \tag{7.9}$$

where we have introduced the shorthand notation

$$(\alpha\beta) \equiv \tfrac{1}{2}\mathrm{Tr}(S^\alpha S^\beta) = \tfrac{1}{4}\delta_{\alpha\beta} \tag{7.10}$$

(2)
$$X_2 = \tfrac{1}{4}\mathrm{Tr}_{12}\big[(\mathbf{S}_1 \cdot \mathbf{S}_2)^3\big]$$
$$= \sum_{\alpha\beta\gamma}(\alpha\beta\gamma)^2 = 6(\tfrac{1}{8}i)^2 = -\tfrac{3}{32} \tag{7.11}$$

since

$$(\alpha\beta\gamma) = \tfrac{1}{2}\mathrm{Tr}(S^\alpha S^\beta S^\gamma) = \tfrac{1}{8}i\epsilon_{\alpha\beta\gamma} \tag{7.12}$$

where $\epsilon_{\alpha\beta\gamma}$ is the usual Levi–Civita symbol.

(3)
$$X_3 = \tfrac{1}{4}\mathrm{Tr}_{12}\big[(\mathbf{S}_1 \cdot \mathbf{S}_2)^4\big]$$
$$= \sum_{\alpha\beta\gamma\delta}(\alpha\beta\gamma\delta)^2 = \tfrac{21}{256} \tag{7.13}$$

where we have used the results

$$(\alpha\alpha\alpha\alpha) = \tfrac{1}{16}, \quad (\alpha\alpha\beta\beta) = \tfrac{1}{16}, \quad (\alpha\beta\alpha\beta) = -\tfrac{1}{16} \tag{7.14}$$

(4) This graph introduces a new feature. There are two distinct operators

$$P_{12} = \mathbf{S}_1 \cdot \mathbf{S}_2, \quad P_{23} = \mathbf{S}_2 \cdot \mathbf{S}_3 \tag{7.15}$$

which do not commute with each other. The graph arises from

$$P^4 = (P_{12} + P_{23})^4 \tag{7.16}$$

and represents six terms.

$$P_{12}P_{12}P_{23}P_{23}, \quad P_{12}P_{23}P_{12}P_{23}, \quad P_{12}P_{23}P_{23}P_{12},$$
$$P_{23}P_{23}P_{12}P_{12}, \quad P_{23}P_{12}P_{23}P_{12}, \quad P_{23}P_{12}P_{12}P_{23}$$

all of which must be included. Using the cyclic property of the trace allows these to be combined to two terms, yielding

$$X_4 = \tfrac{1}{8}\mathrm{Tr}_{123}\left[4(\mathbf{S}_1\cdot\mathbf{S}_2)^2(\mathbf{S}_2\cdot\mathbf{S}_3)^2 + 2(\mathbf{S}_1\cdot\mathbf{S}_2)(\mathbf{S}_2\cdot\mathbf{S}_3)(\mathbf{S}_1\cdot\mathbf{S}_2)(\mathbf{S}_2\cdot\mathbf{S}_3)\right]$$

$$= 4\sum_{\alpha\beta\gamma\delta}(\alpha\beta)(\alpha\beta\gamma\delta)(\gamma\delta) + 2\sum_{\alpha\beta\gamma\delta}(\alpha\gamma)(\alpha\beta\gamma\delta)(\beta\gamma)$$

$$= 4\sum_{\alpha\gamma}(\alpha\alpha)(\alpha\alpha\gamma\gamma)(\gamma\gamma) + 2\sum_{\alpha\beta}(\alpha\alpha)(\alpha\beta\alpha\beta)(\beta\beta)$$

$$= \tfrac{30}{256} \quad \text{using (7.10) and (7.14)} \tag{7.17}$$

(5) ●━━● ●━━● $$X_5 = \tfrac{1}{16}\mathrm{Tr}_{1234}\left[6(\mathbf{S}_1\cdot\mathbf{S}_2)^2(\mathbf{S}_3\cdot\mathbf{S}_4)^2\right]$$

$$= 6\sum_{\alpha\beta\gamma\delta}(\alpha\beta)^2(\gamma\delta)^2 = \tfrac{27}{128} \tag{7.18}$$

Multiplying each term by its lattice constant and including the factor $K^r/r!$ gives

$$\mathcal{Z}_0 = 2^N\left\{1 + \tfrac{3}{32}NK^2 - \tfrac{1}{64}NK^3 + \tfrac{21}{6144}NK^4 + \tfrac{30}{6144}NK^4\right.$$

$$\left. + \tfrac{27}{6144}N(N-3)K^4 + \cdots\right\}$$

$$= 2^N\left\{1 + \tfrac{3}{32}NK^2 - \tfrac{1}{64}NK^3 + \tfrac{1}{2048}(9N^2-10N)K^4 + \cdots\right\} \tag{7.19}$$

Finally, taking the logarithm gives

$$\ln\mathcal{Z}_0 = N\ln 2 + \tfrac{3}{32}NK^2 - \tfrac{1}{64}NK^3 + \tfrac{1}{2048}(9N^2-10N)K^4 - \frac{1}{2}(\tfrac{3}{32}NK^2)^2 + \cdots \tag{7.20}$$

or

$$\frac{1}{N}\ln\mathcal{Z}_0 = \ln 2 + \tfrac{3}{32}K^2 - \tfrac{1}{64}K^3 - \tfrac{5}{1024}K^4 + \cdots \tag{7.21}$$

Note that the terms of order N^2 and higher cancel on taking the logarithm, as in the Ising case in Chapter 2. However disconnected graphs do contribute and must be retained in the expansion.

We now turn to the susceptibility

$$\chi = \beta^{-1}\frac{\partial^2}{\partial B^2}\left(\frac{1}{N}\ln\mathcal{Z}\right)\Big|_{B=0}$$

$$= \beta^{-1}(g\mu_B)^2\frac{\partial^2}{\partial h^2}\left(\frac{1}{N}\ln\mathcal{Z}\right)\Big|_{h=0} \tag{7.22}$$

It is convenient to define a 'reduced susceptibility'

$$\bar{\chi} = k_B T \chi / (g\mu_B)^2$$

$$= \beta^{-2} \frac{\partial^2}{\partial h^2} \left(\frac{1}{N} \ln \mathcal{Z} \right) \Big|_{h=0}$$

$$= \frac{1}{N} \frac{\mathcal{Z}_2}{\mathcal{Z}_0}$$

$$= \frac{1}{N} \frac{\langle M^2 \rangle + \sum_{r=1}^{\infty} \frac{K^r}{r!} \langle P^r M^2 \rangle}{1 + \sum_{r=1}^{\infty} \frac{K^r}{r!} \langle P^r \rangle}$$

$$= \langle M^2 \rangle_N + \sum_{r=1}^{\infty} \frac{K^r}{r!} \langle P^r M^2 \rangle_N \tag{7.23}$$

where the subscript $\langle \cdots \rangle_N$ denotes the coefficient of the term linear in N. The higher powers of N, arising from disconnected graphs, cancel with terms from the denominator.

The graphs now consist of lines, from the P operators, plus two vertex factors from the two M operators which we represent by crosses. As shown by Rushbrooke *et al.* (1974), we only need to consider graphs for which the field operators are located at different sites.

The first term in (7.23) is easily evaluated

$$M^2 = \sum_{ij} S_i^z S_j^z = \sum_i (S_i^z)^2 + \sum_{i \neq j} S_i^z S_j^z \tag{7.24}$$

$$\langle M^2 \rangle = \frac{1}{2} \sum_i \mathrm{Tr}_i (S_i^z S_i^z) = \frac{1}{4} N \implies \langle M^2 \rangle_N = \tfrac{1}{4} \tag{7.25}$$

Let us also evaluate the order K term:

$$\underset{1 \qquad 2}{\overset{\times \qquad \times}{\bullet\!\!-\!\!-\!\!\bullet}} \quad X = 2 \times \tfrac{1}{4}\mathrm{Tr}_{12}[(\mathbf{S}_1 \cdot \mathbf{S}_2)S_1^z S_2^z]$$

$$= 2 \sum_\alpha (\alpha z)^2 = \tfrac{1}{8} \tag{7.26}$$

giving a contribution of $K/8$. Note the factor 2 which comes from $M^2 \to 2M_1 M_2$.

In Table 7.2 we list all of the graphs, their trace factors, and final contributions, up to K^3. The contribution of each graph is the product of lattice constant, trace factor, and $K^r/r!$.

Collecting the data from Table 7.2 gives, to order K^3,

$$\bar{\chi} = \tfrac{1}{4} + \tfrac{1}{8}K - \tfrac{1}{96}K^3 + \cdots \tag{7.27}$$

Table 7.2. *Graphs and their contributions to the susceptibility series for the*
$S = \frac{1}{2}$ *ferromagnetic chain.*

Graph	LC	Trace factor	Contribution
(graph)	1	$(zz) = \frac{1}{4}$	$\frac{1}{4}$
(graph)	1	$2\sum_{\alpha}(\alpha z)^2 = \frac{1}{8}$	$\frac{1}{8}K$
(graph)	1	$2\sum_{\alpha\beta}(\alpha\beta z)^2 = -\frac{1}{16}$	$-\frac{1}{32}K^2$
(graph)	1	$4\sum_{\alpha\beta}(\alpha z)(\alpha\beta)(\beta z) = \frac{1}{16}$	$\frac{1}{32}K^2$
(graph)	2	$\sum_{\alpha\beta}(\alpha\beta)(\alpha\beta zz) = \frac{3}{64}$	$\frac{3}{64}K^2$
(graph)	-2	$\sum_{\alpha\beta}(\alpha\beta)^2(zz) = \frac{3}{64}$	$-\frac{3}{64}K^2$
(graph)	1	$2\sum_{\alpha\beta\gamma}(\alpha\beta\gamma z)^2 = \frac{7}{128}$	$\frac{7}{768}K^3$
(graph)	2	$2\sum_{\alpha\beta\gamma}[(\alpha\beta\gamma z)(\alpha\beta)(\gamma z) + (\alpha\beta\gamma z)(\alpha\gamma)(\beta z)$ $+(\alpha\beta\gamma z)(\beta\gamma)(\alpha z)] = \frac{5}{128}$	$\frac{10}{768}K^3$
(graph)	2	$2\sum_{\alpha\beta\gamma}[(\alpha\beta\gamma)(\alpha\beta z)(\gamma z) + (\alpha\beta\gamma)(\alpha\gamma z)(\beta z)$ $+(\alpha\beta\gamma)(\beta\gamma z)(\alpha z)] = -\frac{4}{256}$	$-\frac{4}{768}K^3$
(graph)	1	$12(zz)^4 = \frac{6}{128}$	$\frac{6}{768}K^3$
(graph)	-3	$6\sum_{\alpha}(\alpha\alpha)^2(zz)^2 = \frac{9}{128}$	$-\frac{27}{768}K^3$

Several points are worth noting. Firstly we have explicitly included, at order K^2, two graphs with field operators at the same site. Inspection shows that these cancel, and it is easy to show that this is a general result. Secondly we note that the coefficient of K^2 is zero. This cancellation between graphs 3 and 4 only occurs for the linear chain, because the lattice constants are equal in this case.

We conclude this section with some general remarks. The moment method is the most obvious and direct one to use. It is easy to generalize to other lattices, since the graph dependent factors remain the same, only the lattice constants need to be changed. It is easy to generalize to higher spin values, by modifying the trace factors appropriately. The early work (Rushbrooke *et al.*, 1974) was for general spin. It would be fairly straightforward to computerize the procedure. However, at least for small spin values, the linked cluster method, which we consider next, appears to be more efficient.

7.2.2 The linked cluster method

The linked cluster approach to high-temperature series in fact predates its application to ground-state problems. It was applied to the $S = \frac{1}{2}$ quantum ferromagnet by Baker and co-workers in the 1960s. The philosophy is essentially identical to the discussion in Chapter 4. The thermodynamic free energy, and its derivatives, are extensive quantities and thus will satisfy the 'cluster-property'. We can thus compute these quantities per site, for an infinite system, by combining the 'reduced' quantities for a sequence of finite connected clusters. We will show how this works for the case of the one-dimensional $S = \frac{1}{2}$ ferromagnect.

The general result is

$$\frac{1}{N} \ln \mathcal{Z}(K, h) = \sum_g c_g^{\mathcal{L}} \Phi^{(g)}(K, h) \tag{7.28}$$

where the variables K, h are defined in Eq. (7.3), $c_g^{\mathcal{L}}$ is the weak lattice constant of cluster g in lattice $\mathcal{L}$, and $\Phi^{(g)}(K, h)$ is the reduced free energy of cluster g, which is independent of the lattice. Expanding this in powers of h

$$\Phi^{(g)}(K, h) = \Phi_0^{(g)}(K) + \frac{1}{2}(\beta h)^2 \Phi_2^{(g)}(K) + \cdots \tag{7.29}$$

yields for the zero-field free energy and susceptibility of the bulk system

$$-\beta f(K) = \sum_g c_g^{\mathcal{L}} \Phi_0^{(g)}(K) \tag{7.30a}$$

$$\chi(K) = \sum_g c_g^{\mathcal{L}} \Phi_2^{(g)}(K) \tag{7.30b}$$

where χ is the reduced susceptibility (7.23). Equations (7.30) are the basic equations which lead to the high-temperature series.

Next we consider the problem of evaluating $\Phi^{(g)}$ for a particular cluster g. The partition function of the cluster can be written as

$$\mathcal{Z}^{(g)} = \text{Tr}_g \{ e^{KP} e^{hM} \}$$
$$= \mathcal{Z}_0^{(g)} + \frac{1}{2}(\beta h)^2 \mathcal{Z}_2^{(g)} + \cdots \tag{7.31}$$

where

$$\mathcal{Z}_0^{(g)} = \sum_{r=0}^{\infty} \frac{\mu_{r0}}{r!} K^r \tag{7.32a}$$

$$\mathcal{Z}_2^{(g)} = \sum_{r=0}^{\infty} \frac{\mu_{r2}}{r!} K^r \tag{7.32b}$$

with

$$\mu_{r0} = \text{Tr}_g(P^r) = \sum_m \langle m | P^r | m \rangle \tag{7.33a}$$

$$\mu_{r2} = \text{Tr}_g(P^r M^2) = \sum_m M_m^2 \langle m | P^r | m \rangle \tag{7.33b}$$

Here P, M are defined as in (7.4), Tr_g denotes the trace over the finite Hilbert space of the cluster, and $\{|m\rangle\}$ is a complete set of states, specified by the quantum numbers S_i^z. M_m is the eigenvalue of the operator M in this state. As a last step we define

$$A^{(g)} \equiv \ln \mathcal{Z}^{(g)}$$

$$= \ln \mathcal{Z}_0^{(g)} + \frac{1}{2}(\beta h)^2 \frac{\mathcal{Z}_2^{(g)}}{\mathcal{Z}_0^{(g)}} + \cdots$$

$$\equiv A_0^{(g)} + \frac{1}{2}(\beta h)^2 A_2^{(g)} + \cdots \tag{7.34}$$

The reduced free energy of cluster g can then be obtained recursively by subtracting the subgraph contributions

$$\Phi_0^{(g)}(K) = A_0^{(g)}(K) - \sum_{g' \in g} c_{g'}^{(g)} \Phi_0^{(g')}(K) \tag{7.35a}$$

$$\Phi_2^{(g)}(K) = A_2^{(g)}(K) - \sum_{g' \in g} c_{g'}^{(g)} \Phi_2^{(g')}(K) \tag{7.35b}$$

The procedure then is as follows.

(i) Construct a list of all simple connected graphs, ordered by number of bonds, up to some maximum order, together with their lattice constants and sub-graph data.
(ii) For each graph/cluster compute $\mathcal{Z}_0^{(g)}$, $\mathcal{Z}_2^{(g)}$ as series in K, up to the maximum order desired.
(iii) Compute the quantities $A_0^{(g)}$, $A_2^{(g)}$ as series in K, from (7.34).
(iv) Compute the reduced quantities $\Phi_0^{(g)}$, $\Phi_2^{(g)}$ from (7.35).
(v) Finally combine the Φ data to obtain the bulk lattice series (7.30).

To illustrate this procedure in more detail we consider, again, the $S = \frac{1}{2}$ ferromagnetic chain

$$H = -J \sum_i \mathbf{S}_i \cdot \mathbf{S}_{i+1} - g\mu_B B \sum_i S_i^z \tag{7.36}$$

which we write as

$$-\beta H = xP + hM \tag{7.37}$$

Table 7.3. 'Input' data for cluster g_3.

Cluster	State	wt	$P\lvert m\rangle$
	$\lvert 1\rangle = \lvert ---\rangle$	2	$P\lvert 1\rangle = 2\lvert 1\rangle$
1 2 3	$\lvert 2\rangle = \lvert +--\rangle$	4	$P\lvert 2\rangle = 2\lvert 3\rangle$
	$\lvert 3\rangle = \lvert -+-\rangle$	2	$P\lvert 3\rangle = 2\lvert 2\rangle - 2\lvert 3\rangle + 2\lvert 4\rangle$
	$\lvert 4\rangle = \lvert --+\rangle$	0	$P\lvert 4\rangle = 2\lvert 3\rangle$
	$\lvert 5\rangle = \lvert ++-\rangle$	0	
	$\lvert 6\rangle = \lvert +-+\rangle$	0	
	$\lvert 7\rangle = \lvert -++\rangle$	0	
	$\lvert 8\rangle = \lvert +++\rangle$	0	

with $x = \frac{1}{4}\beta J = \frac{1}{4}K$, and

$$P = 4\sum_i \mathbf{S}_i \cdot \mathbf{S}_{i+1} = \sum_i \left[4S_i^z S_{i+1}^z + 2S_i^+ S_{i+1}^- + 2S_i^- S_{i+1}^+ \right] \qquad (7.38)$$

The factor 4 is introduced for convenience, to avoid large denominators in the series coefficients.

The list of clusters starts with the following

Note that each of these clusters represents many graphs needed in the moment method, to the same order in K. To compute the coefficients in (7.33) one could, of course, construct the complete matrix P for a cluster and then iteratively raise it to higher powers, computing the trace at each stage. This procedure could be improved by using symmetries to block diagonalize the initial Hamiltonian, for example the conservation of M^z. However this is still grossly inefficient. It is far better to choose each basis state, in turn, as the initial state $\lvert m\rangle$ and to produce a tree structure of 'daughter' states by operating recursively with P, while accumulating the trace data at every step. Even greater efficiency can be achieved by constructing 'half-trees' starting from each end.

We illustrate this with the third cluster ●—●—●. Table 7.3 shows the basis states, their weights, and the effect of P on the basis state. The 'state weight' has been introduced to take account of symmetry. Thus states $\lvert 1\rangle$ and $\lvert 8\rangle$ are clearly equivalent, and we can avoid dealing with state $\lvert 8\rangle$. The other weights are obtained similarly. This will more than halve the work needed. For the remaining

Table 7.4. *'Output' data for cluster* g_3.

$|m\rangle$	wt	M_m^2	$r = 1$	2	3	4	5	6
					$\langle m|P^r|m\rangle$			
$|1\rangle$	2	$9/4$	2	4	8	16	32	64
$|2\rangle$	4	$1/4$	0	4	-8	48	-160	704
$|3\rangle$	2	$1/4$	-2	12	-40	176	-672	2752
$\mathrm{Tr}(P^r)$			0	48	-96	576	-1920	8448
$\mathrm{Tr}(P^r M^2)$			8	28	8	208	-352	2368

states we find

$$P^r|1\rangle = 2^r|1\rangle$$
$$P^2|2\rangle = 2P|3\rangle = 4|2\rangle - 4|3\rangle + 4|4\rangle$$
$$P^3|2\rangle = -8|2\rangle + 24|3\rangle - 8|4\rangle \tag{7.39}$$
$$P^2|3\rangle = -4|2\rangle + 12|3\rangle - 4|4\rangle$$
$$P^3|3\rangle = 24|2\rangle - 40|3\rangle + 24|4\rangle$$

It is helpful to collect these data in table form (Table 7.4).

From Table 7.4 we then obtain

$$Z_0^{(3)} = 8 + 24x^2 - 16x^3 + 24x^4 - 16x^5 + \tfrac{176}{15}x^6 + \cdots \tag{7.40a}$$

$$Z_2^{(3)} = 6 + 8x + 14x^2 + \tfrac{4}{3}x^3 + \tfrac{26}{3}x^4 - \tfrac{44}{15}x^5 + \tfrac{148}{45}x^6 + \cdots \tag{7.40b}$$

The procedure outlined above is easily and efficiently computerized.

For this cluster, and the smaller ones, it is straightforward to obtain closed form results. In fact

$$Z_0^{(3)} = 4e^{2x} + 2 + 2e^{-4x} \tag{7.41a}$$

$$Z_2^{(3)} = \tfrac{1}{2}(10e^{2x} + 1 + e^{-4x}) \tag{7.41b}$$

The reader may wish to verify these. Exact formulae for the general case of (7.39) are also easily obtained. For larger clusters this becomes intractable.

For the benefit of the reader who wishes to obtain the leading terms of the series (7.30) by hand, we provide Table 7.5, which lists the Z, A and Φ series to order x^6 for the first five clusters. Combining these data, for the linear chain, gives in the thermodynamic limit

$$-\beta F = \ln 2 + \tfrac{3}{2}x^2 - x^3 - \tfrac{5}{4}x^4 + 3x^5 + \tfrac{7}{10}x^6 + \cdots \tag{7.42}$$

$$\chi = \tfrac{1}{4} + \tfrac{1}{2}x - \tfrac{2}{3}x^3 + \tfrac{5}{6}x^4 + \cdots \tag{7.43}$$

Table 7.5. $\mathcal{Z}$, A and Φ series for lowest-order clusters

Cluster	Energy and reduced energy
1. •	$\mathcal{Z}_0^{(1)} = 2, \quad \mathcal{Z}_2^{(1)} = \frac{1}{4}$ $A_0^{(1)} = \Phi_0^{(1)} = \ln 2, \quad A_2^{(1)} = \Phi_2^{(1)} = \frac{1}{4}$
2. •—•	$\mathcal{Z}_0^{(2)} = 4 + 6x^2 - 4x^3 + \frac{7}{2}x^4 - 2x^5 + \frac{61}{60}x^6 + \cdots$ $\mathcal{Z}_2^{(2)} = 2 + 2x + x^2 + \frac{1}{3}x^3 + \frac{1}{12}x^4 + \frac{1}{60}x^5 + \frac{1}{360}x^6 + \cdots$ $A_0^{(2)} = 2\ln 2 + \frac{3}{2}x^2 - x^3 - \frac{1}{4}x^4 + x^5 - \frac{13}{30}x^6 + \cdots$ $A_2^{(2)} = \frac{1}{2} + \frac{1}{2}x - \frac{1}{2}x^2 - \frac{1}{6}x^3 + \frac{5}{6}x^4 - \frac{13}{30}x^5 - \frac{77}{90}x^6 + \cdots$ $\Phi_0^{(2)} = \frac{3}{2}x^2 - x^3 - \frac{1}{4}x^4 + x^5 - \frac{13}{30}x^6 + \cdots$ $\Phi_2^{(2)} = \frac{1}{2}x - \frac{1}{2}x^2 - \frac{1}{6}x^3 + \frac{5}{6}x^4 - \frac{13}{30}x^5 - \frac{77}{90}x^6 + \cdots$
3. •—•—•	$\mathcal{Z}_0^{(3)} = 8 + 24x^2 - 16x^3 + 24x^4 - 16x^5 + \frac{176}{15}x^6 + \cdots$ $\mathcal{Z}_2^{(3)} = 6 + 8x + 14x^2 + \frac{4}{3}x^3 + \frac{26}{3}x^4 - \frac{44}{15}x^5 + \frac{148}{45}x^6 + \cdots$ $A_0^{(3)} = 3\ln 2 + 3x^2 - 2x^3 - \frac{3}{2}x^4 + 4x^5 - \frac{8}{15}x^6 + \cdots$ $A_2^{(3)} = \frac{3}{4} + x - \frac{1}{2}x^2 - \frac{4}{3}x^3 + \frac{7}{3}x^4 + \frac{17}{15}x^5 - \frac{617}{90}x^6 + \cdots$ $\Phi_0^{(3)} = -x^4 + 2x^5 + \frac{1}{3}x^6 + \cdots$ $\Phi_2^{(3)} = \frac{1}{2}x^2 - x^3 + \frac{2}{3}x^4 + 2x^5 - \frac{463}{90}x^6 + \cdots$
4. •—•—•—•	$\mathcal{Z}_0^{(4)} = 16 + 72x^2 - 48x^3 + 118x^4 - 104x^5 + \frac{359}{3}x^6 + \cdots$ $\mathcal{Z}_2^{(4)} = 16 + 24x + 64x^2 + 28x^3 + \frac{158}{3}x^4 + \frac{103}{15}x^5 + \frac{854}{45}x^6 + \cdots$ $A_0^{(4)} = 4\ln 2 + \frac{9}{2}x^2 - 3x^3 - \frac{11}{4}x^4 + 7x^5 + \frac{1}{6}x^6 + \cdots$ $A_2^{(4)} = 1 + \frac{3}{2}x - \frac{1}{2}x^2 - 2x^3 + \frac{8}{3}x^4 + \frac{101}{30}x^5 - \frac{977}{90}x^6 + \cdots$ $\Phi_0^{(4)} = \frac{4}{5}x^6 + \cdots$ $\Phi_2^{(4)} = \frac{1}{2}x^3 - \frac{7}{6}x^4 + \frac{2}{3}x^5 + 2x^6 + \cdots$
5. •—•—•—•—•	$\mathcal{Z}_0^{(5)} = 32 + 192x^2 - 128x^3 + 448x^4 - 448x^5 + \frac{10016}{15}x^6 + \cdots$ $\mathcal{Z}_2^{(5)} = 40 + 64x + 224x^2 + \frac{416}{3}x^3 + 320x^4 + \frac{368}{15}x^5 + \frac{12184}{45}x^6 + \cdots$ $A_0^{(5)} = 5\ln 2 + 6x^2 - 4x^3 - 4x^4 + 10x^5 + \frac{13}{15}x^6 + \cdots$ $A_2^{(5)} = \frac{5}{4} + 2x - \frac{1}{2}x^2 - \frac{8}{3}x^3 + \frac{7}{2}x^4 + \frac{64}{15}x^5 - \frac{643}{45}x^6 + \cdots$ $\Phi_0^{(5)} = O(x^8)$ $\Phi_2^{(5)} = \frac{1}{2}x^4 - \frac{4}{3}x^5 + \frac{17}{30}x^6 + \cdots$

which, with the change of variable, agree with (7.21) and (7.27) of the previous section.

The following points can be noted. The lowest power of K (or x) in the free energy contribution of a cluster is K^{2p}, where p is the number of bonds. This is a special feature of the linear chain, where each bond of a graph must be covered at least twice. This will no longer be the case for other lattices, with closed loops, where a loop of length p will give a K^p contribution. For the susceptibility, on the other hand, a chain of p bonds will give a K^p term. This is evident from Table 7.5 by inspection. Thus the data in Table 7.5 only give the χ series correct to order K^4.

7.2.3 The staggered susceptibility

High-temperature series for the simple nearest-neighbour antiferromagnetic Heisenberg model can be obtained from the ferromagnetic series by simply replacing K by $-K$. However the physical susceptibility, which measures the response to a uniform external field, does not diverge at an antiferromagnetic critical point. It is of interest therefore to calculate the so-called *staggered susceptibility*, which is the response to an unphysical *staggered field* that points in opposite directions on the two sub-lattices in a bipartite structure. The calculation of this quantity involves an added complication, which we will explain within the context of the linked cluster method.

To be specific let us consider the spin-$\frac{1}{2}$ antiferromagnetic chain, with Hamiltonian

$$H = J \sum_i \mathbf{S}_i \cdot \mathbf{S}_{i+1} - h_s \sum_i \eta_i S_i^z \qquad (7.44)$$

where $\eta_i = \pm 1$ on alternate sublattices. The staggered susceptibility is then defined as

$$k_B T \chi_s = \lim_{h_s \to 0} \frac{\partial}{\partial h_s^2} \left(\frac{1}{N} \ln \mathcal{Z} \right) \qquad (7.45)$$

Let us write

$$-\beta H = -x P + \beta h_s M_s \qquad (7.46)$$

with $x = \beta J/4$, $P = 4 \sum_i \mathbf{S}_i \cdot \mathbf{S}_{i+1}$, $M_s = \sum_i \eta_i S_i^z$.

This is completely analogous to the ferromagnetic case discussed previously. However the two terms in (7.44) do not commute and so we cannot follow the steps in (7.5), (7.6). Instead we write

$$\mathcal{Z}(x, h_s) = \text{Tr}\left[e^{-xP + \beta h_s M_s} \right]$$

$$= \sum_{r=0}^{\infty} \frac{1}{r!} \text{Tr}\left[(-xP + \beta h_s M_s)^r \right]$$

$$= \mathcal{Z}_0 + \frac{1}{2} (\beta h_s)^2 \mathcal{Z}_{2s} + \cdots \qquad (7.47)$$

where

$$Z_0 = \sum_{r=0}^{\infty} \frac{(-x)^r}{r!} \text{Tr}(P^r) \tag{7.48}$$

as before, and

$$Z_{2s} = \sum_{r=0}^{\infty} \frac{2(-x)^r}{(r+2)!} \text{Tr}\left(\sum_{\Pi} P^r M_s^2\right) \tag{7.49}$$

In (7.49) the summation (Π) is over all $\binom{r+2}{2}$ permutations of the r P-operators and 2 M_s-operators. This can be simplified, using the cyclic property of the trace, by permuting the operators so that one of the M_s is in the rightmost position. This gives a multiplier $\frac{1}{2}(r+2)$, and consequently

$$Z_{2s} = \sum_{r=0}^{\infty} \frac{(-x)^r}{(r+1)!} \text{Tr}\left(\sum_{k=0}^{r} P^{r-k} M_s P^k M_s\right) \tag{7.50}$$

The staggered susceptibility is then $k_B T \chi_s = Z_{2s}/Z_0$.

We can now use the linked cluster method, as in the previous sub-section, with the proviso that we need to consider each operator string in (7.50) in turn. Let us illustrate this by considering, in some detail, the cluster g_3 shown in Table 7.3. If we choose site 1 on the 'up' sub-lattice then the operator M_s has values $-\frac{1}{2}, \frac{1}{2}, -\frac{3}{2}$ in the three states $|1\rangle, |2\rangle, |3\rangle$ which need to be considered. We denote

$$\Sigma_r(m) = \sum_{k=0}^{r} \langle m | P^{r-k} M_s P^k M_s | m \rangle \tag{7.51}$$

Since $P|1\rangle = 2|1\rangle$ we obtain immediately

$$\Sigma_r(1) = \frac{1}{4} \sum_{k=0}^{r} 2^r = 2^{r-2}(r+1) \tag{7.52}$$

Proceeding with state $|2\rangle$

$$\begin{aligned}
\Sigma_1(2) &= \langle 2 | P M_s M_s | 2 \rangle + \langle 2 | M_s P M_s | 2 \rangle \\
&= 2 \times \frac{1}{4} \langle 2 | P | 2 \rangle = 0 \tag{7.53a} \\
\Sigma_2(2) &= \langle 2 | P^2 M_s M_s | 2 \rangle + \langle 2 | P M_s P M_s | 2 \rangle + \langle 2 | M_s P^2 M_s | 2 \rangle \\
&= 2 \times \frac{1}{4} \langle 2 | P^2 | 2 \rangle + \frac{1}{2} \langle 2 | P M_s P | 2 \rangle \\
&= 2 + 2 \langle 3 | M_s | 3 \rangle = -1 \tag{7.53b}
\end{aligned}$$

and so on. Collecting the data in table form, yields Table 7.6

From (7.50) we then obtain

$$Z_{2s}^{(3)} = 6 + 8x + \frac{58}{3}x^2 + 20x^3 + \frac{102}{5}x^4 + \frac{676}{45}x^5 + \cdots \tag{7.54}$$

Table 7.6. *Trace data for* χ_s *for cluster* g_3.

$|m\rangle$	wt	Σ_0	Σ_1	Σ_2	Σ_3	Σ_4	Σ_5
$|1\rangle$	2	1/4	1	3	8	20	48
$|2\rangle$	4	1/4	0	-1	8	-52	272
$|3\rangle$	2	9/4	-9	57	-264	1308	-6000
Trace		6	-16	116	-480	2448	-10816

Table 7.7. *Series data for staggered susceptibility for lowest-order clusters.*

Cluster	Energy and reduced energy
1. ●	$\mathcal{Z}_{2s}^{(1)} = \frac{1}{2}, \quad A_{2s}^{(1)} = \Phi_{2s}^{(1)} = \frac{1}{4}$
2. ●—●	$\mathcal{Z}_{2s}^{(2)} = 2 + 2x + \frac{7}{3}x^2 + \frac{5}{3}x^3 + \frac{61}{60}x^4 + \frac{91}{180}x^5 + \cdots$
	$A_{2s}^{(2)} = \frac{1}{2} + \frac{1}{2}x - \frac{1}{6}x^2 - \frac{5}{6}x^3 - \frac{13}{30}x^4 + \frac{77}{90}x^5 + \cdots$
	$\Phi_{2s}^{(2)} = \frac{1}{2}x - \frac{1}{6}x^2 - \frac{5}{6}x^3 - \frac{13}{30}x^4 + \frac{77}{90}x^5 + \cdots$
3. ●—●—●	$\mathcal{Z}_{2s}^{(3)} = 6 + 8x + \frac{58}{3}x^2 + 20x^3 + \frac{102}{5}x^4 + \frac{676}{45}x^5 + \cdots$
	$A_{2s}^{(3)} = \frac{3}{4} + x + \frac{1}{6}x^2 - 2x^3 - \frac{11}{5}x^4 + \frac{137}{45}x^5 + \cdots$
	$\Phi_{2s}^{(3)} = \frac{1}{2}x^2 - \frac{1}{3}x^3 - \frac{4}{3}x^4 + \frac{4}{3}x^5 + \cdots$
4. ●—●—●—●	$\mathcal{Z}_{2s}^{(4)} = 16 + 24x + 80x^2 + \frac{340}{3}x^3 + \frac{2398}{15}x^4 + \frac{7573}{45}x^5 + \cdots$
	$A_{2s}^{(4)} = 1 + \frac{3}{2}x + \frac{1}{2}x^2 - \frac{8}{3}x^3 - \frac{62}{15}x^4 + \frac{311}{90}x^5 + \cdots$
	$\Phi_{2s}^{(4)} = \frac{1}{2}x^3 - \frac{1}{6}x^4 - \frac{16}{9}x^5 + \cdots$
5. ●—●—●—●—●	$\mathcal{Z}_{2s}^{(5)} = 40 + 64x + \frac{800}{3}x^2 + \frac{1312}{3}x^3 + \frac{11968}{15}x^4 + \frac{5232}{5}x^5 + \cdots$
	$A_{2s}^{(5)} = \frac{5}{4} + 2x + \frac{5}{6}x^2 - \frac{10}{3}x^3 - \frac{167}{30}x^4 + \frac{58}{15}x^5 + \cdots$
	$\Phi_{2s}^{(5)} = \frac{1}{2}x^4 + 0x^5 + \cdots$

From the previous sub-section (Eq. 7.40) we obtain the zero-field partition function for this cluster (remembering to replace $x \to -x$) as

$$\mathcal{Z}_0^{(3)} = 8 + 24x^2 + 16x^3 + 24x^4 + 16x^5 \cdots \tag{7.55}$$

whence

$$A_{2s}^{(3)} = \mathcal{Z}_{2s}^{(3)}/\mathcal{Z}_0^{(3)} = \frac{3}{4} + x + \frac{1}{6}x^2 - 2x^3 - \frac{11}{5}x^4 + \frac{137}{45}x^5 + \cdots \tag{7.56}$$

Table 7.7 gives the series for $\mathcal{Z}_{2s}$, A_{2s} and Φ_{2s} to order x^5 for the first five clusters.

Combining the Φ_{2s} series, for the linear chain, then gives, for the thermodynamic limit

$$\chi_s = \tfrac{1}{4} + \frac{1}{2}x + \tfrac{1}{3}x^2 - \tfrac{2}{3}x^3 - \tfrac{43}{30}x^4 + \cdots \tag{7.57}$$

which is correct to order x^4.

As was the case for the uniform susceptibility considered in Section 7.2.2, a cluster of p bonds has a leading contribution of x^p. Hence to obtain the series to a given order, all clusters with up to that number of bonds must, in principle, be included. For the chain the number of clusters is small; for other lattices, as we have seen elsewhere, the number of clusters increases rapidly. To obtain series of respectable length it is necessary to computerize the entire process. This is rather straightforward to do, following the examples of this section.

7.3 The cubic (SC and BCC) lattices

In the previous section we explained, in some detail, the procedure for computing high-temperature series for the zero-field free energy (and hence the internal energy, entropy, and specific heat) and the uniform and staggered susceptibilities. In the present section we will discuss the most recent results for the spin-$\frac{1}{2}$ Heisenberg ferromagnet and antiferromagnet in three spatial dimensions. We recall that these systems have a finite-temperature critical point, with universal $N = 3$ critical exponents.

Series for the three-dimensional lattices are currently known to order K^{14} (Oitmaa and Bornilla, 1996; Oitmaa and Zheng, 2004a). We present here the series for the body-centred cubic lattice only. A total of 9453 clusters contribute to this order. Table 7.8 lists the coefficients of the series, in integer format, for both the uniform and staggered susceptibility, expressed in the form

$$\chi = \sum_{n=0}^{\infty} \frac{c_n}{4^{n+1}n!} K^n \tag{7.58a}$$

$$\chi_s = \sum_{n=0}^{\infty} \frac{d_n}{4^{n+1}(n+1)!} K^n \tag{7.58b}$$

The series appear to be quite regular, with monotonically increasing positive coefficients. Thus a ratio analysis is worthwhile. In Figure 7.1 we show a ratio plot of both series. On the same plot, we show, for comparison, the susceptibility series for the classical Heisenberg model. For this latter case, we recall that the staggered susceptibility is obtained from the uniform susceptibility via the simple substitution $K \to -K$, which implies that the Curie and Néel temperatures are equal.

Table 7.8. *Coefficients of the uniform and staggered susceptibilities*
(7.58) for the spin-$\frac{1}{2}$ Heisenberg model on the BCC lattice.

n	c_n for χ	d_n for χ_s
0	1	1
1	8	16
2	96	320
3	1664	8192
4	36800	248768
5	1008768	8919296
6	32626560	367854720
7	1221399040	17216475136
8	51734584320	899434884096
9	2459086364672	51925815320576
10	129082499311616	3280345760086016
11	7432690738003968	225270705859919872
12	464885622793134080	16704037174526894080
13	31456185663820136448	133055713552857792512 0
14	228481523821847126016 0	11328264863992151295590 4

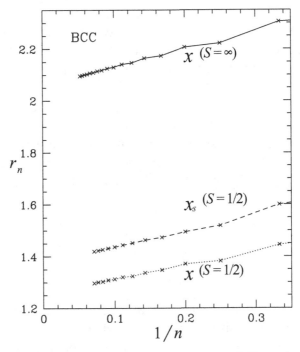

Fig. 7.1. Ratio plot for the uniform and staggered susceptibility series for the $S = \frac{1}{2}$ Heisenberg model on the BCC lattice and, for comparison, the analogous plot for the classical ($S = \infty$) model.

Table 7.9. *Estimates of critical coupling K_C, K_N and exponent γ (in brackets) for the ferromagnet (F) and antiferromagnet (AF) from the $[N, D]$ Padé approximants to $\frac{d}{dK} \ln \chi$ (χ_s).*

	[6,7]	[7,6]	[5,7]	[6,6]	[7,5]	[5,6]	[6,5]
F	0.7935	0.7935	0.7937	0.7936	0.7939	0.7937	0.7936
	(1.416)	(1.416)	(1.419)	(1.417)	(1.423)	(1.418)	(1.418)
AF	0.7266	0.7266	0.7266	0.7264	0.7267	0.7264	0.7264
	(1.436)	(1.435)	(1.434)	(1.431)	(1.436)	(1.432)	(1.432)

A number of points can be noted. Firstly the series are all fairly regular, but show the odd-even oscillation typical of bipartite lattices. Unless something very surprising were to happen at higher order, we can estimate with confidence, from the intercept, that

$$k_B T_C / J \simeq 1.26, \quad k_B T_N / J \simeq 1.38 \tag{7.59}$$

where T_C, T_N are the Curie and Néel temperature respectively. Clearly, in the quantum case, these critical temperatures are unequal, with the Néel temperature the greater. Remembering that the limiting slope of the ratio plot gives $(\gamma - 1)/K_c$, the results appear to confirm the universality prediction that all three quantities diverge with the same exponent γ.

More precise estimates of the critical point and exponent can be obtained from a Padé analysis. Table 7.9 gives estimates of the critical points $K_C = J/k_B T_C$, $K_N = J/k_B T_N$ and the exponent γ from Dlog Padé approximants to the series χ, χ_s.

As is evident, there is good consistency between different approximants and one can reasonably conclude that

$$K_C = 0.7934(2), \quad k_B T_C / J = 1.260(1) \tag{7.60a}$$

$$K_N = 0.7266(2), \quad k_B T_N / J = 1.376(1) \tag{7.60b}$$

consistent with the ratio analysis. The exponent values lie a little above the expected $\gamma = 1.39$. The same effect is seen with series of around 14 terms for the Ising and classical Heisenberg models. Longer series for these models show a regular slow decrease towards the expected values. We would expect the same to occur in the present case.

7.4 Generalizations

This chapter has, so far, focussed on the isotropic spin-$\frac{1}{2}$ Heisenberg model. The same approach can be used for other quantum spin models and a considerable body of work exists. In this section we will summarize this work. The reader interested in these 'special topics' can obtain further details from the literature.

7.4.1 Exchange anisotropy

The Heisenberg model with anisotropic exchange

$$H = -\sum_{\langle ij \rangle} \left(J_x S_i^x S_j^x + J_y S_i^y S_j^y + J_z S_i^z S_j^z \right) \tag{7.61}$$

can be studied via high-temperature expansions, in a similar manner to the previous section. One interesting aspect of this model, as regards its critical point behaviour, is the expected crossover between different universality classes. We have already considered the isotropic O(3) case. However if $J_x = J_y < J_z$ the model will have an O(1) order parameter, and is said to be 'Ising like', while the opposite case $J_x = J_y > J_z$ has O(2) symmetry and is said to be 'XY like'. We are unaware of any series studies of this anisotropic model since the early work of Obokata *et al.* (1967).

The special case, with $J_x = J_y$ and $J_z = 0$, is known as the 'XY model' and has been studied rather extensively, particularly by the Alberta group (Betts, 1974; Rogiers *et al.*, 1978). This model can describe planar magnetic systems, and has also been studied as a model of a quantum lattice fluid. High temperature expansions have been derived for the free energy and the fluctuation of the order parameter, to order K^{13} for all the regular lattices, and analysed to determine the critical behaviour. Undoubtedly, with present techniques and computer power, these series could be extended further if desired.

7.4.2 Further neighbour interactions

We have seen, in Chapter 5, that the inclusion of exchange interactions to further neighbours is often necessary in modelling real materials. Such models may have quite rich phase diagrams, with many different kinds of phases.

The development of high-temperature expansions for such models causes no difficulties apart from increased bookkeeping. Early work, by the Newcastle group, is summarized by Rushbrooke *et al.* (1974). In a recent calculation our group (Oitmaa and Zheng, 2004b) has considered the BCC antiferromagnet with first- and second-neighbour interactions, and substantially extended the previous series. The longer series allow a reasonably precise analysis of the Néel-paramagnetic phase boundary as well as the location of the bicritical point (or perhaps critical end-point) where this boundary meets the antiferromagnetic type 2-paramagnetic phase boundary. The order of the latter phase boundary remains uncertain.

7.4.3 Higher spin

The early work for general S was again from the Newcastle group, and is summarized by Rushbrooke *et al.* (1974). The 'moment method' (Section 7.2.1) can be applied for the general S case with no added difficulty, by generalizing the results for single site traces. As mentioned previously, this is likely to be the most efficient method for larger spin values.

For small S values the linked cluster method is more efficient. We have recently extended the $S = 1$ series for the SC and BCC lattices to order K^{12} for the susceptibility and to order K^{11} for the staggered susceptibility (Oitmaa and Zheng, 2004a). Analysis of these series, together with the $S = \frac{1}{2}$ and $S = \infty$ series, shows clearly that the Néel temperature for the quantum Heisenberg model is higher than the corresponding Curie temperature, with the relative difference decreasing with increasing S, and vanishing in the classical $S = \infty$ limit.

7.5 Perturbation expansions at finite T

The other type of finite-temperature expansion is one in powers of some perturbation parameter λ. In such an expansion the coefficients will be complete functions of $K = J/k_B T$. While such an expansion can be re-expressed as a high-temperature series, of the form considered in previous sections, it contains more 'information' and can be used at lower temperatures, even in an ordered phase and sometime down to $T = 0$.

Let us consider a system with Hamiltonian

$$H = H_0 + \lambda V \tag{7.62}$$

where, in general, H_0 and V do not commute. As shown in Section 1.4.4, the partition function can be written as an expansion in powers of λ, in the form

$$\mathcal{Z}(T) = Z_0 \left\{ 1 + \sum_{n=1}^{\infty} (-\lambda)^n \int_0^{\beta} d\tau_1 \int_0^{\tau_1} d\tau_2 \cdots \int_0^{\tau_{n-1}} d\tau_n \langle \widetilde{V}(\tau_1) \widetilde{V}(\tau_2) \cdots \widetilde{V}(\tau_n) \rangle_0 \right\} \tag{7.63}$$

where Z_0 is the unperturbed partition function,

$$\widetilde{V}(\tau) = e^{\tau H_0} V e^{-\tau H_0} \tag{7.64}$$

and the angular bracket denotes the unperturbed expectation value

$$\langle \widetilde{V}(\tau_1)\widetilde{V}(\tau_2)\cdots\widetilde{V}(\tau_n)\rangle_0 = Z_0^{-1}\mathrm{Tr}\{e^{-\beta H_0}\widetilde{V}(\tau_1)\widetilde{V}(\tau_2)\cdots\widetilde{V}(\tau_n)\} \tag{7.65}$$

To see how to proceed let us assume that the unperturbed Hamiltonian consists of independent 'single site' terms (a 'site' here can be a single spin, a dimer, or some other small complex). Each site i has n quantum states $\{|m_i\rangle, m_i = 1, 2, \cdots, n\}$, assumed to be eigenstates of $H_0(i)$ with energies ϵ_{m_i}. A complete set of states for the system is then given by the direct-product basis

$$|m\rangle = |m_1, m_2, \cdots, m_N\rangle = |m_1\rangle|m_2\rangle\cdots|m_N\rangle \tag{7.66}$$

Then

$$Z_0 = z_0^N \tag{7.67}$$

where z_0 is the single site partition function

$$z_0 = \sum_{r=1}^{n} e^{-\beta\epsilon_r} \tag{7.68}$$

The trace in (7.65) is taken with respect to the product basis (7.66). For simplicity let us consider the second order term

$$\mathrm{Tr}\{e^{-\beta H_0}\widetilde{V}(\tau_1)\widetilde{V}(\tau_2)\} = \sum_{m,p} e^{-\beta E_m}\langle m|\widetilde{V}(\tau_1)|p\rangle\langle p|\widetilde{V}(\tau_2)|m\rangle$$

$$= \sum_{m,p} e^{-\beta E_m} e^{\tau_1(E_m-E_p)} e^{\tau_2(E_p-E_m)}\langle m|V|p\rangle\langle p|V|m\rangle \tag{7.69}$$

and the entire term is

$$\lambda^2 \sum_{m,p} e^{-\beta E_m}\langle m|V|p\rangle\langle p|V|m\rangle \int_0^\beta d\tau_1 e^{\tau_1(E_m-E_p)} \int_0^{\tau_1} d\tau_2 e^{\tau_2(E_p-E_m)}$$

$$= \lambda^2 \sum_{m,p} e^{-\beta E_m}\langle m|V|p\rangle\langle p|V|m\rangle I(\Delta, -\Delta) \tag{7.70}$$

where $\Delta = E_m - E_p$ and

$$I(\Delta, -\Delta) = \int_0^\beta d\tau_1 e^{\Delta\tau_1} \int_0^{\tau_1} d\tau_2 e^{-\Delta\tau_2} = \begin{cases} \frac{1}{\Delta^2}(e^{\beta\Delta} - 1 - \beta\Delta); & \Delta \neq 0 \\ \frac{1}{2}\beta^2; & \Delta = 0 \end{cases} \tag{7.71}$$

Higher-order terms will involve more matrix elements and more complicated integrals, but the steps are the same.

In summary, the procedure is as follows:

(i) Choose an initial state $|m\rangle$.
(ii) Compute all possible paths in state space leading from $|m\rangle$ by the repeated action of V, each path being weighted by a product of matrix elements, and ending in the same state $|m\rangle$.
(iii) For each complete path evaluate the energy differences Δ and evaluate the multiple τ-integral.
(iv) Multiply each term by the Boltzmann factor $e^{-\beta E_m}$, and accumulate results.
(v) Return to step (i), and repeat until all initial states have been used.

There are several 'tricks' that can be used to speed up the computation. In any particular application the energy differences are restricted to a small set. Consequently the number of separate integrals needed is finite, and usually not too large. They can be computed at the outset and stored in a lookup table. Secondly, rather than attempting to compute all $(n-1)$ step paths from state $|m\rangle$, most of which will not reach $|m\rangle$ in the next step, and hence represent wasted effort, it is better to compute 'half-paths' of $n/2$ steps and combine these. We will illustrate this in the next subsection, with the aid of a simple example.

7.5.1 The alternating $S = \frac{1}{2}$ antiferromagnetic chain

To illustrate the general method described above we consider an example: the alternating chain, considered in Section 5.2.1. Let us write the Hamiltonian as

$$H = J \sum_i (\mathbf{S}_{2i} \cdot \mathbf{S}_{2i+1} + \tfrac{3}{4}) + 4\lambda J \sum_i \mathbf{S}_{2i-1} \cdot \mathbf{S}_{2i} \tag{7.72}$$

where we have added a constant term, and introduced a factor of 4, for convenience. Furthermore we will set $J = 1$. The uniform chain then corresponds to $\lambda = 1/4$. This is not the simplest example we could have chosen, as the 'sites' are dimers with four states, and the matrix elements are not as simple as between single spins. However it shows some important features which are worth illustrating.

The 'single site' states and their energies are

$$
\begin{array}{ll}
|1\rangle = \frac{1}{\sqrt{2}}[|+-\rangle - |-+\rangle] & \epsilon_1 = 0 \\
|2\rangle = |++\rangle & \epsilon_2 = 1 \\
|3\rangle = \frac{1}{\sqrt{2}}[|+-\rangle + |-+\rangle] & \epsilon_3 = 1 \\
|4\rangle = |--\rangle & \epsilon_4 = 1
\end{array}
\tag{7.73}
$$

where the dimer is labelled $\overset{\bullet\;\;\;\bullet}{1\;\;\;2}$ and the dimer states are ordered as $|s_1, s_2\rangle$. Because of the minus sign in the singlet wavefunction it is necessary to be careful with this ordering. The notation in (7.73) is a little different from that used in the previous

Table 7.10. *Input data on dimer states.*

State $|m\rangle$	Energy	$V|m\rangle$
$|1\rangle = |1, 1\rangle$	0	$|8\rangle - |11\rangle + |14\rangle$
$|2\rangle = |1, 2\rangle$	1	$-|5\rangle + |7\rangle - |10\rangle$
$|3\rangle = |1, 3\rangle$	1	$|8\rangle - |9\rangle - |14\rangle$
$|4\rangle = |1, 4\rangle$	1	$|12\rangle - |13\rangle - |15\rangle$
$|5\rangle = |2, 1\rangle$	1	$-|2\rangle + |7\rangle - |10\rangle$
$|6\rangle = |2, 2\rangle$	2	$|6\rangle$
$|7\rangle = |2, 3\rangle$	2	$|2\rangle + |5\rangle + |10\rangle$
$|8\rangle = |2, 4\rangle$	2	$|1\rangle + |3\rangle - |8\rangle + |9\rangle + |11\rangle$
$|9\rangle = |3, 1\rangle$	1	$-|3\rangle + |8\rangle - |14\rangle$
$|10\rangle = |3, 2\rangle$	2	$-|2\rangle - |5\rangle + |7\rangle$
$|11\rangle = |3, 3\rangle$	2	$-|1\rangle + |8\rangle + |14\rangle$
$|12\rangle = |3, 4\rangle$	2	$|4\rangle + |13\rangle + |15\rangle$
$|13\rangle = |4, 1\rangle$	1	$-|4\rangle + |12\rangle - |15\rangle$
$|14\rangle = |4, 2\rangle$	2	$|1\rangle - |3\rangle - |9\rangle + |11\rangle - |14\rangle$
$|15\rangle = |4, 3\rangle$	2	$-|4\rangle + |12\rangle - |13\rangle$
$|16\rangle = |4, 4\rangle$	2	$|16\rangle$

chapter. The single site partition function is

$$z_0 = 1 + 3e^{-\beta} \tag{7.74}$$

The next step is to compute the matrix elements $\langle p|V|m\rangle$. Since each operator in V couples a pair of adjacent dimers it suffices to compute the effect of the operator $V_p = 4\mathbf{S}_1 \cdot \mathbf{S}_2$ on all possible pair states $|m_1, m_2\rangle$ where the labelling is ●——● ‑‑ ●——●
           1   2   . In Table 7.10 we list the 16 pair states, their energies, and the effects of V_p on each.

We will use the linked cluster approach, as in Section 7.2.2. The list of clusters is, as before,

●   ●——●   ●——●   ●——●——● $\cdots$
1   2   3   4

We illustrate the calculation of the zero-field free energy to order λ^4.

For cluster 1 we have simply

$$A^{(1)} = \Phi^{(1)} = \ln z_0 \tag{7.75}$$

For cluster 2 we write

$$\mathcal{Z}^{(2)} = z_0^{-2} \left\{ 1 + \sum_{n=1}^{\infty} (-\lambda)^n T_n^{(2)} \right\} \tag{7.76}$$

where

$$T_n^{(2)} = z_0^2 \sum_{m, p_1, \cdots, p_{n-1}} e^{-E_m} I(l_1, l_2, \cdots, l_n) \times$$
$$\langle m|V|p_1\rangle\langle p_1|V|p_2\rangle \cdots \langle p_{n-1}|V|m\rangle \tag{7.77}$$

and $l_k = E_{p_k} - E_{p_{k+1}}$.

The λ term in (7.76) vanishes because V has no diagonal matrix elements. The λ^2 term is given by (7.70). Writing out all the possible paths $|m\rangle \to |p\rangle \to |m\rangle$, and collecting terms, gives

$$z_0^{-2}T_2^{(2)} = 3I(-2, 2) + 6e^{-\beta}I(0, 0) + 12e^{-\beta}I(-1, 1)$$
$$+ 12e^{-2\beta}I(0, 0) + 12e^{-2\beta}I(1, -1) + 3e^{-2\beta}I(2, -2)$$
$$= \tfrac{3}{2}\beta(1 + 8e^{-\beta} - 9e^{-2\beta}) + 3\beta^2(e^{-\beta} + 2e^{-2\beta}) \tag{7.78}$$

It is convenient to introduce the new variable $y = e^{-\beta}/(1 + 3e^{-\beta})$. Then

$$T_2^{(2)} = \tfrac{3}{2}\beta(1 + 2y - 24y^2) + 3\beta^2(y - y^2) \tag{7.79}$$

Continuing in the same way, we obtain

$$T_3^{(2)} = -[\tfrac{3}{2}\beta(1 - 6y + 8y^2) + 3\beta^2(4y - 17y^2) + \beta^3 y^2] \tag{7.80a}$$
$$T_4^{(2)} = \tfrac{3}{8}\beta(1 - 134y + 520y^2) + \tfrac{3}{8}\beta^2(3 + 46y - 106y^2)$$
$$+ 3\beta^3(2y - 9y^2) + \tfrac{1}{4}\beta^4(y + y^2) \tag{7.80b}$$

Taking the logarithm gives

$$\ln \mathcal{Z}^{(2)} = 2\ln z_0 + T_2^{(2)}\lambda^2 - T_3^{(2)}\lambda^3 + \left(T_4^{(2)} - \tfrac{1}{2}T_2^{(2)}T_2^{(2)}\right)\lambda^4 + \cdots \tag{7.81}$$

Subtracting off the sub-graph term $(2\Phi^{(1)})$ then gives for the reduced free energy of cluster 2

$$\Phi^{(2)} = [\tfrac{3}{2}\beta(1 + 2y - 24y^2) + 3\beta^2(y - y^2)]\lambda^2$$
$$+ [\tfrac{3}{2}\beta(1 - 6y + 8y^2) + 3\beta^2(4y - 17y^2) + \beta^3 y^2]\lambda^3$$
$$+ [\tfrac{3}{8}\beta(1 - 134y + 520y^2) + \tfrac{3}{4}\beta^2(17y + 13y^2 + 144y^3 - 864y^4)$$
$$+ \tfrac{3}{2}\beta^3(y - 21y^2 + 78y^3 - 72y^4) + \tfrac{1}{4}\beta^4(y - 17y^2 + 36y^3 - 18y^4)]\lambda^4$$
$$+ \cdots \tag{7.82}$$

The reader who has checked this will appreciate that it is already a very tedious calculation. To proceed to higher order for this cluster, or even to this order for larger clusters requires the process to be computerized. This is quite straightforward to do.

For cluster 3, the free energy is found to be

$$
\begin{aligned}
\ln \mathcal{Z}^{(3)} = {} & 3 \ln z_0 + [3\beta(1 + 2y - 24y^2) + 6\beta^2(y - y^2)]\lambda^2 \\
& + [3\beta(1 - 6y + 8y^2) + 6\beta^2(4y - 17y^2) + 2\beta^3 y^2]\lambda^3 \\
& + [\tfrac{1}{2}\beta(4 - 135y + 330y^2 + 584y^3) + 3\beta^2(7y - 34y^2 + 333y^3 - 864y^4) \\
& + 3\beta^3(2y - 33y^2 + 121y^3 - 144y^4) + \beta^4(y - 15y^2 + 34y^3 - 18y^4)]\lambda^4 \\
& + \cdots
\end{aligned}
\tag{7.83}
$$

Then

$$
\begin{aligned}
\Phi^{(3)} = {} & \ln \mathcal{Z}^{(3)} - 3\Phi^{(1)} - 2\Phi^{(2)} \\
= {} & [\tfrac{1}{4}\beta(5 + 132y - 900y^2 + 1168y^3) - \tfrac{9}{2}\beta^2(y + 27y^2 - 174y^3 + 288y^4) \\
& + 3\beta^3(y - 12y^2 + 43y^3 - 72y^4) + \tfrac{1}{2}\beta^4(y - 13y^2 + 32y^3 - 18y^4)]\lambda^4 \\
& + \cdots
\end{aligned}
\tag{7.84}
$$

We note that the leading term in $\Phi^{(3)}$ is of order λ^4 i.e. each bond in the cluster must be covered twice by a V operator.

Consequently, to obtain the bulk free energy to order λ^4 we need go no further. We have

$$
\begin{aligned}
\frac{1}{N} \ln \mathcal{Z} = {} & \Phi^{(1)} + \Phi^{(2)} + \Phi^{(3)} + \cdots \\
= {} & \ln z_0 + [\tfrac{3}{2}\beta(1 + 2y - 24y^2) + 3\beta^2(y - y^2)]\lambda^2 \\
& + [\tfrac{3}{2}\beta(1 - 6y + 8y^2) + 3\beta^2(4y - 17y^2) + \beta^3 y^2]\lambda^3 \\
& + \Big[\tfrac{1}{8}\beta(13 - 138y - 240y^2 + 2336y^3) + \tfrac{3}{4}\beta^2(11y - 149y^2 \\
& + 1188y^3 - 2592y^4) + \tfrac{3}{2}\beta^3(3y - 45y^2 + 164y^3 - 216y^4) \\
& + \tfrac{1}{4}\beta^4(3y - 43y^2 + 100y^3 - 54y^4)\Big]\lambda^4 + \cdots
\end{aligned}
\tag{7.85}
$$

It is instructive to look at two limiting cases of this general result. Firstly let us consider the high-temperature limit and expand everything in powers of β. This gives

$$
\begin{aligned}
\frac{1}{N} \ln \mathcal{Z} = {} & 2\ln 2 - \tfrac{3}{4}\beta + (\tfrac{3}{32} + \tfrac{3}{2}\lambda^2)\beta^2 + (\tfrac{1}{64} + \lambda^3)\beta^3 \\
& - (\tfrac{1}{1024} + \tfrac{1}{8}\lambda^2 + \tfrac{1}{4}\lambda^4)\beta^4 + \cdots
\end{aligned}
\tag{7.86}
$$

which is a high-temperature expansion for the alternating chain. This could have been derived directly, with less effort, using the methods of Section 7.2. The coefficients are complete functions of λ. Evaluating these for the uniform chain ($\lambda = 1/4$)

gives $\ln \mathcal{Z}$ per site as

$$\frac{1}{2N} \ln \mathcal{Z} = \ln 2 - \tfrac{3}{8}\beta + \tfrac{3}{32}\beta^2 + \tfrac{1}{64}\beta^3 - \tfrac{5}{1024}\beta^4 + \cdots \qquad (7.87)$$

which agrees with the previous result (7.42), with the necessary change of variables. The linear term $-\tfrac{3}{8}\beta$ comes from the constant term $\tfrac{3}{4}J$ in (7.72).

The other interesting limit is $T \to 0$ (or $y \to 0$). Keeping the terms independent of y gives

$$\frac{1}{N} \ln \mathcal{Z} = \beta \left(\tfrac{3}{2}\lambda^2 + \tfrac{3}{2}\lambda^3 + \tfrac{13}{8}\lambda^4 + \cdots \right) \qquad (7.88)$$

or

$$E_0/N = \lim_{\beta \to \infty} \left(-\frac{\beta}{N} \ln \mathcal{Z} \right)$$
$$= -\tfrac{3}{2}\lambda^2 - \tfrac{3}{2}\lambda^3 - \tfrac{13}{8}\lambda^4 + \cdots \qquad (7.89)$$

which agrees with the $T = 0$ expansion for the ground-state energy in Section 5.2.1. In this case, then, the finite-T perturbation expansion can be used down to $T = 0$. We expect this to be the case for any model whose unperturbed ground state is unique.

A program *tpert.f* is provided (www.cambridge.org/9780521842426), which was used to obtain the results of this section, and which can be easily adapted to other systems.

7.6 Further applications

In the previous section we developed the technique of finite temperature perturbation theory for quantum spin models, using as an example the $S = \tfrac{1}{2}$ one-dimensional antiferromagnet. This approach, or variations of it, have been used on many different models and lattices. Some of this work is summarized here, although we do not attempt to be exhaustive.

In the late 1970s Wang and coworkers (Wang and Lee, 1977; Rauchwarger *et al.*, 1979; Wentworth and Wang, 1989) studied various models with crystal field terms, albeit with rather short series. Using present day methods it should be possible to extend this work significantly. Pan (2000) has obtained eighth order series for the Heisenberg antiferromagnet in a magnetic field, by taking a mean-field part as the unperturbed Hamiltonian. This allows the critical line in the field–temperature plane to be obtained.

Thermodynamic perturbation expansions have been carried out on several systems with dimerized ground states. Elstner and Singh (1998a,b) have obtained eighth order series for both the free energy and susceptibility of ladder and bilayer systems (cf. Section 5.2.2). Zheng *et al.* (1999) have obtained finite-temperature dimer expansions for the SCBO lattice (cf. Section 5.4.2) and Rosner *et al.* (2003) have obtained high-temperature expansions for the J_1–J_2 model (cf. Section 5.3) about the limit of decoupled sub-lattices.

A goal of much of this later work has been to fit experimental specific heat and susceptibility data for real materials. This will be discussed in the next section.

7.7 Fitting to experimental data

One of the goals of deriving high-temperature series for lattice models is to attempt to determine, firstly, the applicability of a particular model to some real material and, secondly, the best values of the model parameters. The experimental data at hand usually include the specific heat and uniform zero-field susceptibility, as functions of temperature, and perhaps also field dependent data. We should remark, at the outset, that thermodynamic data, on their own, are relatively insensitive to model details and precise parameter values and it is advisable, wherever possible, to also fit zero-temperature quantities such as excitation energies and structure factor data. This has been discussed previously, in Chapters 5 and 6.

We have also mentioned previously that finite length high-temperature expansions cannot be extrapolated to arbitrarily low temperatures, and certainly not past a phase transition. The only exception, with thermodynamic perturbation theory, is when the system has a simple ground state, as described in Section 7.5.1. Here it is possible to evaluate the series coefficients at arbitrary temperature but, even so, the convergence at low temperatures may be poor. Bernu and Misguich (2001) have proposed a clever interpolation scheme for high-temperature specific heat series, which makes use of the total entropy of the system, the ground state energy at $T = 0$, and the expected low-temperature behaviour of the specific heat, and can obtain very good convergence down to zero temperature.

Let us consider, as an example, the applicability of the J_1–J_2 square lattice antiferromagnet (Section 5.3) to the material Li_2VOSiO_4. The experimental data have been obtained by Melzi *et al.* (2000). High-temperature series have been derived by Rosner *et al.* (2003), and independently to one higher order by Misguich *et al.* (2003).

We allow for independent variation of J_1, J_2 and g, the effective electron g-factor, and define a 'goodness of fit' parameter

$$P = \sum_{T_i} |A^{\text{exp}}(T_i) - A^{\text{theo}}(T_i)| \qquad (7.90)$$

where A is either χ or C_v. The summation is over the experimental points, starting

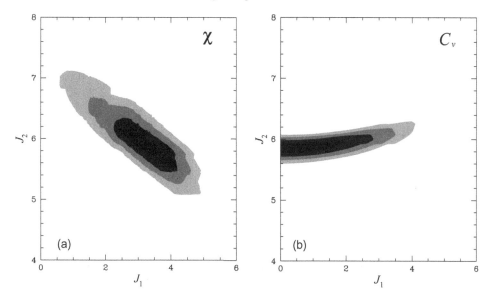

Fig. 7.2. Contour plot for P for (a) susceptibility χ (with g fixed to be 2, and the shaded areas having $P = 2, 3, 4, 5$, respectively, in arbitrary unit) and (b) specific heat C_v (the shaded areas having $P = 6, 8, 10, 12$, respectively).

from a value sufficiently high to achieve good series convergence. In the present calculation we have chosen the region of summation to be $T \gtrsim 3K$ for χ and $T \gtrsim 6K$ for C_v. Choosing the square rather than the absolute value of the deviates in (7.90) makes little difference to the analysis. Figure 7.2(a,b) show contours of P, in arbitrary units, in the J_1–J_2 plane, assuming $g = 2$.

As can be seen, the specific heat fit is relatively insensitive to the value of J_1 in the approximate range $(0 < J_1 < 2)$, and indicates a best $J_2 \sim 5.9K$. On the other hand the susceptibility fit is more sensitive and gives $J_1 \sim 2.6K$, $J_2 \sim 5.85K$. By allowing a variation in g as well, our best numerical fit gives

$$J_1 \sim 1.8K, \quad J_2 \sim 5.9K, \quad g \sim 1.93 \tag{7.91}$$

Figure 7.3 shows the fit of the theoretical curves, using these parameter values, to the experimental results. The susceptibility fit is excellent down to around $T_{\text{peak}}/2$. Above the peak different approximants are indistinguishable, while below the peak a small variation between different approximants is seen. For the specific heat we have used the extrapolation method of Bernu and Misguich (2001). The fit is good in the range $(T_{\text{peak}}, 3T_{\text{peak}})$, but less good at higher temperatures. The experimental values near $2.8K$ reflect the onset of three-dimensional magnetic order, which is not accounted for in the model.

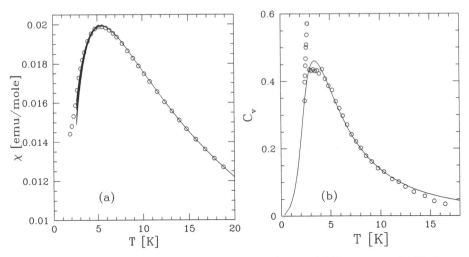

Fig. 7.3. Susceptibility (a) and specific heat (b) for Li_2VOSiO_4 compared with the J_1–J_2 model results for $J_1 = 1.8K$, $J_2 = 5.9K$, and $g = 1.93$.

Misguich *et al.* (2003) use a slightly more involved, and presumably more accurate, procedure for fitting the specific heat curves, and conclude

$$J_1 = 1.25 \pm 0.5K, \quad J_2 = 5.95 \pm 0.2K \tag{7.92}$$

which are consistent with our estimates. They do not allow for a variation in g.

We stress, however, our earlier caution about determining model parameters from high-temperature data alone.

8

Electronic models

8.1 Introduction

In the preceding chapters we have developed a number of series expansion methods, and shown how to use these to study lattice spin models of various kinds. While this encompasses a great deal of interesting physics it does not include perhaps the richest, most interesting, and most challenging area of all: strongly correlated electron systems. Of particular interest here are the cuprate high T_c superconductors and the manganite 'colossal magnetoresistance' (CMR) materials. These systems are far from fully understood but it seems clear that a complex and subtle interplay between charge, spin, and perhaps other degrees of freedom plays a vital role.

A number of generic lattice models for strongly correlated electron systems exist. Besides the Hubbard model, which was briefly introduced in Chapter 1, there is its derivative the 't–J model', the Anderson model, the Kondo lattice model, the Falicov–Kimball model, and others. These all present difficult problems, with little in the way of exact or rigorous results. Many approximate approaches, both analytical and numerical, have been used. In this chapter we will discuss the application of series methods, both at $T = 0$ and at finite temperatures, to some of these models. Some successes have been achieved but, generally speaking, the series are shorter and the analysis more problematic than for spin models. A particular difficulty, for ground-state calculations, is that series are not well suited to studying continuously variable electron density, although we will illustrate one attempt to overcome this. In experiments on real materials, of course, the electron density can often be varied by doping and the variation in physical properties with doping can often be the most striking feature.

Other approaches also have their difficulties. Exact diagonalizations are limited to very small systems and have large finite size effects while quantum Monte Carlo studies of fermionic systems are plagued by the notorious *sign problem*. Thus series methods do have an important place in this area.

8.2 The Hubbard model

We will start our discussion with the Hubbard model, defined by the Hamiltonian

$$H = -t \sum_{\langle ij \rangle, \sigma} (c_{i\sigma}^{\dagger} c_{j\sigma} + c_{j\sigma}^{\dagger} c_{i\sigma}) + U \sum_{i} n_{i\uparrow} n_{i\downarrow} \tag{8.1}$$

As already mentioned in Chapter 1, this is arguably the simplest model to introduce Coulomb repulsion (the U-term) into a single band of free electrons. The operator $c_{i\sigma}^{\dagger} c_{j\sigma}$ transfers an electron of spin $\sigma (= \uparrow, \downarrow)$ from a localized (Wannier) orbital at site j to one at site i. The 't-term' in (8.1) thus describes non-interacting electrons, in a real-space representation. This term can be written, in **k**-space, as

$$\sum_{\mathbf{k}, \sigma} \epsilon_{\mathbf{k}} c_{\mathbf{k}\sigma}^{\dagger} c_{\mathbf{k}\sigma} \tag{8.2}$$

where $\epsilon_{\mathbf{k}} = -t \sum_{\delta} e^{i \mathbf{k} \cdot \delta}$, the sum being over nearest-neighbour lattice vectors δ. This is just the usual band description. The U-term represents an energy cost if two electrons of opposite spin occupy the same orbital. The Pauli principle, of course, forbids double occupancy for electrons of the same spin. It is this innocent-looking term which introduces strong correlations into the motion of the electrons, and makes this simple model difficult to analyse.

We can qualitatively understand some of the physics of the model by considering a variation in electron density n. For small $n \ll 1$ the probability of double occupancy will be small and we would expect metallic behaviour. On the other hand, near and at 'half-filling' ($n = 1$) for large U the electron motion will be strongly suppressed. At exactly $n = 1$ and $U = \infty$ all sites will be occupied by a single electron, and no hopping will occur. Virtual transitions will, as we will show, favour a state in which electrons of opposite spin occupy neighbouring sites – the system will be an antiferromagnetic insulator.

These conclusions are demonstrated rigorously in one dimension, via the exact solution of Lieb and Wu (1968). At half-filling the ground state energy per site, in the thermodynamic limit, is given by the integral

$$E_0/N = -4t \int_0^{\infty} \frac{J_0(\omega) J_1(\omega)}{\omega (1 + e^{\omega U/2t})} d\omega \tag{8.3}$$

where J_0 and J_1 are Bessel functions. Takahashi (1971) showed that the energy is non-analytic at the point $U = 0$. Thus the weak-coupling limit is not amenable to perturbation analysis. Presumably the same is true in higher dimensions, although we are unaware of any rigorous analysis. However the opposite limit, $t = 0$, (or more correctly $U = \infty$) is an analytic point and we will consider such 'strong coupling' expansions in the next subsection.

Before embarking on this, a few more general remarks are worth making. We consider only the case $U > 0$ here – the 'attractive' ($U < 0$) Hubbard model has also been considered and also has rich physical properties. For a bipartite lattice the model is invariant under a sign change of the hopping parameter t. Thus all physical properties will be even functions of t. This is not true for non-bipartite lattices. Another general symmetry property follows from a 'particle-hole' transformation $c_{i\sigma} \rightarrow c_{i\sigma}^\dagger$, which leads to

$$H(n, t, U) \rightarrow (N - N_e)U + H(2 - n, -t, U) \qquad (8.4)$$

which, apart from a trivial constant, relates the models below and above half-filling. For a bipartite lattice this implies that all properties, and in particular the phase diagram, are symmetric about the line $n = 1$.

Readers who wish to learn more about the Hubbard model and, in particular, the results obtained by other methods are referred to several good sources (Rasetti, 1991; Baeriswyl *et al.*, 1995). We will now move on to series work.

8.2.1 t/U expansions at zero temperature

We remarked above that the strong coupling limit $U = \infty$ is an analytic point and thus it seems natural to think of expansions about this point, in powers of t/U.

For the one-dimensional lattice at half-filling the following expansion for the ground-state energy can be derived from the exact result (8.3) (Takahashi, 1971)

$$-E_0/4Nt = (\ln 2)\frac{t}{U} + \sum_{r=2}^\infty (-1)^{r-1} a_{2r-1} \left(\frac{t}{U}\right)^{2r-1} \qquad (8.5)$$

with

$$a_r = \left(\frac{r!!}{(r+1)!!}\right)^2 \frac{2^{2r}}{r}\left(1 - \frac{1}{2^{r-1}}\right)\zeta(r) \qquad (8.6)$$

where $\zeta(r)$ is the Reimann zeta function and the double factorial notation is standard. Explicitly

$$\frac{E_0}{N} = -4\ln 2\frac{t^2}{U} + 9\zeta(3)\frac{t^4}{U^3} - 75\zeta(5)\frac{t^6}{U^5} + \frac{11025}{16}\zeta(7)\frac{t^8}{U^7} + \cdots \qquad (8.7)$$

Takahashi (1971) has shown that the expression (8.3), while analytic on the positive real axis, has logarithmic branch points at $t/U = \pm in/4$ ($n = 1, 2 \ldots$), and hence the series (8.7) has a radius of convergence $|t/U| = 1/4$.

If the exact result (8.3) were not known but, by some means, we were able to derive a reasonable number of terms in the series (8.5) we could use Padé or IDA methods to estimate the ground state energy. As an illustration, we have done this

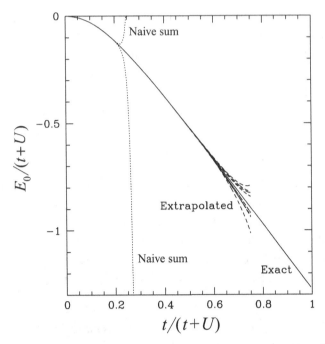

Fig. 8.1. The rescaled ground state energy $E_0/(t + U)$ versus $t/(t + U)$ for the Hubbard chain. Shown are the exact result, the extrapolated result for a series to order t^{20} by IDA methods, and also the naive sum to the series (to order t^{18} and t^{20}).

by taking the series (8.5) to order t^{20} (10 non-zero coefficients). Figure 8.1 shows the comparison of the series estimates with the exact value. To show the whole range $0 < t/U < \infty$ we choose the variable $t/(t + U)$ as the abscissa.

As can be seen, the agreement is good to $t/U \simeq 1.5$, well beyond the radius of convergence, but eventually the estimates from this short series fail.

Unfortunately there is no known way of computing these series coefficients directly, even for the one-dimensional case. This is due to the infinite degeneracy of the $t = 0$ ground state. As we will show shortly, the coefficients of the series (8.5), or corresponding series for other higher dimensional lattices, can be expressed in terms of correlators for an equivalent spin model, but the exact calculation of these is a formidable and unsolved task. Away from half-filling the situation is even more complex. For the one-dimensional case the first five terms have been obtained (Carmelo and Baeriswyl, 1988), and give an accurate estimate of the ground-state energy for small t/U.

An alternative approach to t/U expansions for the Hubbard model is to work at the Hamiltonian level, and to seek an effective Hamiltonian, valid for small t/U, via a series of canonical transformations. This has a long history and many people

Electronic models

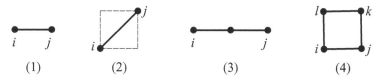

Fig. 8.2. Lattice configurations for the spin Hamiltonian (8.8).

have made important contributions. Our discussion follows closely the recent and comprehensive treatment of MacDonald *et al.* (1988, 1990). The basic idea is the following. The ground state for $t = 0$ (for $n \le 1$, which we assume hereafter) lies in the sector with no doubly occupied sites. The hopping term, acting on a given state, will produce states for which the change in double occupancy $\Delta D = 0, \pm 1$. It is possible to successively eliminate terms in H which couple sectors with different numbers of double-occupied sites, and the result is an effective Hamiltonian for the ground state sector, in the form of a t/U expansion. The same result can be obtained, somewhat differently, via degenerate state perturbation theory and projection operators (Takahashi, 1977).

At half-filling the ground state sector has dimension 2^N, and the effective Hamiltonian can be mapped onto an equivalent spin-$\frac{1}{2}$ Hamiltonian. To order t^4 the result is

$$
\begin{aligned}
H_{\text{eff}} = &-\frac{t^2}{U} \sum_{\langle ij \rangle}^{(1)} (1 - \sigma_i \cdot \sigma_j) + \frac{t^4}{U^3} \Bigg\{ 4 \sum_{\langle ij \rangle}^{(1)} (1 - \sigma_i \cdot \sigma_j) \\
&- 2 \sum_{\langle ij \rangle}^{(2)} (1 - \sigma_i \cdot \sigma_j) - \sum_{\langle ij \rangle}^{(3)} (1 - \sigma_i \cdot \sigma_j) \\
&+ \sum_{\langle ijkl \rangle}^{(4)} [1 - (\sigma_i \cdot \sigma_j + \sigma_j \cdot \sigma_k + \sigma_k \cdot \sigma_l + \sigma_l \cdot \sigma_i) \\
&- (\sigma_i \cdot \sigma_k + \sigma_j \cdot \sigma_l) + 5((\sigma_i \cdot \sigma_j)(\sigma_k \cdot \sigma_l) + (\sigma_i \cdot \sigma_l)(\sigma_j \cdot \sigma_k) \\
&- (\sigma_i \cdot \sigma_k)(\sigma_j \cdot \sigma_l))] \Bigg\} + \cdots
\end{aligned}
\tag{8.8}
$$

where $\sigma_i = 2\mathbf{S_i}$ are the Pauli operators and the sums labelled $1, \ldots, 4$ are over the configurations shown in Figure 8.2.

For the one-dimensional case configurations (2) and (4) do not occur, and we can write explicitly

$$
\begin{aligned}
H_{\text{eff}} = &-\frac{t^2}{U} \left(N - \sum_i \sigma_i \cdot \sigma_{i+1} \right) \\
&+ \frac{t^4}{U^3} \left(3N - 4 \sum_i \sigma_i \cdot \sigma_{i+1} + \sum_i \sigma_i \cdot \sigma_{i+2} \right) + \cdots
\end{aligned}
\tag{8.9}
$$

from which we obtain the ground-state energy per site as

$$E_0/N = -\frac{t^2}{U}(1 - \langle \boldsymbol{\sigma}_0 \cdot \boldsymbol{\sigma}_1 \rangle_0)$$

$$+ \frac{t^4}{U^3}(3 - 4\langle \boldsymbol{\sigma}_0 \cdot \boldsymbol{\sigma}_1 \rangle_0 + \langle \boldsymbol{\sigma}_0 \cdot \boldsymbol{\sigma}_2 \rangle_0) + \cdots \quad (8.10)$$

where $\langle\ \rangle_0$ denotes a ground-state expectation value. From the known solution of the one-dimensional Heisenberg antiferromagnet we have

$$\langle \boldsymbol{\sigma}_0 \cdot \boldsymbol{\sigma}_1 \rangle_0 = 1 - 4\ln 2 \quad (8.11)$$

which gives the leading term of (8.7). The second-neighbour correlator was not known at the time but Takahashi (1977), by exploiting (8.7) and (8.10), deduced the exact results

$$\langle \boldsymbol{\sigma}_0 \cdot \boldsymbol{\sigma}_2 \rangle_0 = 1 - 16\ln 2 + 9\zeta(3) \quad (8.12)$$

This has recently been confirmed by an exact Bethe ansatz calculation (Boos and Korepin, 2001; see also Kato *et al.*, 2004).

For the square lattice, the general result (8.8) can be written as

$$H_{\text{eff}} = -\frac{t^2}{U}\left(2N - \sum_{\langle ij \rangle}^{(1)} \boldsymbol{\sigma}_i \cdot \boldsymbol{\sigma}_j\right)$$

$$+ \frac{t^4}{U^3}\left\{3N - 6\sum_{\langle ij \rangle}^{(1)} \boldsymbol{\sigma}_i \cdot \boldsymbol{\sigma}_j + \sum_{\langle ij \rangle}^{(2)} \boldsymbol{\sigma}_i \cdot \boldsymbol{\sigma}_j + \sum_{\langle ij \rangle}^{(3)} \boldsymbol{\sigma}_i \cdot \boldsymbol{\sigma}_j\right.$$

$$+ 5\sum_{\square}[(\boldsymbol{\sigma}_i \cdot \boldsymbol{\sigma}_j)(\boldsymbol{\sigma}_k \cdot \boldsymbol{\sigma}_l) + (\boldsymbol{\sigma}_i \cdot \boldsymbol{\sigma}_l)(\boldsymbol{\sigma}_j \cdot \boldsymbol{\sigma}_k)$$

$$\left.- (\boldsymbol{\sigma}_i \cdot \boldsymbol{\sigma}_k)(\boldsymbol{\sigma}_j \cdot \boldsymbol{\sigma}_l)]\right\} + \cdots \quad (8.13)$$

and a similar expression can be deduced for the simple cubic lattice. The ground-state energy can then be written formally as a t/U expansion, but the coefficients, expressed in terms of spin correlators, cannot be obtained exactly. Takahashi (1977) evaluated these approximately, using linear spin–wave theory, and obtained

$$\text{SQ:} \quad E_0/N \simeq -4.63t^2/U + 34.6t^4/U^3 + \cdots \quad (8.14a)$$

$$\text{SC:} \quad E_0/N \simeq -6.58t^2/U + 65.6t^4/U^3 + \cdots \quad (8.14b)$$

An alternative approach might be to use Ising expansions to evaluate the correlators, as discussed in Section 6.2.

The terms in the effective Hamiltonian (8.8) become increasingly complicated, and require a computer to keep track of them. MacDonald *et al.* (1988) have carried the expansion to order t^8, but, together with the approximate evaluation of coefficients, this is still a very short series.

An effective Hamiltonian for large U can also be derived away from half-filling. The resulting expression contains electron-hopping terms as well as spin interactions. The lowest order result was obtained by Hirsch (1985), and is

$$
\begin{aligned}
H_{\text{eff}} = -t \sum_{\langle ij\rangle,\sigma} (\tilde{c}_{i\sigma}^\dagger \tilde{c}_{j\sigma} + \text{h.c.}) + J \sum_{\langle ij\rangle} \left(\mathbf{S}_i \cdot \mathbf{S}_j - \frac{1}{4} n_i n_j \right) \\
- \frac{J}{4} \sum_{\langle i,j,k\rangle,\sigma} (\tilde{c}_{i\sigma}^\dagger n_{j\bar{\sigma}} \tilde{c}_{k\sigma} + \text{h.c.}) \\
- \frac{J}{4} \sum_{\langle i,j,k\rangle,\sigma} (\tilde{c}_{i\sigma}^\dagger \tilde{c}_{j\bar{\sigma}}^\dagger \tilde{c}_{j\sigma} \tilde{c}_{k\bar{\sigma}} + \text{h.c.}) + O(t^4)
\end{aligned}
\tag{8.15}
$$

where $\tilde{c}_{i\sigma} = c_{i\sigma}(1 - n_{i\bar{\sigma}})$ and its Hermitian conjugate transfer electrons from site to site but do not allow double occupancy, $n_{i\sigma} = \tilde{c}_{i\sigma}^\dagger \tilde{c}_{i\sigma}$, $\bar{\sigma} = -\sigma$ and $n_i = n_{i\uparrow} + n_{i\downarrow}$. The effective exchange parameter $J = 4t^2/U$. At half-filling electron hopping is completely suppressed, and (8.15) reduces to the leading term in (8.8). We will return to this 't–J' Hamiltonian in a later section.

8.2.2 Ising expansions

In the previous subsection we identified a difficulty in deriving direct perturbation expansions for ground-state properties in powers of t/U. Other approaches are possible, and have been used with some success. Recall that in order to derive linked cluster perturbation series we need an unperturbed Hamiltonian with a simple ground state. Shi and Singh (1995) suggested a modification of the Hamiltonian, by including an artificial symmetry breaking term in H_0 and subtracting it off in the perturbation V. Following their idea, we write the Hamiltonian as $H = H_0 + \lambda V$, with

$$
H_0 = U \sum_i n_{i\uparrow} n_{i\downarrow} + J \sum_{\langle ij\rangle} (\sigma_i \sigma_j + 1)
\tag{8.16a}
$$

$$
V = -t \sum_{\langle ij\rangle,\sigma} (c_{i\sigma}^\dagger c_{j\sigma} + \text{h.c.}) - J \sum_{\langle ij\rangle} (\sigma_i \sigma_j + 1)
\tag{8.16b}
$$

Table 8.1. *Data for perturbation expansion for cluster g_2.*

State $|i\rangle$	e_i	$V|i\rangle$
$|1\rangle = |+-\rangle$	0	$-t|3\rangle - t|4\rangle$
$|2\rangle = |-+\rangle$	$4J$	$t|3\rangle + t|4\rangle - 4J|2\rangle$
$|3\rangle = |\pm 0\rangle$	$U + 3J$	$-t|1\rangle + t|2\rangle - 3J|3\rangle$
$|4\rangle = |0\pm\rangle$	$U + 3J$	$-t|1\rangle + t|2\rangle - 3J|4\rangle$

Here $\sigma_i = n_{i\uparrow} - n_{i\downarrow}$ is an effective Ising spin and J is a parameter which can be tuned to obtain best convergence in the series. Clearly at $\lambda = 1$ the original Hubbard model is recovered.

The unperturbed ground state, on a bipartite lattice at half-filling, is now a simple Néel antiferromagnetic state, with a trivial two-fold degeneracy. It is then possible to carry out cluster expansions, along the same line as described in earlier chapters. The only complication is due to the algebra of the fermion operators. We will illustrate this, in some detail, for the linear chain.

The first few clusters are, as usual,

and we need to consider, for each cluster, only the sector with $n = 1$. The first cluster is trivial, with energy zero. Continuing,

$g_2 : \overset{\bullet\!-\!\bullet}{1\quad 2}$ The states, their unperturbed energies e_i, and effect of V are given in Table 8.1.

Using standard matrix perturbation theory gives, for the ground-state energy

$$E_0^{(2)} = -\frac{2t^2}{U + 3J}\lambda^2 - \frac{6t^2 J}{(U + 3J)^2}\lambda^3$$
$$+ \left(-\frac{t^4}{J(U + 3J)^2} - \frac{18t^2 J^2}{(U + 3J)^3} + \frac{4t^4}{(U + 3J)^3}\right)\lambda^4 + \cdots \quad (8.17)$$

Some points should be noted. Firstly, since the unperturbed Hamiltonian contains interaction terms with sites outside the cluster, these must be included in the unperturbed energies. Secondly, note the divergence in the coefficient of λ^4 as $J \to 0$: this expansion needs a finite J.

The reduced energy is $\epsilon_0^{(2)} = E_0^{(2)} - 2\epsilon_0^{(1)} = E_0^{(2)}$.

Table 8.2. *Data for perturbation expansion for cluster* g_3

State $|i\rangle$	e_i	$V|i\rangle$
$\|1\rangle = \|+-+\rangle$	0	$-t\|2\rangle - t\|3\rangle + t\|4\rangle + t\|5\rangle$
$\|2\rangle = \|0\pm+\rangle$	$U+3J$	$-t\|1\rangle + t\|6\rangle + t\|7\rangle - 3J\|2\rangle$
$\|3\rangle = \|\pm 0+\rangle$	$U+3J$	$-t\|1\rangle + t\|6\rangle - t\|8\rangle - 3J\|3\rangle$
$\|4\rangle = \|+0\pm\rangle$	$U+3J$	$t\|1\rangle - t\|7\rangle - t\|9\rangle - 3J\|4\rangle$
$\|5\rangle = \|+\pm 0\rangle$	$U+3J$	$t\|1\rangle + t\|8\rangle - t\|9\rangle - 3J\|5\rangle$
$\|6\rangle = \|-++\rangle$	$4J$	$t\|2\rangle + t\|3\rangle - 4J\|6\rangle$
$\|7\rangle = \|0+\pm\rangle$	$U+4J$	$t\|2\rangle - t\|4\rangle - 4J\|7\rangle$
$\|8\rangle = \|\pm+0\rangle$	$U+4J$	$-t\|3\rangle + t\|5\rangle - 4J\|8\rangle$
$\|9\rangle = \|++-\rangle$	$4J$	$-t\|4\rangle - t\|5\rangle - 4J\|9\rangle$

g_3 $\underset{1 \quad 2 \quad 3}{\bullet\!\!-\!\!\bullet\!\!-\!\!\bullet}$ Without loss of generality we consider the sector with $n_\uparrow = 2$, $n_\downarrow = 1$, with nine states. The data needed for the perturbation expansion are given in Table 8.2.

Carrying out the perturbation expansion by hand to order λ^4 gives

$$
E_0^{(3)} = -\frac{4t^2}{U+3J}\lambda^2 - \frac{12t^2 J}{(U+3J)^2}\lambda^3 + \left(-\frac{36t^2 J^2}{(U+3J)^3}\right.
$$
$$
\left. -\frac{2t^4}{J(U+3J)^2} - \frac{8t^4}{(U+4J)(U+3J)^2} + \frac{16t^4}{(U+3J)^3}\right)\lambda^4 + \cdots
$$

$$(8.18)$$

The reduced energy is

$$
\epsilon_0^{(3)} = E_0^{(3)} - 3\epsilon_0^{(1)} - 2\epsilon_0^{(2)}
$$
$$
= \left(\frac{8t^4}{(U+3J)^3} - \frac{8t^4}{(U+3J)^2(U+4J)}\right)\lambda^4 + \cdots
$$

$$(8.19)$$

Note that the λ^2, λ^3 terms cancel, as they should, and the leading term is of order λ^4, arising from four V processes.

g_4 $\underset{1 \quad 2 \quad 3 \quad 4}{\bullet\!\!-\!\!\bullet\!\!-\!\!\bullet\!\!-\!\!\bullet}$ A rather tedious hand calculation gives

$$
E_0^{(4)} = -\frac{6t^2}{U+3J}\lambda^2 - \frac{18t^2 J}{(U+3J)^2}\lambda^3 + \left(-\frac{3t^4}{J(U+3J)^2} + \frac{36t^4}{(U+3J)^3}\right.
$$
$$
\left. -\frac{16t^4}{(U+3J)^2(U+4J)} - \frac{16t^4}{(U+3J)^2(2U+5J)} - \frac{54t^2 J^2}{(U+3J)^3}\right)\lambda^4
$$
$$
+ \cdots
$$

$$(8.20)$$

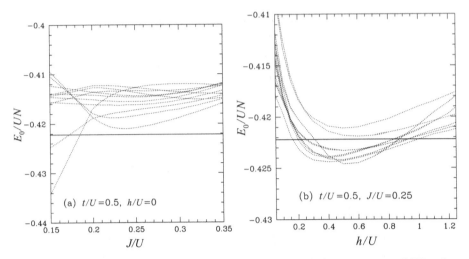

Fig. 8.3. Variation of ground-state energy estimates with the parameters J/U and h/U for $t/U = 0.5$. The horizontal solid line is the exact result.

The reduced energy is

$$\epsilon_0^{(4)} = E_0^{(4)} - 4\epsilon_0^{(1)} - 3\epsilon_0^{(2)} - 2\epsilon_0^{(3)}$$

$$= \left(-\frac{16t^4}{(U+3J)^2(2U+5J)} + \frac{8t^4}{(U+3J)^3} \right)\lambda^4 + \cdots \qquad (8.21)$$

The leading term is again of order λ^4, corresponding to two hops on each of links (12) and (34).

Combining these results gives the expansion for the ground-state energy of the bulk lattice, correct to order λ^4,

$$\frac{E_0}{N} = -\frac{2t^2}{U+3J}\lambda^2 - \frac{6t^2 J}{(U+3J)^2}\lambda^3 + \left(-\frac{t^4}{J(U+3J)^2} - \frac{18t^2 J^2}{(U+3J)^3} \right.$$

$$\left. -\frac{8t^4}{(U+3J)^2(U+4J)} - \frac{16t^4}{(U+3J)^2(2U+5J)} + \frac{20t^4}{(U+3J)^3} \right)\lambda^4$$

$$+ \cdots \qquad (8.22)$$

It is not practicable to go much further by hand.

We have used a computer program to obtain series for the linear chain to order λ^{15}, for any choice of parameters t, U, J. Recall that the J term in (8.16) was included to break the degeneracy. For a sufficiently long series, or for sufficiently small t/U, the values of J will be immaterial. In practice this is not so and the resulting energy estimates will depend on the choice of the parameter J. In Figure 8.3(a) we investigate this by plotting the values of different integrated

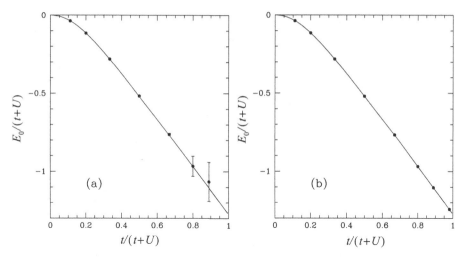

Fig. 8.4. Ground-state energy versus t/U for the Hubbard chain obtained from Ising expansions (a) and dimer expansions (b). Also shown as the full line is the exact result (8.3).

differential approximants to the energy series versus J/U. The known exact result is shown for comparison. As is evident for $J/U \gtrsim 0.2$ all approximants are reasonably consistent, with a spread of around 2%, but all underestimate the correct value. It is possible to do better by also including a staggered field term $h \sum_i (-1)^i \sigma_i /2$ in H_0 and subtracting it off in V. As can be seen in Figure 8.3(b) there is good consistency between different approximants near $h/U = 0.2$ and near $h/U = 0.9$, and both of these regions agree with the exact energy.

In this way we have estimated the ground-state energy in one dimension, as a function of t/U. The results, and a comparison with the exact result, are shown in Figure 8.4(a). It is clear that the order 15 Ising expansion gives accurate results to $t/U \sim 1.5$, but beyond this point longer series would be needed.

Shi and Singh (1995) have used this procedure to obtain series for the ground-state energy, local moment, sub-lattice magnetization, spin stiffness and uniform susceptibility for the Hubbard model on the square lattice. It would be straightforward to extend this approach to other bipartite lattices, in particular in three dimensions. A recent application to the two-dimensional honeycomb lattice (Paiva *et al.*, 2005) has identified a quantum phase transition at $U/t \simeq 5$ for half-filling, between an antiferromagnetic insulating phase and a semi-metallic phase.

We have recently used the Ising expansion to compute the spin-wave dispersion for the Hubbard model on a square lattice. The results for various values of U/t are shown in Figure 8.5, together with the dispersion for the Heisenberg

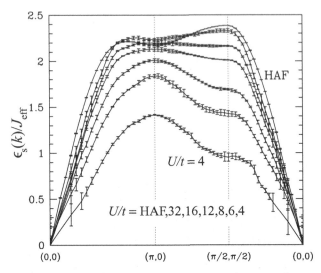

Fig. 8.5. The spin-wave dispersion $\epsilon_s(\mathbf{k})/J_{\text{eff}}$ (scaled by $J_{\text{eff}} = 4t^2/U$) along symmetry lines in the Brillouin zone for the half-filled Hubbard model on a square lattice for $U/t = 32, 16, 8, 6, 4$. Also shown for comparison is the dispersion relation for the Heisenberg antiferromagnet ($U = \infty$). From Zheng *et al.* (2005b).

antiferromagnet. It can be seen that the Hubbard model results approach towards the Heisenberg form at large U/t, as expected. It can also be seen that as U/t decreases, the shape of the curve changes, so that the peak at wave vector $(\pi, 0)$ becomes distinctly higher than that at $(\pi/2, \pi/2)$. This agrees nicely with recent experimental data for the materials $Cu(DCOO)_2 \cdot 4D_2O$ (CFTD) (Ronnow *et al.*, 2001) and $Sr_2Cu_3O_4Cl_2$ (Kim *et al.*, 1999, 2001), and La_2CuO_4 (Coldea *et al.*, 2001).

8.2.3 Dimer expansions

An alternative series approach for ground-state properties of the Hubbard model is via dimer expansions. That is, we start with an unperturbed Hamiltonian for independent dimers, and couple the dimers via the perturbation. For the case of one dimension, we can write $H = H_0 + \lambda V$, with

$$H_0 = U \sum_i n_{i\uparrow} n_{i\downarrow} - t \sum_{i,\sigma} (c^\dagger_{2i,\sigma} c_{2i+1,\sigma} + \text{h.c.}) \tag{8.23a}$$

$$V = -t \sum_{i,\sigma} (c^\dagger_{2i-1,\sigma} c_{2i,\sigma} + \text{h.c.}) \tag{8.23b}$$

The 16 eigenstates of a single dimer can be obtained exactly, and are given in Table 8.3.

Table 8.3. *The sixteen dimer states and their eigen-energies (e_i), where*
$u_1 = \frac{1}{2}\sqrt{1 + U/\sqrt{U^2 + 16t^2}}$, $u_2 = \frac{1}{2}\sqrt{1 - U/\sqrt{U^2 + 16t^2}}$,
$\lambda_1 = \frac{1}{2}(U - \sqrt{U^2 + 16t^2})$, $\lambda_2 = \frac{1}{2}(U + \sqrt{U^2 + 16t^2})$. 0 *represents a*
hole, $+$ ($-$) represent up (down)-spin electrons, $\pm$ represents an
electron pair.

State	Eigenstate	e_i	Name
1	$u_1(\| +-\rangle - \| -+\rangle)$ $-u_2(\| \pm 0\rangle + \| 0\pm\rangle)$	λ_1	singlet
2	$\frac{1}{\sqrt{2}}(\| 0-\rangle + \| -0\rangle)$	$-t$	electron-hole even ($S_{\text{tot}}^z = -\frac{1}{2}$)
3	$\frac{1}{\sqrt{2}}(\| 0+\rangle + \| +0\rangle)$	$-t$	electron-hole even ($S_{\text{tot}}^z = \frac{1}{2}$)
4	$\frac{1}{\sqrt{2}}(\| \pm -\rangle - \| -\pm\rangle)$	$U - t$	three-electron odd ($S_{\text{tot}}^z = -\frac{1}{2}$)
5	$\frac{1}{\sqrt{2}}(\| \pm +\rangle - \| +\pm\rangle)$	$U - t$	three-electron odd ($S_{\text{tot}}^z = \frac{1}{2}$)
6	$\| --\rangle$	0	triplet ($S_{\text{tot}}^z = -1$)
7	$\frac{1}{\sqrt{2}}(\| +-\rangle + \| -+\rangle)$	0	triplet ($S_{\text{tot}}^z = 0$)
8	$\| ++\rangle$	0	triplet ($S_{\text{tot}}^z = 1$)
9	$\| 00\rangle$	0	hole-pair singlet
10	$\frac{1}{\sqrt{2}}(\| \pm 0\rangle - \| 0\pm\rangle)$	U	electron pair and hole singlet
11	$\| \pm\pm\rangle$	$2U$	electron-pair singlet
12	$\frac{1}{\sqrt{2}}(\| 0-\rangle - \| -0\rangle)$	t	electron-hole odd ($S_{\text{tot}}^z = -\frac{1}{2}$)
13	$\frac{1}{\sqrt{2}}(\| 0+\rangle - \| +0\rangle)$	t	electron-hole odd ($S_{\text{tot}}^z = \frac{1}{2}$)
14	$\frac{1}{\sqrt{2}}(\| \pm -\rangle + \| -\pm\rangle)$	$U + t$	three-electron even ($S_{\text{tot}}^z = -\frac{1}{2}$)
15	$\frac{1}{\sqrt{2}}(\| \pm +\rangle + \| +\pm\rangle)$	$U + t$	three-electron even ($S_{\text{tot}}^z = \frac{1}{2}$)
16	$u_2(\| +-\rangle - \| -+\rangle)$ $+u_1(\| \pm 0\rangle + \| 0\pm\rangle)$	λ_2	mixed singlet

For large U (in particular for $t/U < \frac{1}{3}$) the lowest state of a dimer is a non-degenerate singlet, and hence the unperturbed ground state is a direct product of dimer singlets. The dimer expansion then proceeds along similar lines to our earlier example (Section 8.2.2). The dimers become single 'objects' or sites for our clusters, the V terms induce transitions to higher dimer states, and perturbation expansions are developed in powers of λ. Again care must be taken to preserve the correct phase factors arising from fermion operators. At the end the limit $\lambda = 1$ is taken to regain the full Hubbard Hamiltonian.

Because of the large number of states and resulting terms it is hardly feasible to proceed by hand. We have written a computer program and obtained expansions for the linear chain up to order λ^{14}. For example, the series for the ground-state energy for $U = t = 1$ is

$$
\begin{aligned}
E_0/N = & -7.807764064 \times 10^{-1} - 2.321098855 \times 10^{-1}\lambda^2 \\
& -2.114784210 \times 10^{-2}\lambda^4 - 1.237851610 \times 10^{-3}\lambda^6 \\
& -3.155856521 \times 10^{-3}\lambda^8 + 1.120883080 \times 10^{-4}\lambda^{10} \\
& -1.018507218 \times 10^{-3}\lambda^{12} + 4.047217476 \times 10^{-5}\lambda^{14} + \cdots \quad (8.24)
\end{aligned}
$$

The estimated ground-state energy, from the dimer expansion, is shown in Figure. 8.4(b). As is apparent from the figure the agreement with the exact result is excellent, even for small t/U. Note that the dimer expansion gives better convergence for larger t/U, while the Ising expansion is better for small t/U.

The dimer expansion has been used to study the two-leg Hubbard ladder (Zheng *et al.*, 2001c). As well as obtaining the ground-state energy, and spin excitation spectrum at half-filling, results were obtained for one- and two-hole charge excitations.

8.2.4 t/U expansions at finite temperature

In Section 8.2.1 we outlined the difficulty of obtaining perturbation expansions in t/U for ground-state properties of the Hubbard model. No such difficulty exists at finite temperature and thermodynamic perturbation expansions about the 'atomic limit' ($t = 0$) have a long history, going back to the 1970s. We refer the interested reader to more recent papers (Henderson *et al.*, 1992; ten Haaf *et al.*, 1995) for the latest results and references to earlier work.

The calculations follow the approach outlined in Section 7.5. We write the Hamiltonian as $H = H_0 + V$, with

$$
H_0 = U \sum_i n_{i\uparrow} n_{i\downarrow} - \mu \sum_i (n_{i\uparrow} + n_{i\downarrow}) \tag{8.25a}
$$

$$
V = -t \sum_{\langle ij \rangle, \sigma} (c^\dagger_{i\sigma} c_{j\sigma} + c^\dagger_{j\sigma} c_{i\sigma}) \tag{8.25b}
$$

We have included a chemical potential term $-\mu N_e$, and work in the grand ensemble, with variable electron density. Thus the difficulty, in ground-state calculations, of dealing with continuous electron density is overcome. On the other hand, as we shall show, extrapolation of our finite-temperature results to low temperature becomes problematic. To compute the magnetic susceptibility we can include a field term in H_0. A staggered field can be included to obtain the antiferromagnetic susceptibility.

Following Section 7.5, we write the grand partition function as a perturbation expansion

$$\mathcal{Z}^{(g)} = \mathcal{Z}_0^{(g)} \left\{ 1 + \sum_{n=1}^{\infty} (-1)^n \int_0^\beta d\tau_1 \int_0^{\tau_1} d\tau_2 \cdots \int_0^{\tau_{n-1}} d\tau_n \langle \widetilde{V}(\tau_1) \widetilde{V}(\tau_2) \cdots \widetilde{V}(\tau_n) \rangle_0 \right\}$$

(8.26)

with

$$\widetilde{V}(\tau) = e^{\tau H_0} V e^{-\tau H_0}$$

(8.27)

and

$$\mathcal{Z}_0^{(g)} = \text{Tr}\{e^{-\beta H_0}\} = z_0^N$$

(8.28)

with

$$z_0 = 1 + 2e^{\beta\mu} + e^{2\beta\mu - \beta U}$$
$$= 1 + 2\zeta + \zeta^2 w$$

(8.29)

where $\zeta = e^{\beta\mu}$, the fugacity, and $w = e^{-\beta U}$. The angular bracket denotes, as usual, the unperturbed expectation value.

Rather than use the linked-cluster approach we will follow the alternative approach, mentioned briefly in Section 7.2.1, in which disconnected clusters also occur. For this problem this alternative method is computationally more efficient, and no sub-graph subtraction is needed. Taking the logarithm of (8.26) yields the grand potential

$$-\beta\Omega = N \ln z_0 + \sum_{r=1}^{\infty} (-1)^r \int_0^\beta d\tau_1 \int_0^{\tau_1} d\tau_2 \cdots \int_0^{\tau_{r-1}} d\tau_r \langle \widetilde{V}(\tau_1) \widetilde{V}(\tau_2) \cdots \widetilde{V}(\tau_r) \rangle_N$$

(8.30)

where the subscript N signifies that only the order N term is to be included. As in previous examples, higher powers of N cancel on taking the logarithm. The various terms in (8.30) can, as usual, be represented by cluster diagrams or graphs, in which each application of V is represented by a bond. For bipartite lattices the graphs which contribute up to order 4 are shown in Table 8.4.

The contribution to (8.30) from each particular graph is computed along the same lines as in Section 7.5, the only complicating factor being the phase arising from commutation of fermion operators. The resulting contribution from graph g is conveniently written in the form

$$z_0^{-s} c_g X_g(\zeta, \beta, U)(\beta t)^l$$

(8.31)

where s, l are the numbers of sites and bonds respectively, c_g is the weak lattice constant and X_g is a 'graph weight'. The lowest few can be obtained by hand,

Table 8.4. *Graphs for the grand potential through order 4 and their weak lattice constants for the linear chain (LC), square (SQ) and simple cubic (SC) lattices.*

			Lattice constant	
Graph	Order	LC	SQ	SC
1.	2	1	2	3
2.	4	1	2	3
3.	4	1	6	15
4.	4	$-3/2$	-7	$-33/2$
5.	4	0	1	3

and are

$$X_1 = 2\zeta(1 + \zeta^2 w) + \frac{4\zeta^2}{\beta U}(1 - w) \tag{8.32a}$$

$$X_2 = \frac{1}{6}\zeta(1 + \zeta^2 w) + \frac{8\zeta^2}{(\beta U)^2}(1 + w) - \frac{16\zeta^2}{(\beta U)^3}(1 - w) \tag{8.32b}$$

$$X_3 = \frac{1}{3}\zeta(1 + \zeta^2 w)(1 + 2\zeta + \zeta^2 w) + \frac{6\zeta^2}{\beta U}(1 - 2\zeta w + \zeta^2 w)$$

$$- \frac{4\zeta^2}{(\beta U)^2}[2 - w - 2\zeta(1 - 2w) + \zeta^2 w(2 - w)]$$

$$+ \frac{4\zeta^2}{(\beta U)^3}(1 - w)(1 + 2\zeta + \zeta^2 w) \tag{8.32c}$$

$$X_4 = 4\zeta^2(1 + \zeta^2 w)^2 + \frac{16\zeta^3}{\beta U}(1 - w)(1 + \zeta^2 w)$$

$$+ \frac{16\zeta^4}{(\beta U)^2}(1 - w)^2 \tag{8.32d}$$

$$X_5 = \frac{2}{3}\zeta(1 - 4\zeta + \zeta^2 + \zeta^4 w - 4\zeta^5 w^2 + \zeta^6 w^3)$$

$$+ \frac{8\zeta^2}{\beta U}[1 - \zeta + \zeta(1 - 2\zeta - \zeta^3)w + 2\zeta^3(1 + \zeta)w^2]$$

$$- \frac{8\zeta^2}{(\beta U)^2}[2 - \zeta - \zeta(1 + 8\zeta - \zeta^2)w + \zeta^3(1 - 2\zeta)w^2]$$

$$+ \frac{16\zeta^2}{(\beta U)^3}[1 - \zeta - \zeta^2 - (1 - \zeta + 2\zeta^2 + \zeta^3)w$$

$$+ \zeta^2(3 + \zeta + \zeta^2)w^2 - \zeta^4 w^3] \tag{8.32e}$$

The reader may wish to verify these. We note that $X_4 = X_1^2$. The weights of discon-nected graphs can always be obtained directly from their components, and hence the inclusion of disconnected graphs causes no great difficulties. The advantage is that each bond of a graph is covered only once by a V operator, and this makes the computation of graph weights much more efficient. It is also apparent that the graph weights are becoming rather complex and tedious to calculate, and further terms will require a computer.

Before considering further terms it is as well to consider analysis of the results. We write the grand potential in the form

$$-\beta\Omega/N = \ln z_0 + \sum_{r=2}^{\infty} z_0^{-r} A_r(\zeta, \beta U)(\beta t)^r \tag{8.33}$$

with, from the discussion above for bipartite lattices,

$$A_2(\zeta, \beta U) = c_1 X_1 \tag{8.34}$$

and

$$A_4(\zeta, \beta U) = c_2 z_0^2 X_2 + c_3 z_0 X_3 + c_4 X_4 + c_5 X_5 \tag{8.35}$$

While (8.33) looks like a high-temperature expansion, the coefficients $A_n(\zeta, \beta U)$ are themselves temperature dependent. An alternative expansion

$$-\beta\Omega/N = \ln z_0 + \sum_{r=2}^{\infty} z_0^{-r} \tilde{A}_r(\zeta, \beta U)(t/U)^r \tag{8.36}$$

can sometimes prove more useful. In either case, our series coefficients are still in terms of the fugacity ζ, whereas it is usually desirable to evaluate thermodynamic quantities for fixed electron density n. The connection is

$$n = \zeta \frac{\partial}{\partial \zeta}(-\beta\Omega/N)$$

$$= \frac{2\zeta(1 + \zeta w)}{1 + 2\zeta + \zeta^2 w} + \sum_{r=2}^{\infty} z_0^{-(r+1)} Y_r(\zeta, \beta U)(\beta t)^r \tag{8.37}$$

with

$$Y_r(\zeta, \beta U) = \zeta z_0 \frac{\partial A_r}{\partial \zeta} - 2r\zeta(1 + \zeta w)A_r \tag{8.38}$$

The task then is to invert the series (8.37) to obtain ζ in terms of n, βt and to substitute in (8.33). In the general case it is not possible to do this analytically and a numerical procedure must be used (Henderson *et al.*, 1992).

A substantial simplification occurs in the 'strong correlation' limit $U \to \infty$. Introducing, for convenience, the variable $p = \zeta/(1 + 2\zeta)$, the grand potential becomes

$$-\beta\Omega/N = -\ln(1-2p) + 2c_1 p(1-2p)(\beta t)^2$$
$$+ \frac{1}{6}p(1-2p)\big[c_2 + 2c_3 + 24p(1-2p)c_4$$
$$+ 4(1-8p+13p^2)c_5\big](\beta t)^4 + \cdots \qquad (8.39)$$

whence we obtain

$$n = p(1-2p)\frac{\partial}{\partial p}(-\beta\Omega/N)$$

$$= 2p + 2c_1 p(1-2p)(1-4p)(\beta t)^2 + \frac{1}{6}p(1-2p)[(1-4p)c_2 + 2(1-4p)c_3$$
$$+ 48p(1-2p)(1-4p)c_4 + 4(1-20p+87p^2-104p^3)c_5](\beta t)^4 + \cdots$$

$$(8.40)$$

If we now write

$$p = p_0 + p_2(\beta t)^2 + p_4(\beta t)^4 + \cdots \qquad (8.41)$$

we obtain, by iteration,

$$p_0 = n/2$$
$$p_2 = -\frac{1}{2}n(1-n)(1-2n)c_1 \qquad (8.42)$$
$$p_4 = \frac{1}{2}n(1-n)(1-2n)(1-6n+6n^2)c_1^2 - \frac{1}{24}n(1-n)(1-2n)c_2$$
$$- \frac{1}{12}n(1-n)(1-2n)c_3 - n^2(1-n)^2(1-2n)c_4$$
$$- \frac{1}{24}n(1-n)(4-40n+87n^2-52n^3)c_5$$

These results can then be substituted into (8.39) to obtain the grand potential, and from it the free energy, internal energy, entropy and specific heat. These are high-temperature expansions in (βt) with the coefficients dependent only on the parameter n. Actually it is not essential to set $U = \infty$. For large βU we can drop all exponential terms but keep inverse powers of (βU) in (8.32). It is still possible to do the series inversion analytically in this case, although the resulting expressions are more complicated. We leave this as an exercise for the reader.

Of greater interest are the uniform and staggered susceptibilities as their divergence can, at least in principle, be used to identify regions of magnetic order in the Hubbard model phase diagram. Derivation of these series proceeds along standard lines. In the strong correlation limit the results, to order $(\beta t)^4$, are found to be

$$\chi = n + 0(\beta t)^2 - \frac{1}{6}n^2(1-n)(8-11n)c_5(\beta t)^4 + \cdots \tag{8.43}$$

$$\chi_s = n - \frac{1}{3}n(1-n)c_1(\beta t)^2 + \frac{1}{30}n(1-n)[20(1-2n)^2c_1^2 - 2c_2$$
$$- 2(2+n)c_3 + 40n(1-n)c_4 - (8-48n+47n^2)c_5](\beta t)^4 + \cdots \tag{8.44}$$

Pan and Wang (1991, 1997) also give the additional terms for large (βU), neglecting the exponential terms.

Fourth-order series are much too short to shed any light on the Hubbard phase diagram. A number of authors have derived further terms. For $\beta U = \infty$ Kubo and Tada (1983, 1984) obtained the grand potential and uniform susceptibility to order $(\beta t)^9$. These were subsequently extended to $(\beta t)^{12}$, but these data remain unpublished. For the general case Henderson *et al.* (1992) obtained the grand potential and uniform susceptibility to order $(\beta t)^{10}$, ten Haaf and van Leeuwen (1992) and ten Haaf *et al.* (1995) obtained the nearest-neighbour spin correlator to order $(\beta t)^8$ and Pan and Wang (1997) have obtained the staggered susceptibility to order $(\beta t)^8$, but only at half filling. This has also been obtained to order $(\beta t)^8$ for general n by our group, but has not been published. Attempts to analyse these series have had limited success. The series are irregular and even at tenth order (five terms for bipartite lattices) are too short for a proper Padé analysis. Two main approaches have been used to locate phase boundaries. One is to look for zeros in the series for χ^{-1}, the other is to use the sign of the nearest-neighbour correlator as an indicator. Both methods give indications of antiferromagnetic order near half filling and ferromagnetism at large U, with maximum T_c at around $n = 0.85$. However, in our view, the analysis remains far from conclusive.

8.3 The *t–J* model

The *t-J* model is another much studied model for strongly correlated electrons on a lattice. In its simplest form it is defined by the Hamiltonian

$$H = -t \sum_{\langle ij \rangle, \sigma} (c_{i\sigma}^\dagger c_{j\sigma} + c_{j\sigma}^\dagger c_{i\sigma}) + J \sum_{\langle ij \rangle} \left(\mathbf{S}_i \cdot \mathbf{S}_j - \frac{1}{4} n_i n_j \right) \tag{8.45}$$

where the $c_{i\sigma}^\dagger$, $c_{i\sigma}$ are 'constrained' creation and destruction operators for electrons, acting in a Hilbert space where double occupancy is forbidden, $n_i = 0, 1$ is the number operator for site i and $\mathbf{S}_i = (S_i^x, S_i^y, S_i^z)$ is the corresponding spin operator. Note that the J-term gives zero except when operating on a pair of electrons with opposite spin. The parameters t, J are assumed independent. From the earlier discussion leading to (8.15) it is evident that the $t - J$ Hamiltonian can be 'derived' as an effective Hamiltonian for the large U Hubbard model, with $J = 4t^2/U$, provided that we drop the 3-site terms in (8.15). However it is now generally regarded as an interesting model in its own right and is studied in the entire (t, J) parameter space, where the mapping from the Hubbard model may no longer be valid. The Hamiltonian (8.45) is often written with the inclusion of projection operators to ensure no double occupancy, or using the operators $\tilde{c}_{i\sigma}$ as in (8.15). We will use the simpler notation (8.45), but it must be remembered that the c variables are not true fermion operators.

Quite a lot is known about the t–J model in one dimension. At the special 'supersymmetric' point $J = 2t$ an exact solution is possible via the Bethe ansatz approach (Bares *et al.*, 1991). This result, together with numerical studies (Ogata *et al.*, 1991) confirm that the one dimensional t–J model lies in the same universality class as the corresponding repulsive Hubbard model, referred to as 'Luttinger liquids'. Spin and charge excitations 'separate', each propagating with its own characteristic velocity. The phase diagram in the space of J/t and electron density n has been mapped out. At large J/t phase separation occurs, and the ground state is non-uniform, consisting of coexisting hole-rich and hole-poor phases. There is numerical evidence for a phase of singlet-bound electron pairs for $J/t > 2$ and low n, which will have a gap for spin excitations, and evidence for enhanced superconducting correlations in the region of $n \sim 1$, just below phase separation.

In higher dimensions there are no exact results known, but some of the features of the phase diagram are expected to persist. For $n = 1$, which is usually referred to as 'half-filling' although it has the maximum electron density, the charge motion is completely frozen out. The Hamiltonian reduces to the antiferromagnetic Heisenberg model, which has been discussed in previous chapters. The system is a gapless antiferromagnetic insulator. The physics of a small number of holes in an antiferromagnetic background remains an interesting and much studied problem. For large J we expect phase separation, as in one dimension. There are predictions of superconductivity, particularly when longer range hopping terms are included (Kotov and Sushkov, 2004). It is possible, but by no means proven, that the two-dimensional t–J model can provide a basis for a comprehensive theory of high-temperature superconductivity.

We turn now to the application of series expansion methods to this model.

8.3.1 Expansions at zero temperature

As already mentioned, the undoped ($n = 1$) t–J model is just the Heisenberg antiferromagnet and extensive series work exists, as described in Chapter 5.

A single hole is able to propagate through the lattice as a well-defined quasiparticle. Linked cluster series, using the methods of Section 4.5, can yield the dispersion relation for this excitation. We will illustrate this, in some detail, for the linear chain. To allow for an Ising expansion we introduce anisotropy into the spin exchange term in (8.45), and write

$$H = H_0 + \lambda V \tag{8.46}$$

with

$$H_0 = J \sum_{\langle ij \rangle} \left(S_i^z S_j^z - \frac{1}{4} n_i n_j \right) + r \sum_i \eta_i S_i^z \tag{8.47a}$$

$$V = -t \sum_{\langle ij \rangle, \sigma} (c_{i\sigma}^\dagger c_{j\sigma} + c_{j\sigma}^\dagger c_{i\sigma}) + \frac{1}{2} J \sum_{\langle ij \rangle} (S_i^+ S_j^- + S_i^- S_j^+)$$
$$- r \sum_i \eta_i S_i^z \tag{8.47b}$$

where an additional staggered field term of strength r, with $\eta_i = \pm 1$ on alternate sub-lattices, has been added to improve convergence. At $\lambda = 1$ the original t–J Hamiltonian is recovered. It is convenient to set $J = 1$ to fix the energy scale. We also set $r = 0$ in the following example.

The total numbers of spin up and spin down electrons, N_+ and N_-, are separately conserved. The ground state has $N_+ = N_- = N/2$, while there are two degenerate one-hole sectors with $N_+ = N_- \pm 1$. We consider the sector where a hole is introduced on the A (up) sub-lattice. Table 8.5 shows the clusters which must be included to obtain the dispersion relation correct to order λ^5.

To illustrate the calculation in greater detail, let us consider the cluster 3A. ●——○——● We first consider the ground-state sector. The three states which contribute, their unperturbed energies, and the effect of V on each are

state $|i\rangle$	e_i	$V|i\rangle$
$|1\rangle = |+-+\rangle$	-2	$V|1\rangle = \frac{1}{2}|2\rangle + \frac{1}{2}|3\rangle$
$|2\rangle = |-++\rangle$	-1	$V|2\rangle = \frac{1}{2}|1\rangle$
$|3\rangle = |++-\rangle$	-1	$V|3\rangle = \frac{1}{2}|1\rangle$

Note that the unperturbed energies include interactions between the cluster and the unperturbed background. We have set the auxiliary field to zero. The 3×3 Hamiltonian matrix can be diagonalized exactly and yields

$$E_g = -\frac{3}{2} - \frac{1}{2}\sqrt{1 + 2\lambda^2} = -2 - \frac{1}{2}\lambda^2 + \frac{1}{4}\lambda^4 + \cdots \tag{8.48}$$

The one-hole sector has six states, the two unperturbed states with the hole at

Table 8.5. *Low-order clusters and their reduced transition amplitudes needed to compute the one-hole dispersion relation for the linear chain t-J model to order λ^5. Filled and empty circles denote A and B sites, respectively.*

1.	●	$d_0^{(1)} = 1$
2.	●—○	$d_0^{(2)} = (\frac{1}{4} - 2t^2)\lambda^2 + (-\frac{1}{16} + 8t^4)\lambda^4 + \cdots$
3A.	●—○—●	$d_0^{(3A)} = -\frac{1}{2}\lambda^2 + (\frac{5}{8} + 4t^2 - 16t^4)\lambda^4 + \cdots$
		$d_2^{(3A)} = 8t^2\lambda^3 + (-24t^2 - 64t^4)\lambda^5 + \cdots$
3B.	○—●—○	$d_0^{(3B)} = (-\frac{1}{8} + 16t^4)\lambda^4 + \cdots$
4.	●—○—●—○	$d_0^{(4)} = (\frac{1}{2} - \frac{19}{4}t^2)\lambda^4 + \cdots$
		$d_2^{(4)} = (-2t^2 - 16t^4)\lambda^5 + \cdots$
5A.	●—○—●—○—●	$d_0^{(5A)} = -\frac{5}{4}\lambda^4 + \cdots$
		$d_2^{(5A)} = 34t^2\lambda^5 + \cdots$
		$d_4^{(5A)} = O(\lambda^6)$
5B.	○—●—○—●—○	$d_0^{(5B)} = O(\lambda^6)$
		$d_2^{(5B)} = O(\lambda^7)$

sites 1,3 respectively and the other states generated by V. The data for these are

state $|i\rangle$	e_i	$V|i\rangle$
$|1\rangle = |0 - +\rangle$	-1	$V|1\rangle = -t|3\rangle + \frac{1}{2}|5\rangle$
$|2\rangle = |+ - 0\rangle$	-1	$V|2\rangle = -t|4\rangle + \frac{1}{2}|6\rangle$
$|3\rangle = |- 0 +\rangle$	$-\frac{1}{2}$	$V|3\rangle = -t|1\rangle - t|6\rangle$
$|4\rangle = |+ 0 -\rangle$	$-\frac{1}{2}$	$V|4\rangle = -t|2\rangle - t|5\rangle$
$|5\rangle = |0 + -\rangle$	$-\frac{1}{2}$	$V|5\rangle = \frac{1}{2}|1\rangle - t|4\rangle$
$|6\rangle = |- + 0\rangle$	$-\frac{1}{2}$	$V|6\rangle = \frac{1}{2}|2\rangle - t|3\rangle$

The Hamiltonian matrix is then

$$H = \begin{pmatrix} -1 & 0 & -t & 0 & \frac{1}{2} & 0 \\ 0 & -1 & 0 & -t & 0 & \frac{1}{2} \\ -t & 0 & -\frac{1}{2} & 0 & 0 & -t \\ 0 & -t & 0 & -\frac{1}{2} & -t & 0 \\ \frac{1}{2} & 0 & 0 & -t & -\frac{1}{2} & 0 \\ 0 & \frac{1}{2} & -t & 0 & 0 & -\frac{1}{2} \end{pmatrix} \tag{8.49}$$

This must now be transformed to block diagonal form to yield the 2×2 effective Hamiltonian

$$H_{\text{eff}} = \begin{pmatrix} h_{11} & h_{12} \\ h_{21} & h_{22} \end{pmatrix} \tag{8.50}$$

This can be done using any of the methods described in Section 4.5.1. Using the TBOT approach yields

$$h_{11} = h_{22} = -1 - \left(\tfrac{1}{2} + 2t^2\right)\lambda^2 + \left(\tfrac{1}{2} + 2t^2\right)\lambda^4 + \cdots$$
$$h_{12} = h_{21} = 4t^2\lambda^3 + (-12t^2 - 32t^4)\lambda^5 + \cdots \tag{8.51}$$

The reduced transition amplitudes are then

$$d_0^{(3A)} = h_{11} + h_{22} - 2E_g - 2d_0^{(1)} - 2d_0^{(2)}$$
$$= -\tfrac{1}{2}\lambda^2 + \left(\tfrac{5}{8} + 4t^2 - 16t^4\right)\lambda^4 + \cdots$$
$$d_2^{(3A)} = h_{12} + h_{21} = 8t^2\lambda^3 + (-24t^2 - 64t^4)\lambda^5 + \cdots \tag{8.52}$$

as shown in Table 8.5.

The results given in Table 8.5 can then be combined to yield the transition amplitudes and dispersion relation for the infinite chain

$$\Delta_0 = d_0^{(1)} + 2d_0^{(2)} + d_0^{(3A)} + d_0^{(3B)} + 2d_0^{(4)} + d_0^{(5A)} + d_0^{(5B)} + \cdots$$
$$= 1 - 4t^2\lambda^2 + \left(\tfrac{1}{8} - \tfrac{11}{2}t^2 + 16t^4\right)\lambda^4 + \cdots$$
$$\Delta_2 = d_2^{(3A)} + 2d_2^{(4)} + d_2^{(5A)} + d_2^{(5B)} + \cdots$$
$$= 8t^2\lambda^3 + (6t^2 - 96t^4)\lambda^5 + \cdots$$
$$\Delta_4 = O(\lambda^6) \tag{8.53}$$

To obtain the λ^6 term correctly requires clusters with up to seven sites. The one-hole energy is then

$$\epsilon(k) = \Delta_0 + \Delta_2 \cos(2k) + \Delta_4 \cos(4k) + \cdots \tag{8.54}$$

and can be evaluated for any k by extrapolation of the series to $\lambda = 1$. To achieve this requires considerably longer series, which must be obtained by computerizing the process. It is also much more efficient to choose fixed values of t, rather than keeping it is a general parameter, as above.

We have computed series to order λ^{21} and show, in Figure 8.6, the resulting dispersion curve. It can be seen that the single-hole energy gap is a minimum at $k = \pi/2$, and the bandwidth decreases with decreasing J/t.

Our group has used the approach described above to study one- and two-hole states on the square lattice (Hamer *et al.*, 1998). Figure 8.7 shows the one-hole

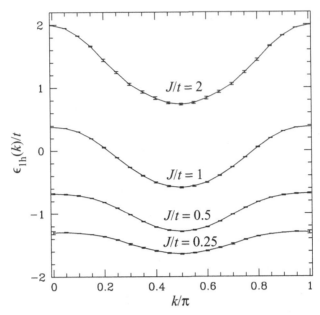

Fig. 8.6. One-hole excitation spectrum for the 1-D *t–J* model.

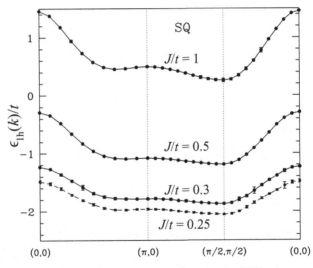

Fig. 8.7. One-hole energies along symmetry lines in the Brillouin zone for the *t–J* model on the square lattice.

energies along symmetry lines in the antiferromagnetic Brillouin zone for various ratios t/J. The single-hole dispersion relation is anomalously flat near the degenerate points $(0, \pm\pi)$, $(\pm\pi, 0)$ of the Brillouin zone, with the minimum of the dispersion at $(\pm\pi/2, \pm\pi/2)$. The bandwidth for small J/t appears to decrease with J/t, an anomalous behaviour which is a result of the coupling to spin waves (Martinez and Horsch, 1991). We also obtained energies for two-hole states of s-, p-, and d-symmetry and found that for small t all states are bound, with the strongest binding for d-states. However, for $t/J \sim 2.5$ there is a crossover with d-states becoming unbound but the p-state remaining weakly bound. The interested reader is referred to the original paper for further details and for references to other related work.

8.3.2 Variable electron density

We have mentioned previously the difficulty of treating systems with finite doping or, in other words, incommensurate electron density via zero-temperature series methods. The difficulty arises because one cannot identify a simple regular unperturbed ground state. One scheme to circumvent this difficulty is to introduce an artificial term in the unperturbed Hamiltonian H_0 which is not particle conserving, and remove it again in the perturbation V. This approach has been explored recently (Zheng *et al.*, 2002a) within the context of a t–J ladder system. The Hamiltonian is written as $H = H_0 + \lambda V$ with

$$H_0 = J_\perp \sum_i \left(\mathbf{S}_{i,1} \cdot \mathbf{S}_{i,2} - \frac{1}{4} n_{i,1} n_{i,2} \right) - t_\perp \sum_{i,\sigma} P(c^\dagger_{i,1,\sigma} c_{i,2,\sigma} + \text{h.c.})P$$

$$+ \frac{h}{\sqrt{2}} \sum_i P(c^\dagger_{i,1,\uparrow} c^\dagger_{i,2,\downarrow} - c^\dagger_{i,1,\downarrow} c^\dagger_{i,2,\uparrow} + \text{h.c.})P - \mu \sum_i (n_{i,1} + n_{i,2})$$

$$V = J \sum_{i,a} \left(\mathbf{S}_{i,a} \cdot \mathbf{S}_{i+1,a} - \frac{1}{4} n_{i,a} n_{i+1,a} \right) - t \sum_{i,a,\sigma} P(c^\dagger_{i,a,\sigma} c_{i+1,a,\sigma} + \text{h.c.})P$$

$$- \frac{h}{\sqrt{2}} \sum_i P(c^\dagger_{i,1,\uparrow} c^\dagger_{i,2,\downarrow} - c^\dagger_{i,1,\downarrow} c^\dagger_{i,2,\uparrow} + \text{h.c.})P \tag{8.55}$$

In other words, this is a 'dimer' expansion on the rungs. H_0 describes the rung terms (with parameter $t_\perp$, $J_\perp$) and V describes the coupling between rungs (with parameter t, J). The particle non-conserving term creates or destroys a pair of electrons on a particular rung, creating a spin singlet on an empty rung or destroying a spin singlet to create an empty rung. Other kinds of non-conserving terms could also be used. A chemical potential term is included to control the total electron density. When we extrapolate to $\lambda = 1$, we regain the (anisotropic) t–J Hamiltonian.

This approach naturally works best when the rung parameters are dominant. Zheng *et al.* (2002a) give a range of results for the case $J/J_\perp = t/t_\perp = 0.25$. Figure 8.8 shows the ground-state energy as a function of electron density for various t/J ratios. Figure 8.9 shows the ground-state energy versus electron density at $t/J = 2$, compared with two approximate methods and with a numerical density-matrix renormalization (DMRG) calculation. Our results are in excellent agreement with the DMRG calculation, which is expected to be very accurate. The approximate methods clearly give too high an energy.

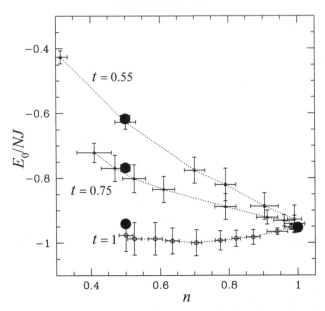

Fig. 8.8. Ground-state energy versus electron density for the *t–J* ladder. The full circles are previous series estimates for half and quarter filling. From Zheng *et al.* (2002a).

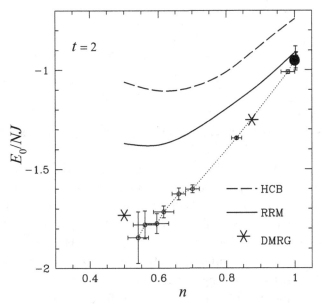

Fig. 8.9. Ground-state energy versus electron density for the *t–J* ladder, compared with results by other methods. From Zheng *et al.* (2002a).

The interested reader can find further details and references to related work in our paper (Zheng *et al.*, 2002a). There is clearly much more that can be done along similar lines.

8.3.3 High-temperature expansions

Derivation of high-temperature expansions for the *t–J* model follows the same general approach as for the Hubbard model, described in Section 8.2.4. There are, however, some differences and we provide a fairly detailed description, taking the linear chain as an example.

The Hamiltonian is taken to be

$$H = H_0 + \lambda V \tag{8.56}$$

with

$$H_0 = -\mu \sum_i n_i - h \sum_i (n_{i\uparrow} - n_{i\downarrow}) \tag{8.57}$$

and

$$V = -t \sum_{\langle ij \rangle \sigma} (c_{i\sigma}^\dagger c_{j\sigma} + \text{h.c.}) + J \sum_{\langle ij \rangle} \left(\mathbf{S}_i \cdot \mathbf{S}_j - \frac{1}{4} n_i n_j \right) \tag{8.58}$$

Note that we work in the grand ensemble and include a uniform external field coupling to the magnetization. Our aim is to derive an expansion for the uniform susceptibility.

Since H_0 and V commute we can express the grand partition function as

$$\mathcal{Z}_g = \mathcal{Z}_{g0} \left\{ 1 + \sum_{r=1}^\infty \frac{(-\beta)^r}{r!} \langle V^r \rangle \right\} \tag{8.59}$$

where

$$\mathcal{Z}_{g0} = \text{Tr}\{e^{-\beta H_0}\} = z_0^N \tag{8.60}$$

and

$$z_0 = 1 + \zeta(y + y^{-1}) \tag{8.61}$$

with $\zeta = e^{\beta\mu}$, the fugacity, and $y = e^{\beta h}$. The angular bracket denotes an unperturbed average

$$\langle A \rangle = \mathcal{Z}_{g0}^{-1} \text{Tr}\{e^{-\beta H_0} A\} \tag{8.62}$$

Note that there are no τ-integrals here, and this is a true high-temperature expansion. At this point there are again two ways to proceed. In the linked cluster approach we need to included connected clusters only, evaluate the grand partition function for

Table 8.6. *Cluster contributions to t–J linear chain to order β^4.*

Cluster	LC	$X_g = z_0^r W_g$
1.	1	$\gamma \zeta^2 (\beta t) + [\zeta y_1 + \frac{1}{2}\gamma^2 \zeta^2](\beta t)^2$ $+ (\frac{1}{6}\gamma^3 \zeta^2)(\beta t)^3 + [\frac{1}{12}\zeta y_1 + \frac{1}{24}\gamma^4 \zeta^2](\beta t)^4 + \cdots$
2.	1	$\frac{1}{4}\gamma^2 \zeta^3 y_1 (\beta t)^2 + [\gamma \zeta^2 + \frac{1}{4}\gamma^3 \zeta^3 y_1](\beta t)^3$ $+ \frac{1}{24}[4\zeta y_1 + 4\zeta^2(2 + y_2 + 2\gamma^2) + \frac{25}{8}\gamma^4 \zeta^3 y_1](\beta t)^4 + \cdots$
3.	$-\frac{3}{2}$	X_1^2
4.	1	$(\frac{1}{4}\gamma^3 \zeta^4)(\beta t)^3 + [\frac{7}{24}\gamma^2 \zeta^3 y_1 + \frac{1}{96}\gamma^4 \zeta^4 (y_2 + 34)](\beta t)^4 + \cdots$
5.	-4	$X_2 X_1$
6.	$\frac{10}{3}$	X_1^3
7.	1	$(\frac{1}{16}\gamma^4 \zeta^5 y_1)(\beta t)^4 + \cdots$
8.	-5	$X_4 X_1$
9.	$-\frac{5}{2}$	X_2^2
10.	15	$X_2 X_1^2$
11.	$-\frac{35}{4}$	X_1^4

each cluster, compute the logarithm to get the contribution to the grand potential, subtract off sub-graph contributions and finally combine the cluster contributions with their lattice constants to obtain the series for the bulk lattice.

The approach we follow is to take the logarithm of the infinite lattice partition function, giving the grand potential per site

$$-\beta \Omega / N = \frac{1}{N} \ln \mathcal{Z}_g = \ln z_0 + \sum_{r=1}^{\infty} \frac{(-\beta)^r}{r!} \langle V^r \rangle_N \tag{8.63}$$

and identifying the various contributions to $\langle V^r \rangle_N$ with connected and disconnected clusters. Alternatively we can write

$$-\beta \Omega / N = \ln z_0 + \sum_{\{g\}} c_g W_g(\zeta, y, \beta) \tag{8.64}$$

where the sum is over all clusters, c_g is the usual weak lattice constant and W_g is the 'cluster weight' which must be calculated for each cluster. A computer program can be written to do this, as the number of terms grows rapidly.

Table 8.6 lists all of the clusters which contribute to order β^4, for the linear chain, together with their lattice constants and cluster weights (to order β^4). Unlike the Hubbard case (Table 8.4) we use a cluster symbol to include multiple usage of any bond, with each bond being used at least once. We use the notation $\gamma = J/t$, $y_n = y^n + y^{-n}$, and write $W_g = z_0^{-r} X_g$ where r is the number of sites in the cluster.

To analyse the resulting series it is convenient to introduce a new parameter $p = \zeta/(1 + 2\zeta)$. The data of Table 8.6 can then be combined to yield, for the linear chain, the zero-field grand potential

$$
\begin{aligned}
-\beta\Omega/N = & -\ln(1-2p) + (\gamma p^2)(\beta t) + [2p(1-2p) \\
& + \frac{1}{2}\gamma^2 p^2(1+p-3p^2)](\beta t)^2 + [\gamma p^2(1-8p+12p^2) \\
& + \frac{1}{12}\gamma^3 p^2(2+6p-15p^2-24p^3+40p^4)](\beta t)^3 \\
& + \Big[\frac{1}{2}p(1-14p+48p^2-48p^3) + \frac{1}{12}\gamma^2 p^2(4-37p-38p^2 \\
& + 432p^3 - 480p^4) + \frac{1}{96}\gamma^4 p^2(4+25p-48p^2-276p^3+300p^4 \\
& + 720p^5 - 840p^6)\Big](\beta t)^4 + \cdots
\end{aligned}
\tag{8.65}
$$

and for the susceptibility

$$
\begin{aligned}
\chi = & \frac{1}{\beta^2}\frac{\partial^2}{\partial h^2}(-\beta\Omega/N) \\
= & 2p - (4\gamma p^3)(\beta t) + \frac{1}{2}(1-2p)[4-16p-3\gamma^2 p^2(1+4p)](\beta t)^2 \\
& - [12\gamma p^3(1-6p+8p^2) + \frac{1}{6}\gamma^3 p^3(1+18p-48p^2-120p^3+240p^4)](\beta t)^3 \\
& + \Big[\frac{1}{2}p(1-30p+200p^2-480p^3+384p^4) - \frac{1}{12}\gamma^2 p^3(53-242p \\
& - 928p^2 + 5280p^3 - 5760p^4) + \frac{1}{96}\gamma^4 p^3(9-142p+108p^2+2640p^3 \\
& - 2880p^4 - 10080p^5 + 13440p^6)\Big](\beta t)^4 + \cdots
\end{aligned}
\tag{8.66}
$$

To express things in terms of the electron density n we proceed as in Section 8.2.4 and compute the elctron density

$$
n = \zeta\frac{\partial}{\partial\zeta}(-\beta\Omega/N) = p(1-2p)\frac{\partial}{\partial p}(-\beta\Omega/N)
$$

$$
= 2p + \sum_{r=1}^{\infty}Q_r(\gamma, p)(\beta t)^r
\tag{8.67}
$$

with

$$
Q_1 = 2\gamma p^2(1-2p)
\tag{8.68a}
$$

$$
Q_2 = 2p(1-2p)(1-4p) + \frac{1}{2}\gamma^2 p^2(1-2p)(2+3p-12p^2)
\tag{8.68b}
$$

$$
\begin{aligned}
Q_3 = & 2\gamma p^2(1-2p)(1-12p+24p^2) \\
& + \frac{1}{6}\gamma^3 p^2(1-2p)(2+9p-30p^2-60p^3+120p^4)
\end{aligned}
\tag{8.68c}
$$

$$
\begin{aligned}
Q_4 = & \frac{1}{2}p(1-2p)(1-4p)(1-24p+48p^2) + \frac{1}{12}\gamma^2 p^2(1-2p)(8-111p \\
& - 152p^2 + 2160p^3 - 2880p^4) + \frac{1}{96}\gamma^4 p^2(1-2p)(8+75p \\
& - 192p^2 - 1380p^3 + 1800p^4 + 5040p^5 - 6720p^6)
\end{aligned}
\tag{8.68d}
$$

The series can now be inverted analytically by writing

$$p = p_0 + p_1(\beta t) + p_2(\beta t)^2 + p_3(\beta t)^3 + p_4(\beta t)^4 + \cdots \qquad (8.69)$$

and solving by iteration to give $p_0, p_1, \ldots$ in terms of n. The results are

$$p_0 = n/2 \qquad (8.70a)$$

$$p_1 = \tfrac{1}{4}\gamma(n-1)n^2 \qquad (8.70b)$$

$$p_2 = \left[-\tfrac{1}{2}(1-2n) - \tfrac{1}{32}\gamma^2 n(4 - 5n + 6n^2)\right](1-n)n \qquad (8.70c)$$

$$p_3 = \left[\tfrac{1}{4}(1-2n)(2-3n)\right.$$
$$\left. - \tfrac{1}{192}\gamma^2(8 - 30n + 57n^2 - 51n^3 + 30n^4)\right]\gamma(1-n)n^2 \qquad (8.70d)$$

$$p_4 = \left[\tfrac{3}{8}(1-2n)^3 + \tfrac{1}{192}\gamma^2 n(56 - 273n + 592n^2\right.$$
$$- 708n^3 + 360n^4) - \tfrac{1}{1536}\gamma^4 n(16 - 149n + 420n^2$$
$$\left. - 738n^3 + 819n^4 - 543n^5 + 210n^6)\right](1-n)n \qquad (8.70e)$$

Finally we obtain the high-temperature series for thermodynamic quantities in terms of electron density

$$-\beta\Omega/N = -\ln(1-n) - \tfrac{1}{4}\gamma n^2(\beta t) + n^2\left[1 - \tfrac{1}{32}\gamma^2(4 - 4n + 3n^2)\right](\beta t)^2$$
$$+ \gamma n^2\left[\tfrac{1}{4}(1-n)(1-3n) - \tfrac{1}{192}\gamma^2(8 - 24n + 33n^2 - 24n^3 + 10n^4)\right](\beta t)^3$$
$$+ n^2\left[-\tfrac{1}{4}(3 - 8n + 6n^2) + \tfrac{1}{96}\gamma^2(16 - 82n + 147n^2 - 144n^3 + 60n^4)\right.$$
$$\left. - \tfrac{1}{3072}\gamma^4(32 - 248n + 528n^2 - 744n^3 + 660n^4 - 360n^5 + 105n^6)\right](\beta t)^4$$
$$+ \cdots \qquad (8.71a)$$

$$\beta\mu = \ln(n/(2-2n)) - \tfrac{1}{2}\gamma n(\beta t) + \left[-1 + 2n - \tfrac{1}{16}\gamma^2 n(4 - 3n + 2n^2)\right](\beta t)^2$$
$$+ \left[\tfrac{1}{2}\gamma(1-n)n(1-2n) - \tfrac{1}{96}\gamma^3 n(8 - 18n + 22n^2 - 15n^3 + 6n^4)\right](\beta t)^3$$
$$+ \left[\tfrac{1}{4}(1-2n)^3 + \tfrac{1}{96}\gamma^2 n(32 - 123n + 196n^2 - 180n^3 + 72n^4)\right.$$
$$\left. - \tfrac{1}{1536}\gamma^4 n(32 - 186n + 352n^2 - 465n^3 + 396n^4 - 210n^5 + 60n^6)\right](\beta t)^4$$
$$+ \cdots \qquad (8.71b)$$

$$\chi = n - \tfrac{1}{2}\gamma n^2(\beta t) - \tfrac{1}{8}\gamma^2(2-n)(1-n)n^2(\beta t)^2$$
$$+ \left[\tfrac{1}{2}\gamma(1-n)^2 n^2 - \tfrac{1}{48}\gamma^3 n^2(4 - 18n + 21n^2 - 12n^3 + 3n^4)\right](\beta t)^3$$
$$+ \left[\tfrac{1}{48}\gamma^2(1-n)n^2(16 - 49n + 60n^2 - 24n^3) - \tfrac{1}{384}\gamma^4 n^2(8\right.$$
$$\left. - 87n + 206n^2 - 249n^3 + 177n^4 - 75n^5 + 15n^6)\right](\beta t)^4 + \cdots \qquad (8.71c)$$

Several groups have studied the square lattice *t–J* model via high-temperature expansions. Putikka *et al.* (1992) and Putikka and Luchini (2000) have computed tenth-order series for the free energy and susceptibility and, by extrapolation of the series to low temperatures, have attempted to investigate the conditions for phase separation. Singh and Glenister (1992a, b) computed ninth-order series for the static structure factor, and eighth-order series [subsequently extended to twelfth-order (Putikka *et al.*, 1998)] for the distribution function $n(\mathbf{k})$, and have investigated the development of a Fermi surface as the temperature is reduced.

8.4 Further topics and possibilities

The bulk of this chapter has focused on the two most studied lattice models of strongly correlated electrons, the Hubbard and *t–J* models, and it has been shown how series expansions, both at zero and finite temperatures, can yield information about the models. To conclude the chapter we will summarize existing work on several other models and attempt to identify promising areas for further study.

One direction of work is to generalize the Hubbard and *t–J* models in some fairly obvious ways, for example by including further neighbour-hopping terms. Such terms are potentially important in modelling real materials and may lead to new physics. There is little, if any, series work in this area. Another generalization is to include a first-neighbour Coulomb repulsion term V into the Hubbard model. This is known as the 'extended Hubbard model' and has been studied via short finite-temperature series by Bartkowiak *et al.* (1995). There is, again, an opportunity for further series work.

A new model is the Falicov–Kimball model which, in its simplest form, is described by the Hamiltonian

$$H = -t \sum_{\langle ij \rangle} (c_i^{\dagger} c_j + c_j^{\dagger} c_i) + E \sum_i n_i^{(f)} + U \sum_i n_i^{(f)} n_i^{(c)} \qquad (8.72)$$

There are two species of spinless fermions, one species which can hop (represented by operators $c^{\dagger}$, c) and one which cannot (represented by $f^{\dagger}$, f), with occupation numbers $n_i^{(f)} = f_i^{\dagger} f_i, n_i^{(c)} = c_i^{\dagger} c_i$. There is a repulsion between a c and an f fermion on the same site and it is this term which leads to strong correlations. The model was originally introduced to study mixed valence systems and metal–insulator transitions in rare-earth compounds but other interpretations are also possible, and of considerable interest. There is a substantial literature on the model, and a number of exact results in both one and two dimensions. A variety of generalizations have also been studied including the addition of spin and hybridization between the two species. The reader is referred to a paper by Czycholl (1999) as a lead into the literature. Interest in the model has increased recently in connection with a conjectured electronic mechanism for ferroelectricity (Portengen *et al.*, 1996; Batista *et al.*, 2004). The only published series work on this model, of which we are aware, is Yang *et al.* (1992), who studied the infinite U model in infinite spatial dimension via high-temperature series. Clearly there is much scope for further work.

Another model of interest in strongly correlated electron physics is the so-called Kondo lattice model. This model describes a band of conduction electrons interacting, via a spin-exchange term, with a set of immobile $S = \frac{1}{2}$ spins $\mathbf{S}_i$ (representing f electrons). The Hamiltonian is

$$H = -t \sum_{\langle ij\rangle\sigma} (c^{\dagger}_{i\sigma} c_{j\sigma} + \text{h.c.}) + J \sum_i \mathbf{S}_i \cdot \mathbf{s}_i \qquad (8.73)$$

where $\mathbf{s}_i$ is the conduction electron spin operator. This model has been extensively studied in connection with a class of materials known as Kondo insulators ($J > 0$), and in connection with the manganites ($J < 0$). Despite the apparent simplicity of the model, in which neither the conduction electrons nor the localized spins interact directly among themselves, the spin exchange leads to a strongly correlated, many-body system. The model incorporates two competing physical processes. In the large J limit the conduction electrons are frozen out, via the formation of local singlets or triplets. In this case there will be a gap to spin excitations and spin correlations will be short ranged. On the other hand at weak coupling, the conduction electrons can induce the usual RKKY interaction between localized spins, giving rise to possible magnetically ordered phases with no spin gap and long-range correlations. Thus one should expect a quantum phase transition at some intermediate t/J.

Series methods have been used, with some success, to study this model. Early studies by Shi *et al.* (1995) (for $T = 0$) and by Röder *et al.* (1997) (for higher temperatures) have been recently extended by our group (Zheng and Oitmaa, 2003; Oitmaa and Zheng, 2003). We refer the reader to these papers for details of the methods and results, as well as for a lead into the large body of literature which exists. There remains scope for further series work.

This brief discussion by no means exhausts the range of models which have been proposed and studied in this very fertile field. There are, of course, more realistic multi-band models which have hardly been touched by series methods. On the other hand, generally speaking, the more complex the model the shorter are the series which can be obtained, and the more uncertain is the analysis.

9

Review of lattice gauge theory

The prime motivation for the study of lattice gauge theory, as outlined in Chapter 1, is to provide a non-perturbative method of calculating the properties of quantum chromodynamics (QCD) at low energies and long distances. To a theorist, however, lattice gauge models are a fascinating topic of study in their own right.

In this chapter we shall briefly review the essential elements of quantum chromodynamics and lattice field theory, before moving in the next chapter to a discussion of series calculations in the area. More detailed discussions of this background material can be found in any of several books on lattice gauge theory (Montvay and Münster, 1994; Rothe, 1997; Smit, 2002).

9.1 Quantum chromodynamics

The theory of quantum chromodynamics is now universally recognized as part of the *Standard Model* in particle physics, describing the strong interactions among hadrons. The theory is invariant under gauge transformations belonging to the non-Abelian group SU(3). The quarks in QCD belong to the fundamental representation of SU(3) and come in three 'colours', while the gluons or gauge bosons belong to the adjoint representation, forming an octet.

Let us recall the formalism of QCD (Itzykson and Zuber, 1980). The eight generators of SU(3), $\{t^a, a = 1, \cdots, 8\}$, obey the Lie algebra

$$[t^a, t^b] = i f_{abc} t^c \tag{9.1}$$

where the f_{abc} are the completely antisymmetric structure constants of the group. In the fundamental representation we may choose

$$t^a = \tfrac{1}{2}\lambda^a, \qquad a = 1, \cdots, 8 \tag{9.2}$$

211

where the λ^a are the 3×3 Gell–Mann matrices, normalized so that

$$\text{Tr}(t^a t^b) = \tfrac{1}{2}\delta_{ab} \tag{9.3}$$

Any element lying in the Lie algebra can be written as a combination of generators, e.g.

$$A_\mu = \sum_a A_{\mu a} t^a, \qquad F_{\mu\nu} = \sum_a F_{\mu\nu a} t^a, \tag{9.4}$$

where A_μ, $F_{\mu\nu}$ are the gauge field and field strength tensors, respectively.

Then the action for QCD is

$$\begin{aligned}
S &= \int d^4x \left\{ -\frac{1}{2g^2} \text{Tr}(F_{\mu\nu} F^{\mu\nu}) + \bar\psi [i D_\mu \gamma^\mu - M]\psi \right\} \\
&= \int d^4x \left\{ -\frac{1}{4g^2} \sum_a F_{\mu\nu a} F^{\mu\nu a} + \bar\psi [i D_\mu \gamma^\mu - M]\psi \right\}
\end{aligned} \tag{9.5}$$

where $\psi(x)$ is the fermion (quark) field of mass M, with spin and colour indices suppressed, $A_\mu(x)$ is the gauge (gluon) field, the field strength tensor is

$$F_{\mu\nu} = \partial_\mu A_\nu - \partial_\nu A_\mu - i[A_\mu, A_\nu], \tag{9.6}$$

and the covariant derivative is defined

$$D_\mu = \partial_\mu - i A_\mu, \tag{9.7}$$

obeying the identity

$$[D_\mu, D_\nu] = -i F_{\mu\nu} \tag{9.8}$$

The dimensionless parameter g is the coupling constant; one may rescale $A \to gA$ to regain the more usual form of the theory.

Under a gauge transformation $g(x)$, where $g(x)$ is an element of the group SU(3) in the fundamental representation which varies with the space–time point x, the fields transform as

$$\begin{aligned}
\psi &\to g\psi, \\
D_\mu \psi &\to g D_\mu \psi, \\
F_{\mu\nu} &\to g F_{\mu\nu} g^{-1}
\end{aligned} \tag{9.9}$$

while the gauge field A_μ transforms as

$$A_\mu \to g A_\mu g^{-1} - i(\partial_\mu g)g^{-1} \tag{9.10}$$

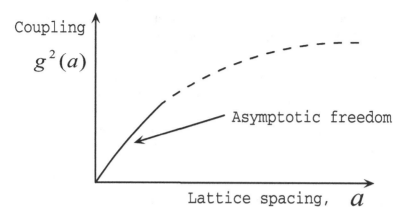

Fig. 9.1. Sketch of the effective coupling $g^2(a)$ as a function of the short-distance cutoff a for SU(3).

The 'parallel transporter' between points x_1 and x_2 is

$$P(x_1, x_2) = \exp\left[i \int_{x_1}^{x_2} dx^\mu A_\mu\right] = \exp\left[\frac{i\lambda^a}{2} \int_{x_1}^{x_2} dx^\mu A_{\mu a}\right] \qquad (9.11)$$

which corresponds to the 'path integral phase factor' of Schwinger in electrodynamics, and transforms as

$$P(x_1, x_2) \rightarrow g(x_1)P(x_1, x_2)g^{-1}(x_2) \qquad (9.12)$$

Using standard perturbative techniques, QCD can provide an excellent description of 'deep-inelastic' phenomena involving large momentum transfers; and it convincingly explains the pattern of quantum numbers in the hadronic spectrum. Unfortunately, however, the vast bulk of hadronic phenomena do not involve large momentum transfers, and *cannot* be treated by perturbation theory. The reason for this lies in renormalization theory. A renormalization group analysis of QCD by Politzer (1973) and Gross and Wilczek (1973) revealed an unexpected behaviour of the effective coupling constant. In the familiar case of quantum electrodynamics, the polarization of the vacuum into virtual e^+e^- pairs 'shields' any introduced charge, and causes the effective charge to decrease with distance. In the case of QCD, however, the gluons themselves carry colour charge, and produce an 'anti-shielding' effect, so that the effective charge *increases* with distance (Figure 9.1).

Thus at short distances or large momenta the effective coupling is small, so that quarks and gluons behave almost like free particles, and can be treated by weak-coupling perturbation theory. This is the phenomenon of *asymptotic freedom*: it explains why perturbative QCD may be applied to deep inelastic scattering processes. At large distances or small momenta, on the other hand, the effective coupling grows large, and traditional methods break down. Furthermore, it is known that non-perturbative 'instanton' effects become important in this theory – they are completely missed by perturbation theory.

It therefore becomes necessary to develop new methods to deal with the large distance, low-energy regime of QCD. Of particular interest is the question of 'confinement': do free quarks exist, or not? Can we calculate the low-energy, bound-state spectrum of QCD, i.e. the masses of all the hadrons observed in nature? Does the spectrum exhibit spontaneous breakdown of chiral symmetry?

Several non-perturbative approaches have been proposed to attack these problems, but the most successful so far has undoubtedly been the lattice technique. The introduction of a space–time lattice was suggested by Wilson (1974) as an intermediate 'scaffolding' when doing calculations in QCD. Advantages of the approach include the following.

(1) The lattice provides an ultraviolet (short-distance) cutoff, which regularizes the theory automatically.
(2) Gauge invariance is explicitly maintained on the lattice, which is crucial for renormalizability.
(3) The Euclidean lattice formalism is mathematically equivalent to that of statistical mechanics (as we shall see in the next section), which allows the importation of many techniques from that field, including Monte Carlo simulations and strong-coupling (equivalent to high-temperature) series expansions.

9.2 The path integral approach to field theory

The modern approach to quantum field theory starts from the Feynman path integral

$$Z = \int D\phi\, e^{iS/\hbar} \tag{9.13}$$

where $\int D\phi$ denotes a functional integral over the fields in the theory at every point in space–time, and S is the action. The path integral acts as a generating functional for the theory: by adding appropriate source terms (analogous to the external fields in statistical mechanics) and taking the appropriate functional derivatives, one may generate any desired Green's function or propagator in the theory (see e.g. Zinn-Justin, 1996).

To give a concrete mathematical definition to the path integral, we make a Wick rotation to imaginary time, when the metric becomes Euclidean, and replace continuous space–time by a discrete lattice of points, e.g. a hypercubic lattice with spacing a:

$$t \rightarrow -i\tau \tag{9.14a}$$

$$iS\{\phi\} \rightarrow -S_E\{\phi\} \tag{9.14b}$$

$$Z \rightarrow \prod_i \left\{ \int d\phi_i \right\} e^{-S_E\{\phi_i\}/\hbar} \tag{9.14c}$$

where ϕ_i denotes the fields at site i of the space–time lattice, and S_E is the Euclidean action. For example, in the case of a free scalar field with mass m

$$S = \frac{1}{2} \int dt d^3x [\partial^\mu \phi \partial_\mu \phi - m^2 \phi^2] \tag{9.15a}$$

$$S_E = \frac{1}{2} \int d^4x [(\partial_\mu \phi)^2 + m^2 \phi^2] \quad (\tau = x_4) \tag{9.15b}$$

and the transcription to the lattice may be made (for instance)

$$x_\mu \rightarrow n_\mu a, \tag{9.16a}$$

$$\phi(x) \rightarrow \phi(na), \tag{9.16b}$$

$$\int d^4x \rightarrow a^4 \sum_n, \tag{9.16c}$$

$$\partial_\mu \phi \rightarrow \frac{1}{2a} [\phi((n + \mu)a) - \phi((n - \hat{e}_\mu)a)], \tag{9.16d}$$

where derivatives of the fields are replaced by the corresponding finite-difference expressions.

The fields are now integrated at a denumerable set of points, and the Euclidean action is positive definite, and supplies a Gaussian convergence factor at large values of ϕ, so that the theory is mathematically well-defined.

In the form (9.14), it can be seen that the path integral has the same mathematical structure as the *partition function* in lattice statistical mechanics (Zinn-Justin, 1996). It involves an integral over all possible configurations of the system in four-dimensional Euclidean space–time, weighted by an exponential factor corresponding to the Maxwell–Boltzmann factor in statistical mechanics. This has led to the recognition that the formalisms of quantum field theory and statistical mechanics are in fact equivalent, and the same theoretical techniques can be freely used in both disciplines. In general, a quantum field theory in one time and $(d - 1)$ space dimensions is equivalent to a classical statistical system in d Euclidean space dimensions. For convenience, we present in Appendix 8 a list of some quantities in the two disciplines which correspond to each other. The first obvious correspondence following from equation (9.14) is that the Euclidean action in field theory

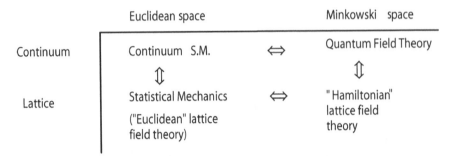

Fig. 9.2. 'Road map' of space–time environments.

plays the same role as the Hamiltonian in statistical mechanics:

$$\frac{1}{\hbar} S_E \leftrightarrow \beta H_{SM} \tag{9.17}$$

Let us pause to draw a 'road-map' of the different space–time arenas in which a field theory may be formulated (Figure 9.2).

At the top right, we start with quantum field theory in its normal setting of continuous Minkowski space–time. By a Wick rotation to Euclidean space–time, followed by a discretization to the lattice, one arrives at Euclidean lattice field theory in the bottom left-hand corner, whose formalism is equivalent to that of statistical mechanics. Alternatively, one may retain the original Minkowski metric, and replace only the three space dimensions by a discrete lattice, arriving at the so-called 'Hamiltonian lattice' formulation of field theory discussed by Kogut and Susskind (1975). This will be discussed further in Section 9.8.

At the end of the day, as the last stage in a lattice model calculation, one would like to regain the continuum theory by taking the continuum limit $a \to 0$. In general, one expects the continuum limit will correspond to a *critical point* of the lattice model. The reason is simple: the physical correlation length ξ can be written.

$$\xi = n_\xi a \tag{9.18}$$

where n_ξ is the equivalent quantity in lattice units. If ξ is to remain non-zero as $a \to 0$ (i.e. if the continuum field theory is non-trivial), then n_ξ must diverge to infinity as $a \to 0$, which indicates critical behaviour.

Note that the discretization process going from the continuum to the lattice is not in general unique. One might choose a square or triangular lattice in two dimensions, for instance, or employ one of several different finite-difference approximations to a derivative. The predictions of all these different lattice models should be the same in the continuum limit, however. This is due to the hypothesis of *universality*, which tells us that the microscopic details of a lattice model do not affect its behaviour at a critical point, where the correlation length diverges. This hypothesis can be demonstrated by renormalization group arguments, and proved for soluble models in two dimensions (see e.g. Zinn-Justin, 1996).

The introduction of the lattice also breaks the continuum Lorentz symmetry down to its discrete lattice analogue (displacements in units of the lattice spacing a; rotations by $90°$, etc). It is assumed that these discrete lattice symmetries will be 'promoted' back to their continuum analogues as the critical point corresponding to the continuum limit is approached. This phenomenon can be demonstrated in the soluble case of the two-dimensional Ising model, for instance (Wu *et al.*, 1976).

9.3 Euclidean lattice gauge theory

Transcription of the QCD action (9.5) to the Euclidean regime makes use of the same correspondences as equations (9.14), plus the additional rule

$$A^0 \to i A_4 \tag{9.19}$$

It is also convenient to define a new set of γ-matrices in Euclidean space γ_μ^E $(\mu = 1, \cdots, 4)$ satisfying

$$\{\gamma_\mu^E, \gamma_\nu^E\} = 2\delta_{\mu\nu} \tag{9.20}$$

which may be implemented by the Hermitian choice

$$\gamma^0 \to \gamma_4^E, \tag{9.21a}$$

$$\gamma^j \to i\gamma_j^E \quad (j = 1, \cdots, 3) \tag{9.21b}$$

The Euclidean action then takes the form

$$S_E = \int d^4x \left\{ \frac{1}{4g^2} F_{\mu\nu a} F_{\mu\nu a} + \bar{\psi}[\gamma_\mu^E D_\mu + M]\psi \right\} \tag{9.22}$$

where the metric is now Euclidean.

To set up a lattice version of pure gauge QCD (i.e. SU(3) Yang–Mills theory), we merely have to recognize that the 'link variable' U_{ij} discussed in Section 1.2.5 plays the role of the 'parallel transporter' between neighbouring sites (i, j) – the transformation law (1.9a) is identical with (9.12). Thus the link variable $U(n, n + \hat{\mu}) \equiv U_\mu(n)$, obeying

$$U(n + \hat{\mu}, n) = U^\dagger(n, n + \hat{\mu}), \tag{9.23}$$

can be written

$$U_\mu(n) = e^{ia A_\mu(n)} \tag{9.24}$$

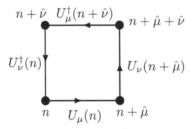

Fig. 9.3. An elementary plaquette lying in the $\mu\nu$ plane, based at site n.

where $A_\mu(n)$ is the gauge field on the lattice, defined on links.

The plaquette variable corresponding to Figure 9.3 is then

$$U_{\mu\nu}(n) = U_\mu(n)U_\nu(n+\hat{\mu})U_\mu^\dagger(n+\hat{\nu})U_\nu^\dagger(n) \tag{9.25}$$

If one takes the 'naive' continuum limit $a \to 0$ (neglecting renormalization effects), letting

$$A_\mu(n) \to A_\mu(x), \tag{9.26a}$$

$$(A_\mu(n+\hat{\nu}) - A_\mu(n))/a \to \partial_\nu A_\mu(x) \tag{9.26b}$$

and uses the Baker–Campbell–Haussdorf identity

$$e^x e^y = e^{x+y+[x,y]/2+\cdots}, \tag{9.27}$$

then one can show that

$$U_{\mu\nu}(n) \to \exp(ia^2 F_{\mu\nu}) \qquad \text{as} \quad a \to 0, \tag{9.28}$$

where $F_{\mu\nu}$ is given by (9.6) – this is left as an exercise for the reader.

Wilson's form of the Euclidean lattice action for the pure gauge theory is

$$S = \beta \sum_P \left[1 - \frac{1}{3}\mathrm{ReTr}\{U_{\mu\nu}(n)\} \right] \tag{9.29}$$

where the sum runs over all plaquettes of the lattice (the factor $1/3$ appears because $U_{\mu\nu}(n)$ is a 3×3 matrix in colour space). Expanding the exponential in (9.28) and taking the real part, we obtain

$$S \to \frac{\beta}{6} \int d^4x \frac{1}{2}\mathrm{Tr}(F_{\mu\nu}F_{\mu\nu}) + O(a^6) \tag{9.30}$$

which is the standard Euclidean action of pure Yang–Mills gauge theory in the continuum (9.22), provided we set the coupling

$$\beta = \frac{6}{g^2} \tag{9.31}$$

Geometrically, the Wilson action is a measure of the effect of parallel transport around a closed plaquette, i.e. the 'curvature'.

The integration measure in the path integral for pure Yang–Mills theory is written

$$\int D\phi \rightarrow \int DU = \prod_l \left\{ \int dU_l \right\} \tag{9.32}$$

where U_l is the link variable on link l. The precise meaning of this expression will be discussed in the next chapter.

9.4 Confinement and phase structure on the lattice

In a lattice spin model such as the Ising model, the phase structure is characterized by an order parameter. One looks for a disordered phase at high temperatures, and an ordered phase at low temperatures, with all the spins aligned towards the same direction. The order parameter is the spontaneous magnetization: a non-zero value signals spontaneous breakdown of rotational symmetry in spin space.

For a gauge model, however, Elitzur's theorem (Elitzur, 1975) states that there is *no* spontaneous breakdown of local gauge invariance. In the statistical mechanics analogy, the local value of a gauge-non-invariant order parameter can always be changed by a local gauge transformation at small cost in 'energy'; thus 'thermal' fluctuations will always restore the gauge symmetry as a symmetry-breaking field is removed, even in the bulk limit.

For the pure gauge theory without quarks, Wilson (1974) introduced a non-local but gauge-invariant order parameter, the *Wilson loop*

$$W_C = \left\langle \mathrm{Tr} \left\{ \prod_{i,j \in C} U(i, j) \right\} \right\rangle \tag{9.33}$$

which is the expectation value of a product of link variables around a closed curve C. For very large loops C, there are two possible limiting behaviours

$$W_C \sim \exp[-K \times \mathrm{area}(C)] \tag{9.34}$$

– an 'area law', or

$$W_C \sim \exp[-K_1 \times \mathrm{perimeter}(C)] \tag{9.35}$$

– a 'perimeter law'. The area law is characteristic of a strong-coupling phase (small β) with confined quarks (Wilson, 1974). Heuristically, this may be understood as follows.

Consider a large, rectangular 'timelike' loop as shown in Figure 9.4. Choose the 'timelike' gauge $A_\tau = 0$; then the Wilson loop is

$$W = \exp \left\{ i \left[\int_A^B dx \, A_x(T_1) + \int_B^A dx \, A_x(T_2) \right] \right\} \tag{9.36}$$

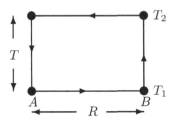

Fig. 9.4. A large, rectangular 'timelike' Wilson loop.

which is the correlation function between a 'string operator' between points A and B at time T_1, and its conjugate at time T_2. The string operators might be connecting a static $q\bar{q}$ pair at points A and B, for instance. By general arguments (Appendix 8), the asymptotic decay of the correlation function is

$$W(R, T) \sim \exp[-V(R)T], \tag{9.37}$$

where $V(R)$ is the energy gap corresponding to states which overlap with the string operator: in this case, the field energy or potential energy of a fictitious, static $q\bar{q}$ pair at distance R apart.

If the theory is unconfined, $V(R) \to$ constant at large R, and the exponent is proportional to T, the linear dimension (perimeter) of the loop. If the theory is confining, we expect the potential difference to rise linearly with R, so that the exponent is proportional to the area of the loop. In fact the constant K in equation (9.34) is just the 'string tension' of the theory, which we picture as the energy per unit length of a 'string' of colour flux joining the $q\bar{q}$ pair.

In the strong-coupling limit $\beta \to 0$ (or $g \to \infty$), a lattice gauge theory with compact gauge group such as SU(3) is automatically confining. This will be shown by strong-coupling perturbation theory in the next chapter. The upshot is that the string tension in this limit is

$$Ka^2 \sim -\ln\left(\frac{\beta}{18}\right), \tag{9.38}$$

which actually diverges logarithmically in the limit.

Renormalization group theory predicts, however, that the continuum limit of the theory occurs at *weak* coupling, $g \to 0$ or $\beta \to \infty$, as we shall see in the next section. Thus the question of confinement is determined by the phase diagram of the theory. If the theory remains in a confining phase, with a finite string tension at all couplings, then the continuum theory is confining. If there is a phase transition at finite coupling where the string tension goes to zero, then the corresponding continuum theory is not confining. We expect the first scenario to apply to QCD, and the second to apply to QED (quantum electrodynamics).

9.5 Renormalization theory and the continuum limit

In the renormalization procedure, we let the lattice spacing $a \to 0$, while varying the coupling g in such a way as to maintain a physical observable, such as the string tension, fixed at its 'measured' value. The induced dependence of g on a is given by the renormalization group function

$$\gamma(g) = a \frac{d}{da} g(a). \tag{9.39}$$

As $a \to 0$, the lattice coupling approaches a fixed point

$$\lim_{a \to 0} g = g^*, \tag{9.40}$$

and equation (9.39) then implies

$$\gamma(g^*) = 0. \tag{9.41}$$

For an ultraviolet fixed point such as that considered here, we expect the slope of γ to be positive at g^*.

At weak coupling, one can use perturbation theory to compute the renormalization group function, and for a non-Abelian gauge theory one finds an ultraviolet fixed point at $g = 0$, as illustrated in Figure 9.1. If the theory remains confining, and no phase transition at finite coupling supervenes, then we expect this fixed point to be the continuum limit we are looking for.

For example, in the case of SU(3) Yang–Mills theory we find (Politzer, 1973; Gross and Wilczek, 1973)

$$\gamma(g) = \gamma_0 g^3 + \gamma_1 g^5 + \cdots$$
$$\gamma_0 = 11/(16\pi^2). \tag{9.42}$$

Integrating equation (9.39) in the vicinity of $g = 0$, we obtain:

$$\frac{1}{g^2} = \gamma_0 \log\left(\frac{1}{a^2 \Lambda_0^2}\right) + \left(\frac{\gamma_1}{\gamma_0}\right) \log\left(\log \frac{1}{a^2 \Lambda_0^2}\right), \tag{9.43}$$

where the 'scale factor' Λ_0 appears as a constant of integration. It is a remarkable feature of pure gauge theory that it contains no dimensionful parameter (the 'fine structure' constant $g^2/(4\pi \hbar c)$ is dimensionless), but dimensional dependence is 'smuggled in' via the parameter Λ_0. All the dimensional quantities in the theory such as string tension, masses, etc. may be expressed in units of Λ_0. Inverting equation (9.43), for instance, the lattice spacing can be expressed:

$$a = \frac{1}{\Lambda_0} (g^2 \gamma_0)^{-\gamma_1/2\gamma_0^2} \exp\left(-\frac{1}{2\gamma_0 g^2}\right) (1 + O(g^2)). \tag{9.44}$$

– note the essential singularity at weak coupling. Multiplying by a corresponding mass, we can obtain the weak-coupling dependence of a lattice correlation length:

$$Ma = n_\xi^{-1} = \left(\frac{M}{\Lambda_0}\right)(g^2\gamma_0)^{-\gamma_1/2\gamma_0^2}\exp\left(-\frac{1}{2\gamma_0 g^2}\right)(1 + O(g^2)). \tag{9.45}$$

or for the string tension

$$Ka^2 = \left(\frac{K}{\Lambda_0^2}\right)(g^2\gamma_0)^{-\gamma_1/\gamma_0^2}\exp\left(-\frac{1}{\gamma_0 g^2}\right)(1 + O(g^2)). \tag{9.46}$$

If we measure the correlation length in the lattice model, the result is a value for the ratio (m/Λ_0). The scale factor is not predicted by the theory, and depends on the particular renormalization scheme chosen. Hasenfratz and Hasenfratz (1980) have calculated the ratio of the lattice scale factor Λ_0 and the standard continuum scale factor Λ_{mom} as

$$\Lambda_{\text{mom}}/\Lambda_0 = \begin{cases} 57.5 & SU(2); \\ 83.5 & SU(3). \end{cases} \tag{9.47}$$

for the pure gauge theory.

The scale factor Λ_0 has to be determined by comparison with experiment. Other than this, however, the theory for pure gauge fields has no free parameters. If one takes a dimensionless quantity such as the ratio between two masses, the Λ_0 parameter drops out, and the result is predicted uniquely by the theory, provided the relevant numerical calculations can be done.

9.6 Monte Carlo simulations

Up till now, the most powerful numerical method for investigating the properties of QCD has been Monte Carlo simulation of the Euclidean path integral (see e.g. Creutz, 1979). The technique is the same in principle as the Metropolis Monte Carlo algorithm in statistical mechanics. Successive gauge field configurations of a finite lattice system are generated by a Markov chain process of Metropolis type, so that at equilibrium the probability of any given configuration is that given by the Euclidean path integral, or partition function. Measurements of physical quantities may then be made by averaging over the ensemble of configurations. For further details, we refer the reader to excellent reviews elsewhere (Montvay and Münster, 1994; Rothe, 1997). Here we shall only mention a few illustrative results.

The phase structure of Yang–Mills theory was investigated in the original work of Creutz and collaborators (Creutz, 1983). They measured the value of the 'mean plaquette' $\langle 1 - \text{ReTr}U_P/n\rangle$, analogous to the internal energy in statistical mechanics, as a function of the coupling β (analogous to the inverse temperature), as shown

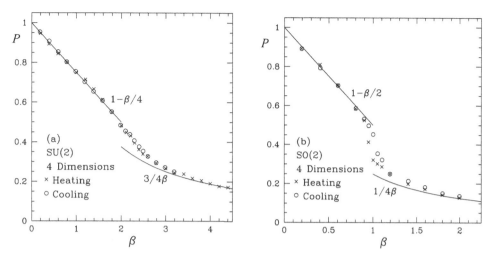

Fig. 9.5. The mean plaquette value as a function of β: (a) SU(2) Yang–Mills theory in four dimensions; (b) SO(2) in four dimensions. From Creutz (1979).

in Figure 9.5. For the Abelian SO(2) (or compact U(1)) model in four dimensions there is a clear 'hysteresis' effect, indicating a phase transition to a non-confining phase at finite coupling, as we expect for QED. Guth (1980) later showed rigorously that there is a non-confining phase at weak couplings for the U(1) theory; but there has been hot debate whether the transition is weakly first-order, or higher order (see e.g. Klaus and Roiesnel, 1998).

For the non-Abelian SU(2) and SU(3) theories in four dimensions, on the other hand, there is no sign of any phase transition at finite coupling, and it is accepted that these theories remain confining in the continuum limit, as we expect for QCD.

By measuring the expectation values of large Wilson loops and fitting them to equation (9.37), one can also extract values for $V(R)$, the potential between static, massive 'quarks' (Stack, 1984). The results (Figure 1.6c) show very clearly the expected linear rise at large R characteristic of confinement. This is pictured as due to the formation of a narrow 'tube' of colour flux between the quark and antiquark sources. At small distances a Coulomb-type $1/R$ behaviour is evident, as expected from asymptotic freedom. Thus the data provide beautiful confirmation of our theoretical expectations.

Making fits to the linear rise of the potential at large distances, one can extract values for the string tension K. A graph of the string tension for SU(3) is given in Figure 9.6.

The results for $\beta \geq 6$ scale in a roughly exponential manner as predicted by equation (9.46); but a more careful look shows that the results are not well fitted even when two-loop corrections are included. Correspondingly, estimates of the

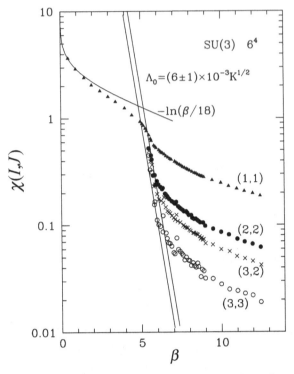

Fig. 9.6. The string tension for SU(3), From Creutz and Moriarty (1982).

string tension in units of Λ_0 have shown a noticeable tendency to decrease with time, from Creutz and Moriarty's early estimate (Creutz and Moriarty, 1982)

$$\Lambda_0 = (6 \pm 1) \times 10^{-3}\sqrt{K} \tag{9.48}$$

to a recent estimate by Edwards *et al.* (1998)

$$\Lambda_0 = 0.01338(12)\sqrt{K}. \tag{9.49}$$

The pure Yang–Mills theory will exhibit a spectrum of excited 'glueball' states: colourless, massive bound states made up of gluons. Their masses may be measured by studying the Euclidean time dependence of the appropriate correlation functions

$$G(T) = \langle O(T)O(0)\rangle - \langle O\rangle^2 \tag{9.50}$$

where $O(T)$ is some suitable operator (e.g. a spacelike plaquette operator) designed to overlap with the glueball wavefunction of interest. The correlation function should decay asymptotically like

$$G(T) \sim \exp(-MT) \qquad \text{as} \quad T \to \infty \tag{9.51}$$

where M is the energy gap to the lowest-lying glueball state with the appropriate quantum numbers. Using refinements of these ideas (Rothe, 1997), recent estimates for the lowest-lying $J^{PC} = 0^{++}$ scalar glueball mass lie in the range 1500–1750 MeV (Bugg *et al.*, 2000). We see that the glueball states are predicted to lie 1–2 Gev above the low-lying $q\bar{q}$ meson states; they have proved difficult to identify experimentally. Calculations including dynamical quarks (Hart and Teper, 2002) typically give masses around 20% lower than the pure gauge results.

9.7 Including fermions on the lattice

So far we have not considered the inclusion of fermion fields representing the quarks in the lattice theory. This question is discussed at some length in the book by Rothe (1997). The simplest option is to transcribe the fermion field directly to sites of the lattice:

$$\psi_\alpha(x) \rightarrow \left(\frac{1}{a^{3/2}}\right) \hat{\psi}_\alpha(n), \tag{9.52}$$

$$\partial_\mu \psi_\alpha(x) \rightarrow \left(\frac{1}{2a^{5/2}}\right) [\hat{\psi}_\alpha(n + \hat{\mu}) - \hat{\psi}_\alpha(n - \hat{\mu})], \tag{9.53}$$

where the powers of a are included so that the lattice field $\hat{\psi}$ is dimensionless. Then the lattice version of the fermionic piece of (9.22), ignoring gauge fields to begin with, reads

$$S_F = \sum_{n,m} \hat{\bar{\psi}}(n) K(n, m) \hat{\psi}(m) \tag{9.54}$$

where

$$K(n, m) = \frac{1}{2} \sum_\mu \gamma_\mu^E [\delta_{m,n+\hat{\mu}} - \delta_{m,n-\hat{\mu}}] + Ma\delta_{n,m} \tag{9.55}$$

with spinor indices suppressed, and where M is the fermion mass.

In a gauge theory, the partial derivative is replaced by the covariant derivative. Correspondingly, in the lattice action $K(n, m)$ is replaced by an expression involving link variables:

$$K(n, m) = \frac{1}{2} \sum_\mu \gamma_\mu^E [\delta_{m,n+\hat{\mu}} U(n, n + \hat{\mu}) - \delta_{m,n-\hat{\mu}} U(n, n - \hat{\mu})] + Ma\delta_{n,m}.$$

$$\tag{9.56}$$

It can easily be checked that the lattice action is then gauge invariant, and reduces in the naive continuum limit to equation (9.22).

The integration measure in the path integral for fermions is written

$$\int D\phi \rightarrow \int D\bar{\psi} D\psi = \prod_n \left\{ \int d\hat{\bar{\psi}}(n) \int d\hat{\psi}(n) \right\} \tag{9.57}$$

where $\hat{\bar{\psi}}, \hat{\psi}$ are Grassmann fields defined on the lattice sites n.

This simple prescription suffers, however, from the notorious problem of 'fermion doubling'. On the lattice, the free fermion momentum is

$$p_\mu \rightarrow \frac{1}{a} \sin(k_\mu a), \tag{9.58}$$

which has zeros not only at $k_\mu = 0$, but also at the edge of the Brillouin zone, $k_\mu = \pm\pi/a$. Hence one discovers that in the continuum limit there are in fact *two* independent fermion degrees of freedom for each direction; or 2^4 species in all. This doubling of fermion species turns out to be an essential mechanism whereby the lattice model avoids the divergences associated with the so-called Adler–Bell–Jackiw anomaly (Karsten and Smit, 1981). There is a 'no-go' theorem due to Nielsen and Ninomiya (1981) which implies that one cannot solve the fermion doubling problem without breaking chiral symmetry for vanishing fermion mass.

There are various ways to remedy this situation. One is to add a term to the action which explicitly gives a large mass to all but one of the fermion species ('Wilson fermions'; Wilson, 1975). Such a term carries the penalty that it explicitly breaks chiral symmetry, and the parameters have to be 'tuned' so that the symmetry is restored. Another is the Kogut–Susskind 'staggered fermion' prescription (Kogut and Susskind, 1975) where the different spinor components of the fermion and anti-fermion fields are distributed individually around the corners of each unit hypercube on the lattice. This reduces the overall number of fermion species by a factor of four, but does not eliminate the problem entirely. One may identify the remaining species with different 'flavours' of the quark field.

A further difficulty for Monte Carlo simulation is that the fermion fields are Grassmann variables (anticommuting), and the path integral is no longer a probabilistic sum over configurations with positive definite weights, but includes both positive and negative weights. The required 'signal' gets swamped by 'noise'.

For these reasons, many Monte Carlo calculations have been done in the 'quenched' approximation, where valence quarks are allowed, but quark loops (virtual $q\bar{q}$ pairs) are forbidden. Then one can calculate many properties of hadrons: for instance, Table 9.1 gives a recent estimate (Aoki *et al.*, 2000) of the masses for some of the common hadronic resonances, along with their experimental values.

It can be seen that the masses broadly agree with experiment down to about the 10% level of accuracy, but beyond that there are some systematic deviations, particularly in the strange quark sector.

In doing calculations complete with 'dynamical' fermions, in which quark loops are allowed, the usual procedure is to integrate out the fermion fields explicitly corresponding to each given gauge field configuration. This involves a heavy computational cost, no matter what clever techniques are involved: one of the favourite

Table 9.1. *Quenched Monte Carlo estimates of the light hadron spectrum. From Aoki* et al. *(2001). Masses* m_π, m_ρ, m_K *taken as input.*

	Mass (GeV)		
	Experiment	Theory	Deviation
K^*	0.8961	0.858(09)	−4.2%
ϕ	1.0194	0.957(13)	−6.1%
N	0.9396	0.878(25)	−6.6%
Λ	1.1157	1.019(20)	−8.6%
Σ	1.1926	1.117(19)	−6.4%
Ξ	1.3149	1.201(17)	−8.7%
Δ	1.2320	1.257(35)	2.0%
Σ^*	1.3837	1.359(29)	−1.8%
Ξ^*	1.5318	1.459(26)	−4.7%
Ω	1.6725	1.561(24)	−6.7%

techniques at present is the 'hybrid molecular dynamics' algorithm (Kennedy, 1989; Gottlieb *et al.*, 1987). Simulations with dynamical fermions are now finding agreement with experiment to within about 3% (Davies *et al.*, 2004).

9.8 The Hamiltonian lattice formulation

An alternative to Wilson's Euclidean lattice formulation of QCD is the 'Hamiltonian lattice' formulation of Kogut and Susskind (1975). Here, the theory is set up in ordinary Minkowski space–time, with a continuous time variable and a discrete lattice in three-dimensional space, corresponding to the lower right corner in Figure 9.2. Most attention has been focused on the quantum Hamiltonian for the system, which for the pure Yang–Mills theory is

$$H = \frac{g^2}{a} \left\{ \frac{1}{2} \sum_{l,a} (E_l^a)^2 + \lambda \sum_P \left[1 - \frac{1}{3} \text{Re} \text{Tr}(U_P) \right] \right\} \tag{9.59}$$

where E_l is a component of the 'electric' field defined on spatial links l, U_P is the plaquette operator on spatial plaquettes P, the coupling $\lambda = 6/g^4$, and the index a runs over the eight generators of SU(3). The electric fields are the conjugate 'momenta' to the gauge fields A_l^a, and their commutation relations with the link variables are:

$$[E_{l'}^a, U_l] = \delta_{ll'} t^a U_l \tag{9.60}$$

The most direct way to derive the Hamiltonian (9.59) is to calculate the continuum field theory Hamiltonian by canonical procedures, and then transcribe the magnetic field term $\mathbf{B}^2$ to a space-like plaquette operator following Wilson's procedure. An alternative is the transfer matrix method of Creutz (1983).

If they can be calculated, the eigenvalues of the Hamiltonian give the ground state energy and the spectrum of excited states in the model directly, without requiring the measurement of correlation functions and their exponential decay rates. There are many techniques borrowed from quantum many-body theory which can be used to calculate the eigenvalues of a quantum Hamiltonian such as (9.59). These include strong-coupling series methods (Banks *et al.*, 1977b), the coupled-cluster method (Guo *et al.*, 1996; Baker *et al.*, 1996), the *t*-expansion (Horn and Weinstein, 1984), the plaquette expansion (Hollenberg, 1994) and the density matrix renormalization group method (Byrnes *et al.*, 2002). Generally speaking, all these techniques converge well and give accurate answers at strong couplings. Unfortunately, they do not converge well in the weak coupling regime, which is the real region of interest as we move towards the continuum limit. Some examples will be seen in the next chapter.

Given the difficulty of obtaining reliable results in the weak coupling regime, there have been a number of attempts to develop Monte Carlo algorithms to treat the quantum Hamiltonian (9.59). These have included quantum Monte Carlo techniques such as variational Monte Carlo, projector Monte Carlo, and diffusion or Green's Function Monte Carlo techniques. In the end, however, it appears that the classical Euclidean path integral Monte Carlo approach reviewed in Section (9.6) is still the best way to proceed. Using this technique on 'anisotropic' lattices where the spacing a_τ in the timelike direction is smaller than the spacing a in the spacelike directions (Morningstar and Peardon, 1997), one may extrapolate to $a_\tau/a \rightarrow 0$, which corresponds to the Hamiltonian limit (after a Wick rotation). This turns out to be the most powerful and convergent technique for calculating dynamical quantities such as the string tension and excited state masses at weak couplings (Byrnes *et al.*, 2004).

9.9 Conclusions

As we have seen, the lattice approach allows one to make non-perturbative calculations in QCD and other field theories, starting from first principles, and introducing no extraneous assumptions or parameters. The Euclidean Monte Carlo technique, in particular, has proved amazingly powerful, allowing accurate estimates for lattices of up to 1 million sites.

The Monte Carlo results are generally speaking in good agreement with theoretical expectations, and even give quantitative agreement with experiment at a level of a few percent in accuracy. It is now possible to make phenomenological predictions of such quantities as decay widths, electromagnetic form factors, and hyperfine splittings in heavy quark systems.

In the next chapter we shall examine the results obtained by the alternative numerical technique of 'strong-coupling' series expansions. So far, this approach has not given nearly such accurate results as Monte Carlo in the weak-coupling or continuum limit, but it gives a very accurate basis for comparison in the strong-coupling regime on the lattice: It may also prove useful in situations where the Monte Carlo approach encounters difficulties, e.g. at finite baryon density, or even when dynamical fermions are included.

10

Series expansions for lattice gauge models

There are many different types of series expansions that have been used in connection with lattice gauge models. These include weak coupling expansions in powers of g^2, strong coupling expansions in powers of $1/g^2$, $1/N$ expansions, hopping parameter expansions, t-expansions, and others. For lack of space, we do not attempt to discuss all of these different expansions here. Weak coupling expansions, for instance, are extensively discussed elsewhere (e.g. Rothe, 1997). We shall confine ourselves to expansions which are related to those we have discussed for spin models, namely the strong coupling and t-expansions.

10.1 Strong coupling expansions for Euclidean lattice Yang–Mills theory

The Euclidean path integral for pure SU(N) Yang–Mills theory with Wilson action is

$$Z = \int DU \, e^{\beta/N \sum_P \text{Re Tr}\{U_P\}} \tag{10.1}$$

(dropping an irrelevant constant term), where $\beta = 2N/g^2$, and U_P is the plaquette variable at plaquette P. In the Wilson action, the link variables are taken as belonging to the fundamental representation of the group. This path integral has an equivalent structure to that of the partition function in statistical mechanics, and so a 'strong coupling' expansion in powers of β is analogous to a high temperature expansion in statistical mechanics, and involves many of the same results and techniques. In particular, one can prove (Osterwalder and Seiler, 1978) that the strong coupling expansion has a finite radius of convergence. In this section we shall review the methods used in calculating strong coupling expansions, and the results that have been obtained to date. Further background on this topic can be found in the reviews by Drouffe and Zuber (1983) or Montvay and Münster (1994). We follow the notation of the latter.

Table 10.1. *Some irreducible representations of SU(3).*

p	q	d_r	C_r
0	0	1	0
1	0	3	4/3
1	1	3	4/3
2	0	6	10/3
2	1	8	3
2	2	6	10/3

10.1.1 Group theory considerations

In order to evaluate the path integral and related observables, a few results from group theory are required. Here we merely collect the most necessary results: a fuller discussion may be found elsewhere [Drouffe and Zuber 1983 (Appendix A), Hamermesh 1962]. As examples, we refer to the Abelian compact U(1) (or SO(2)) group, and the physical colour group SU(3).

   Elements of the semi-simple Lie group $SU(N)$ can be parameterized in terms of N continuous parameters or generalized 'angles', just as an element of U(1) can be parameterized as $\exp(i\theta)$. *Irreducible representations* of $SU(N)$ can be labelled by $(N-1)$ Casimir invariants, e.g. the total spin S in the case of SU(2). Representations of U(1) are also labelled by an integer 'spin' L, and have elements $e^{iL\theta}$. In the case of SU(3), irreducible representations can be labelled by two integers p, q such that

$$p \geq q \geq 0 \tag{10.2}$$

with the dimension of the representation given by

$$d_r = \frac{1}{2}(p+2)(q+1)(p-q+1) = \chi_r(1) \tag{10.3}$$

where $\chi_r(U)$ is the trace or '*character*' of the element U in representation r. The quadratic Casimir invariant of the irreducible representation is given by

$$C_r = \frac{1}{6}[2(p^2+q^2) - 2pq + 6p] \tag{10.4}$$

The first few irreducible representations of SU(3) are listed in Table 10.1.

   The functional integral in equation (10.1) runs over all the link variables U_l:

$$\int DU = \prod_l \left\{ \int dU_l \right\} \tag{10.5}$$

where $\int dU_l$ is the invariant Haar measure, normalized to

$$\int dU = 1 \tag{10.6}$$

and invariant under group transformations

$$dU = dU^{-1} = d(UV) \qquad \forall V \in G \tag{10.7}$$

In the U(1) case, for example, we may take

$$\int dU = \frac{1}{2\pi} \int_0^{2\pi} d\theta \tag{10.8}$$

– we will not need an explicit representation for SU(3).

A group invariant or *class function* obeys

$$f(U) = f(VUV^{-1}) \qquad \forall V \in G. \tag{10.9}$$

Irreducible characters (i.e. traces of irreducible representations) form a complete basis for these class functions. If the character for representation r is:

$$\chi_r(U) = \sum_\alpha D^r_{\alpha\alpha}(U) \tag{10.10}$$

then the class function can be decomposed

$$f(U) = \sum_r f_r \chi_r(U) \tag{10.11}$$

where

$$f_r = \int dU \chi_r^*(U) f(U) \tag{10.12}$$

In the case of U(1), this amounts to nothing more than a Fourier decomposition:

$$\chi_n(U) = e^{in\theta} \tag{10.13}$$

$$f(U) = \sum_n f_n e^{in\theta} \tag{10.14}$$

where

$$f_n = \frac{1}{2\pi} \int_0^{2\pi} d\theta e^{-in\theta} f(U). \tag{10.15}$$

In particular, the Boltzmann weight for a single plaquette is an invariant, and can be decomposed

$$\exp\left[\frac{\beta}{N}\mathrm{Re\,Tr}U_P\right] = \sum_r d_r c_r(\beta)\chi_r(U_P)$$

$$= c_0(\beta)\left\{1 + \sum_{r\neq 0} d_r a_r(\beta)\chi_r(U_P)\right\}$$

$$\equiv c_0(\beta)\{1 + f_P(U_P)\}, \tag{10.16}$$

with

$$a_r(\beta) = \frac{c_r(\beta)}{c_0(\beta)}. \tag{10.17}$$

The coefficients a_r behave like powers of β for small β. For the U(1) case, for instance, (10.16) reads

$$\exp[\beta\cos\theta_P] = \sum_n c_n(\beta)e^{in\theta_P} \tag{10.18}$$

with

$$c_n(\beta) = I_n(\beta) \tag{10.19}$$

where $I_n(\beta)$ is a modified Bessel function. Then

$$a_n(\beta) = \frac{I_n(\beta)}{I_0(\beta)} \sim \frac{1}{n!}\left(\frac{\beta}{2}\right)^n \quad \text{as} \quad \beta \to 0, \ n > 0. \tag{10.20}$$

For SU(3), the equivalent result is given by Drouffe and Zuber (1983).

Finally, we often need to evaluate integrals over link variables of the form

$$\int dU d_f \mathrm{Tr}(UA)d_f\mathrm{Tr}(U^{-1}B) = d_f\mathrm{Tr}(AB) \tag{10.21}$$

for the fundamental representation f. For U(1), an analogous formula to (10.21) is

$$\frac{1}{2\pi}\int_0^{2\pi} d\theta\, A^n e^{in\theta} B^p e^{-ip\theta} = \delta_{n,p}(AB)^n. \tag{10.22}$$

10.1.2 The free energy

The strong coupling expansion for the free energy follows essentially the same procedures as for the high-temperature expansion of the free energy in Chapters 2 and 3. Let us begin by considering the U(1) model on a single plaquette as an example. At large couplings (small β), one can naively expand the Boltzmann

Fig. 10.1. Examples of diagrams contributing to the naive expansion of the free energy for a single plaquette. The oriented squares correspond to plaquette variables U_P or $U_P^\dagger$.

weight

$$\exp(-\beta S_E) = \prod_P \exp\left[\frac{\beta}{2}(U_P + U_P^\dagger)\right]$$

$$= \prod_P \left[1 + \frac{\beta}{2}(U_P + U_P^\dagger) + \frac{1}{2!}\left(\frac{\beta}{2}\right)^2 (U_P + U_P^\dagger)^2 + \cdots\right] \quad (10.23)$$

where

$$U_P = \exp[i\theta_P] = \exp[i(\theta_1 + \theta_2 - \theta_3 - \theta_4)] \quad (10.24)$$

and $\{\theta_i, i = 1, \cdots, 4\}$ are the 'link angles' around the plaquette P. Using (10.6), one easily sees that

$$\prod_{l=1}^{4}\left\{\int dU_l\right\} e^{-\beta S_E\{U_P\}} = \prod_{i=1}^{4}\left\{\frac{1}{2\pi}\int_0^{2\pi} d\theta_i\right\} e^{-\beta S_E\{U_P\}} = \frac{1}{2\pi}\int_0^{2\pi} d\theta_P e^{-\beta S_E\{U_P\}}$$

$$(10.25)$$

One may then integrate each term in equation (10.23) to obtain an expansion of the partition function in powers of β. The only non-zero terms in this expansion are those with equal numbers of U_P and $U_P^\dagger$ variables, e.g.

$$\frac{1}{2\pi}\int_0^{2\pi} d\theta_P U_P U_P^\dagger = 1 \quad (10.26)$$

corresponding to Figure 10.1(a). Hence one finds

$$Z_P = 1 + \tfrac{1}{4}\beta^2 + \tfrac{1}{64}\beta^4 + \cdots = c_0(\beta) = I_0(\beta), \quad (10.27)$$

as given by equation (10.19).

Now let us consider the U(1) model on a lattice in two or more dimensions. The standard procedure is to employ the character expansion (10.16). Then we obtain the Boltzmann weight

$$e^{-S} = [c_0(\beta)]^N \prod_P \left\{1 + \sum_{r \neq 0} d_r a_r(\beta)\chi_r(U_P)\right\} \quad (10.28)$$

where N is the total number of plaquettes in the lattice. Expanding the product over plaquettes gives a sum of terms, each of which is of the form $\prod_P d_{r_P} a_{r_P} \chi_{r_P}(U_P)$.

We define a *graph g* to consist of a set of plaquettes, with a representation $r_P \neq 0$ attached to each plaquette. Then the Boltzmann factor can be written as

$$e^{-S} = c_0^N \sum_g \prod_{P \in g} d_{r_P} a_{r_P} \chi_{r_P}(U_P) \tag{10.29}$$

Integrating over link variables, we get

$$Z_N = c_0^N \sum_g \Phi(g) \tag{10.30}$$

where $\Phi(g)$ is the 'contribution' of the graph g.

To calculate the contribution of a graph, in general one needs to know integrals of the type

$$\int dU \, \chi_{r_1}(V_1 U) \cdots \chi_{r_n}(V_n U).$$

Some simple cases are discussed by Drouffe and Zuber (1983). One way to proceed is to write the integrand in terms of representation matrices and form a Clebsch–Gordan series for the resulting product. Note that if the Clebsch–Gordan series does not contain the singlet representation the integral *vanishes*, since

$$\int dU \, \chi_r(U) = \delta_{r,0} \tag{10.31}$$

This has two important consequences.

(1) The contribution from any graph with a 'free edge', i.e. a link involved with only one plaquette of the graph, must vanish by Eq. (10.31). Thus the only graphs that contribute are those with no free edges, i.e. *closed* graphs.
(2) If just two plaquettes with representations r and r' meet in a link, we must have $r' = r$ or r^*, depending how the plaquettes are oriented, since

$$\int dU \, \chi_r(AU)\chi_{r'}(U^{-1}B) = \delta_{rr'} \frac{1}{d_r} \chi_r(AB). \tag{10.32}$$

Now for a two-dimensional planar lattice, all non-trivial graphs have free edges (e.g. Figure 10.2a), and there are no closed graphs. Therefore the partition function is simply

$$Z_N = c_0(\beta)^N \tag{10.33}$$

and the free energy per plaquette

$$F = -\frac{1}{N} \ln Z = -\ln c_0(\beta) \tag{10.34}$$

(a)   (b)

Fig. 10.2. Examples of graphs contributing to the character expansion of the free energy: (a) a graph in two dimensions with free edges, (b) the cube graph. In the simplest case, each plaquette carries the fundamental representation.

is identical with that of a single plaquette. For U(1),

$$F = - \ln I_0(\beta). \tag{10.35}$$

The two-dimensional model is thus exactly solvable, which corresponds to the fact that gauge fields in two dimensions have no independent dynamical degrees of freedom.

In higher dimensions, then, the only graphs which make non-zero contributions are those with no free edges, i.e. closed graphs such as the cube in Figure 10.2(b). If each face of the cube corresponds to the fundamental representation, the contribution of this graph is

$$\Phi(g) = a_f(\beta)^6 \int \prod_l dU_l \prod_{i=1}^{6} [d_f \mathrm{Tr}(U_{P_i})]. \tag{10.36}$$

Using equations (10.21) or (10.6), successive integrations over link variables yield a final result

$$\Phi(\text{cube}, r = f) = d_r^2 a_f(\beta)^6. \tag{10.37}$$

To carry the expansion to higher orders, one must generate a list of the relevant graphs and their lattice embedding constants, and then calculate the contribution of each graph. The graphs must again be closed: but in this type of expansion one is calculating the partition function directly, and there is no requirement that the graphs must be connected – e.g. a graph consisting of two disconnected cubes will contribute. For the free energy, on the other hand, the result may be expressed in terms of connected clusters alone, as usual (Montvay and Münster, 1994). The list of graphs may be generated using techniques such as those discussed in Section 2.2. For further rules and details of the character expansion method, we refer the reader to Drouffe and Zuber (1983). Using these procedures, the free energy of the U(1) model has been calculated to order β^{16} in arbitrary dimension by Balian *et al.* (1975).

Now let us consider the same problem for the SU(3) case, where the naive expansion reads

$$e^{-\beta S_E} = \prod_P e^{\beta(\mathrm{Tr}U_P + \mathrm{Tr}U_P^\dagger)/6}$$

$$= \prod_P \left[1 + \frac{\beta}{6}(\mathrm{Tr}U_P + \mathrm{Tr}U_P^\dagger) + \frac{1}{2!}\left(\frac{\beta}{6}\right)^2 (\mathrm{Tr}U_P + \mathrm{Tr}U_P^\dagger)^2 + \cdots \right] \quad (10.38)$$

Again, diagrams contributing to the free energy can be built out of oriented pla-quettes, representing either $\mathrm{Tr}U_P$ or $\mathrm{Tr}U_P^\dagger$. Thus, for the single plaquette, using Eq. (10.21), the graph of Figure 10.1a gives a contribution of $\beta^2/36$. At the next order, the triality rule for SU(3) allows an extra graph corresponding to Figure 10.2(b), because the singlet representation appears in the Clebsch–Gordan decom-position of the product of three fundamental representations. This diagram turns out to give a contribution of $\beta^3/648$. So to this order we have

$$Z_P = c_0(\beta) = 1 + \frac{\beta^2}{36} + \frac{\beta^3}{648} + \cdots \quad (10.39)$$

For the two-dimensional problem, using the character expansion approach, the general arguments given above imply once again that the free energy per site is the same as for a single plaquette:

$$F = -\ln c_0(\beta) \quad (10.40)$$

providing us with an exact solution.

In four dimensions, the only non-zero contributions again come from closed graphs. The geometrical problem of enumerating the closed graph topologies is the same as for the U(1) model, or any gauge model. There remains the group theoretic problem of evaluating the contributions from each graph: for instance, at lowest order the cubic graph of Figure 10.2(b) gives $2(\beta/6)^6(1/3)^5$. At higher orders, the problem of combining representations during the successive integrations over links becomes more complex. The free energy of the SU(3) model has been calculated to order β^{22} in four dimensions, and to order β^{16} in arbitrary dimensions – see Drouffe and Zuber (1983).

The series for the free energy can now be analysed to search for any phase transition at finite coupling, which would signal the breakdown of confinement before the continuum limit is reached. In the case of a continuous transition, a standard Dlog Pade analysis of the series for the internal energy

$$E = \frac{\partial F}{\partial \beta} \quad (10.41)$$

will reveal the location and critical index α of the transition. A first-order transition cannot be located from a single series expansion: the series will usually indicate an unphysical singular point beyond the transition.

For the U(1) model in four dimensions, there is a phase transition at finite coupling between the confining phase and a massless Coulomb phase. A Dlog Padé analysis of the series for E indicates a singularity at $\beta \sim 0.98 - 1.01$, with an exponent α ranging from 0.38–0.65 (Falcioni *et al.*, 1981). Monte Carlo simulations find a transition at much the same coupling; but there has been a long-standing controversy as to whether the transition is second order, or weakly first order. A recent study (Klaus and Roiesnel, 1998) favours a weak first-order transition at $\beta_c \simeq 1.011$, which is indeed very close to the series estimate.

For SU(3) gauge theory in four dimensions, the series analyses are again somewhat ambiguous. The specific heat shows a sharp bump in the cross-over region between strong and weak coupling at around $\beta \sim 5.6$, and approximants to the series tend to show a singularity a little beyond that point, at $\beta \sim 6.2$. The singularity structure is generally interpreted as a pair of complex conjugate singularities off the real axis, which do not correspond to a physical phase transition (Drouffe and Zuber, 1983). The behaviour for SU(2) is similar. For SU(N) with $N \geq 4$, a first-order transition is again found.

10.1.3 The string tension

The string tension can be derived from the asymptotic behaviour of Wilson loops – see Section 9.4. Consider the expectation value of an $M \times N$ Wilson loop:

$$\langle W_{M,N} \rangle = \frac{1}{Z} \prod_l \left\{ \int dU_l \right\} W_{M,N} e^{-\beta S_E} \tag{10.42}$$

Naively expanding the Boltzmann weight in powers of β as before, the non-zero contributions will correspond to connected diagrams as usual, where the Wilson loop variable is represented by a large $M \times N$ rectangle. The integral over any link containing only a single-link variable vanishes, by Eq. (10.31); so the first non-zero contribution corresponds to Figure 10.3(a), where the Wilson loop has been 'tiled' completely with plaquette variables. Evaluating this graph, we find that in leading order

$$\langle W_{M,N} \rangle = \begin{cases} (\beta/2)^{MN}, & U(1) \\ 3(\beta/18)^{MN}, & SU(3) \end{cases} \tag{10.43}$$

which by equation (9.35) corresponds to a string tension

$$K = \begin{cases} -\ln(\beta/2), & U(1) \\ -\ln(\beta/18), & SU(3) \end{cases} \tag{10.44}$$

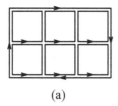 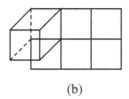

(a)   (b)

Fig. 10.3. (a) The lowest-order diagram contributing to the Wilson loop expectation value in the naive expansion; (b) a graph contributing to the character expansion for the Wilson loop.

Hence, as promised, we find that in the strong-coupling limit the string tension is non-zero, and the theory is confining. This conclusion holds for a finite radius of convergence about the strong-coupling limit (Osterwalder and Seiler, 1978).

At higher orders, a character expansion for the string tension can again be formulated (Münster, 1981a). Any graph with a free edge (included in only one plaquette and not part of the Wilson loop) will vanish. Therefore non-zero contributions correspond to closed graphs connected to the Wilson loop, such as Figure 10.3(b). The contributions of these graphs can be calculated using similar methods to the free energy. Series for the string tension in four dimensions have been calculated to order β^{14} for U(1) and SU(3) (Münster, 1981b).

For the confining theories in four dimensions, such as SU(3), the string tension is expected to remain finite at all couplings, and to scale exponentially in the weak-coupling region as in Eq. (9.46). One would therefore like to extrapolate the series for the string tension from strong coupling into the weak-coupling region, and see if the expected scaling behaviour is reproduced. Figure 10.4 illustrates the results. Approximants to the string tension converge well in the strong-coupling region, as one would expect. In the crossover region, however, the convergence rapidly deteriorates, and the series approximants run away from the Monte Carlo estimates.

The reason for this was quickly identified: it is due to the presence of a 'roughening' transition (Montvay and Münster, 1994). Roughening is a phenomenon well-known in connection with lattice spin models. Consider an Ising lattice in two or three dimensions, with boundary conditions creating an interface between two phases of opposite spin (Figure 10.5). The interfacial energy ΔE is proportional to the area of the interface,

$$\Delta E = -KA, \tag{10.45}$$

where K is the surface tension. At low temperatures the area will be minimized, forming a smooth plane across the lattice. As the temperature increases, thermal fluctuations increase, until at some point they become so large that the surface

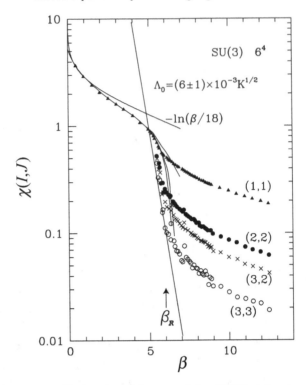

Fig. 10.4. Strong coupling series estimates of the SU(3) string tension, super-imposed on Monte Carlo data. Also indicated is the roughening coupling. From Drouffe and Zuber (1983).

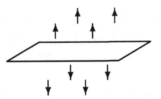

Fig. 10.5. The interface between two phases of opposite spin in the Ising model.

becomes delocalized – the width of the interfacial region diverges. The surface tension at this point does not vanish, but a weak singularity occurs there. The same thing happens here for the string tension, and the strong-coupling series cannot be continued beyond the roughening point. For SU(3), the roughening transition is estimated by Drouffe and Zuber (1981) to occur at $\beta_R \simeq 5.9(4)$. This renders the strong coupling series useless for estimating the weak coupling behaviour of the string tension. The roughening transition does not affect other physical quantities, however.

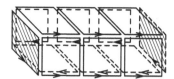

Fig. 10.6. The lowest-order diagram contributing to the plaquette–plaquette correlation function. The correlation is between the shaded plaquettes at either end of the tube.

10.1.4 Glueball masses

Glueball masses can be derived from the asymptotic behaviour of correlation functions, as discussed in Section 9.6. Consider for example the connected correlation function between parallel plaquettes at a distance of N sites along the 'time' axis:

$$G_N = \langle \mathrm{Tr} U_{P_1} \mathrm{Tr} U_{P_2}^{\dagger} \rangle - \langle \mathrm{Tr} U_{P_1} \rangle \langle \mathrm{Tr} U_{P_2}^{\dagger} \rangle, \tag{10.46}$$

where plaquettes P_1 and P_2 are as illustrated in Figure 10.6. At leading order in the strong-coupling expansion, the link variables may all be 'paired off' by 'tiling' the tube joining the two plaquettes with plaquette variables. Hence at leading order

$$G_N = \begin{cases} (\beta/2)^{4N}, & U(1) \\ (\beta/18)^{4N}, & SU(3) \end{cases} \tag{10.47}$$

corresponding to a lowest lying glueball mass

$$ma = \begin{cases} -4\ln(\beta/2), & U(1) \\ -4\ln(\beta/18), & SU(3) \end{cases} \tag{10.48}$$

Higher-order contributions will involve closed 'decorations' added to the tube graph. Some complications then arise. Many different glueball states can contribute to a given correlation function, with different momenta and different properties under lattice rotations and other symmetries. Exactly the same problem is encountered in Monte Carlo simulations (see Rothe, 1997). It then becomes necessary to 'project out' states of the desired quantum numbers by taking the appropriate linear combinations of correlation functions. For instance, to project out a state with zero momentum, we must sum over all spatial locations of the initial and final plaquettes P_1 and P_2 with a fixed time difference. To project out the lowest scalar state with $J^{PC} = 0^{++}$, we sum over all space-like orientations of P_1 and P_2 separately, with equal weight. Using these methods, series expansions for the lowest lying glueball masses have been computed in the form

$$ma = -4\ln u + \sum_{k=1}^{\infty} m_k u^k \tag{10.49}$$

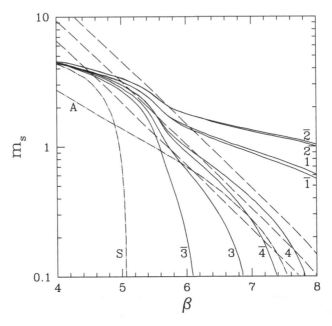

Fig. 10.7. The scalar glueball mass as a function of β. Curve S represents the series truncated at eighth order; the solid and dot-dashed lines are various approximants to the series; and the almost straight dashed curves are scaling curves following the expected weak-coupling behaviour. From Smit (1982).

where instead of β, the 'average character' u has been used as expansion variable:

$$u = \frac{\int dU \, \mathrm{Re} \, \mathrm{Tr}\{U/3\} e^{\beta \mathrm{Re} \, \mathrm{Tr}\{U/3\}}}{\int dU \, e^{\beta \mathrm{Re} \, \mathrm{Tr}\{U/3\}}} = \frac{\beta}{18} + \cdots \quad [SU(3)] \qquad (10.50)$$

The coefficients m_k have been computed up to order $k = 8$ by Münster (1981b) and Seo (1982) for the lowest-lying $J^{PC} = 0^{++}$, 2^{++} and 1^{+-} states in the SU(3) theory. Figure 10.7 shows some series approximants for the scalar mass gap calculated by Smit (1982).

The series is not very well behaved, and it can be seen that the approximants have begun to diverge as the crossover region $\beta = 5 - 6$ is reached; but if they are matched on to the expected scaling behaviour in the weak coupling regime, represented by the dashed straight lines, then a rough estimate of the continuum limit can be made

$$m(0^{++}) \simeq 340(40)\Lambda_0. \qquad (10.51)$$

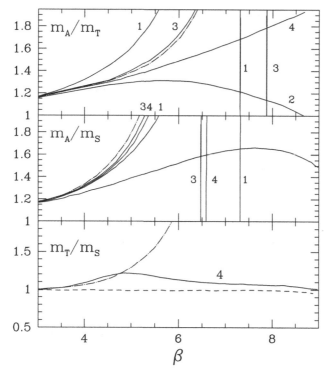

Fig. 10.8. Glueball mass ratios as functions of β. The dot-dashed curves are the truncated series, and the solid lines are [1/1] – [4/4] Padé approximants. From Smit (1982).

If we use the recent estimates (Edwards *et al.*, 1998) $\Lambda_0 = 0.01338\sqrt{K}$ (Eq. 9.49) and $\sqrt{K} = 465$ MeV, this corresponds to

$$m(0^{++}) \simeq 2100(200) \text{ MeV}, \tag{10.52}$$

which is about 30% higher than the best current Monte Carlo estimate of 1500–1750 MeV (Bugg *et al.*, 2000), but within the same ballpark.

The mass ratios $m(2^{++})/m(0^{++})$ and $m(1^{+-})/m(0^{++})$ are expected to approach universal constants in the continuum limit $\beta \to \infty$. Figure 10.8 shows some series approximants for these mass ratios. Again, the results are beginning to diverge as the crossover region is reached, and they allow only rough estimates (Smit, 1982):

$$m(2^{++})/m(0^{++}) \simeq 1.2 \tag{10.53}$$

$$m(1^{+-})/m(0^{++}) \simeq 1.7 - 2 \tag{10.54}$$

These may be compared with recent Monte Carlo estimates (Morningstar and Peardon, 1997)

$$m(2^{++})/m(0^{++}) = 1.46(8) \tag{10.55}$$

$$m(1^{+-})/m(0^{++}) = 1.82(9) \tag{10.56}$$

10.2 Strong coupling expansions in Hamiltonian Yang–Mills theory

Strong coupling expansions have also been a standard tool in the Hamiltonian version of lattice gauge theory, beginning with the original paper of Kogut and Susskind (1975), and a treatment of the Schwinger model by Banks *et al.* (1976). To illustrate the techniques, we shall again use the compact U(1) (or SO(2)) pure gauge theory, and the physical SU(3) theory in three space and one time dimensions.

The primary object of interest is the quantum Hamiltonian, as discussed in the previous chapter. It is convenient to work with the dimensionless operator

$$W = \frac{2a}{g^2} H = \sum_l \mathbf{E}_l^2 - x \sum_P (Z_P + Z_P^\dagger) \tag{10.57}$$

where $Z_P = \text{Tr}[U_P]$, $x = 2/g^2$, and the electric field operator has eight components E_l^a for SU(3) Yang–Mills theory. For the U(1) case, the coupling $x = 1/g^2$, and the electric field only has one component. Many quantities of physical interest can be obtained directly from the eigenvalues of the quantum Hamiltonian operator. The ground-state energy density itself is not a physically observable quantity, but it corresponds to the free energy in the Euclidean formalism (Appendix E), and can be analyzed in the same way to reveal the phase structure of the model. The energy gaps to excited states correspond directly to the masses of observable 'hadrons' in the model, and can also give the string tension directly, as we shall see shortly. In principle, transition matrix elements could also be calculated to give vertex functions and form factors, but little has yet been done in this direction. Strong coupling series expansions in powers of x have been calculated for quantities including the ground-state energy per site, the string tension, and the glueball masses in these models.

10.2.1 Group theory considerations

For the compact U(1) case, the commutation relations between the electric field and its corresponding link field is (Banks *et al.*, 1977a)

$$[E_l, U_{l'}] = U_l \delta_{ll'}, \qquad [E_l, U_{l'}^\dagger] = -U_l^\dagger \delta_{ll'} \tag{10.58}$$

so that U_l ($U_l^\dagger$) acts as a 'raising operator' ('lowering operator') for the electric field, respectively. If we represent $U_l = e^{i\theta_l}$, the gauge field θ_l is an angular variable in the range $[0, 2\pi]$, and the electric field is the conjugate 'spin operator', $-i\partial/\partial\theta_l$. Hence the allowed spectrum of eigenvalues of E_l consists of the integers, $E_l = 0$, $\pm 1, \pm 2, \cdots$ and the corresponding link eigenfunctions are $e^{iL\theta_l}$, for $E_l = L$.

For the SU(3) model, the commutation relations are (Kogut and Susskind, 1975)

$$[E_l^a, U_{l'}] = \delta_{ll'} t^a U_l. \tag{10.59}$$

where the t^a are generators of SU(3). The link eigenfunctions of $\mathbf{E}_l^2$ correspond to irreducible representations of SU(3), specified by a pair of integers p, q as in the previous section. The quadratic Casimir operator of the representation is given by Eq. (10.4). This is the corresponding eigenvalue of $\mathbf{E}_l^2$.

In calculating the perturbation series in x for the energy eigenvalues, one must calculate the effect of multiple link operators U_l acting on a given link, where U_l belongs to the fundamental representation $(1,0)$ of SU(3). Then one meets the Clebsch–Gordan problem of decomposing $(U_l)^n$ into irreducible representations. For a more detailed discussion of this problem in the context of lattice gauge theory, we refer the reader to Hamer *et al.* (1986).

10.2.2 Ground state energy

To perform a strong coupling expansion, we write

$$W = W_0 - xV \tag{10.60}$$

where the unperturbed term W_0 is the electric field term, and the perturbation V consists of the magnetic term, i.e. the spacelike plaquette operators:

$$W_0 = \sum_l \mathbf{E}_l^2 \tag{10.61}$$

$$V = \sum_P (Z_P + Z_P^\dagger) \tag{10.62}$$

In the strong coupling limit $x \to 0$, we have $W = W_0$, and the strong-coupling basis states are eigenstates of the electric field on each link l. The ground state is clearly that for which the electric field vanishes

$$\mathbf{E}_l^2 |0\rangle = 0, \quad \text{all } l \tag{10.63}$$

The strong coupling expansion for the ground-state energy can be calculated using the linked cluster algorithm of Chapter 4. Again, let us start with the U(1) model on a single plaquette. The unperturbed ground state $|0\rangle$ has $\mathbf{E}_l^2 = 0$, all l. The

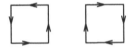

Fig. 10.9. Diagrams representing an excited loop of electric flux on a single plaquette. Arrows up or to the right represent $E_l = +1$; left or downwards represent $E_l = -1$.

effect of U_P acting on the ground state is to excite a 'loop' of electric flux $E_l = \pm 1$ on the plaquette, as shown in Figure 10.9.

Higher excited states can then be labelled $|n\rangle$, where n units of electric flux circulate around the loop. The Hamiltonian matrix is tridiagonal

$$W_{nn} = 4n^2 \tag{10.64}$$

$$W_{n,\pm 1} = -x \tag{10.65}$$

This simple system is equivalent to a lattice harmonic oscillator, and the Schrödinger equation in Fourier space is equivalent to the Mathieu equation. Hence the system can be solved exactly (Robson and Webber, 1980), giving the ground-state energy

$$\omega_0 = a_0(x) = -\frac{x^2}{2} + \frac{7x^4}{128} + \cdots \tag{10.66}$$

where $a_0(x)$ is the lowest characteristic value of the Mathieu equation (see Abramowitz and Stegun, 1965).

Now let us turn to the U(1) model on a cubic three-dimensional lattice. The 'linked clusters' appropriate to this problem will consist of connected sets of plaquettes on the cubic lattice. The single plaquette has a lattice constant of 3, corresponding to its three possible orientations, and hence the ground-state energy per site to leading order is

$$\omega_0/N = -3x^2/2 \tag{10.67}$$

Higher-order terms were calculated by Kogut *et al.* (1980) up to order x^6, and the series has since been extended to order x^{14} by Hamer *et al.* (1994a) using linked cluster methods.

For the SU(3) model, the unperturbed ground state again has $\mathbf{E}_l^2 = 0$, for all l. At first order in V, the plaquette operator Z_P will excite a loop of 'colour' flux on a single plaquette, as illustrated in Figure 10.9, with $\mathbf{E}_l^2 = 4/3$ on every link (corresponding to $p = 1, q = 0$ in Eq. (10.4)). Hence Rayleigh–Schrödinger theory gives

$$\omega_0 = -\tfrac{3}{8}x^2 \quad \text{(single plaquette)} \tag{10.68}$$

Table 10.2. *Coefficients of x^n in series expansions for the vacuum energy per site ω_0/N, the string tension K, and scalar, axial vector and tensor glueball masses M_S, M_A and M_T for the SU(3) Yang–Mills theory in (3+1)D. Data from Hamer, et al. (1986), and Hamer (1989).*

n	ω_0/N	K	M_S
0	0	$4/3$	$16/3$
1	0	0	-1
2	$-9/8$	$-0.718954248366 \times 10^{-1}$	-0.106372549020
3	-0.2109375	$-0.373774509804 \times 10^{-1}$	$-0.259751057286 \times 10^{-1}$
4	$-0.202081294165 \times 10^{-1}$	$-0.127115015989 \times 10^{-1}$	-0.13886954227
5	$0.512047659085 \times 10^{-2}$	$-0.428422611694 \times 10^{-2}$	-0.1343376068
6	$0.17542606536 \times 10^{-2}$	$-0.14802690944 \times 10^{-2}$	$-0.994739930 \times 10^{-1}$
7	$-0.1282518288 \times 10^{-2}$	$-0.9884542510 \times 10^{-3}$	$-0.232246442 \times 10^{-1}$
8	$-0.157335145 \times 10^{-2}$		

n	M_A	M_T
0	$16/3$	$16/3$
1	1	-1
2	$0.179738562092 \times 10^{-1}$	0.109313725490
3	-0.109283889209	$0.501374471357 \times 10^{-1}$
4	$-0.968932328773 \times 10^{-1}$	$0.481356764 \times 10^{-2}$
5	$-0.69813863780 \times 10^{-1}$	$0.1315377296 \times 10^{-1}$
6	$-0.41089676435 \times 10^{-1}$	$0.495936854 \times 10^{-2}$
7	$-0.17154548532 \times 10^{-1}$	$-0.42445283 \times 10^{-3}$

or

$$\omega_0/N = -\tfrac{9}{8}x^2 \quad \text{(cubic lattice)} \tag{10.69}$$

At higher orders, the Clebsch–Gordan decomposition problem arises. One must formulate a set of gauge-invariant basis states on the lattice, and calculate matrix elements of the plaquette operator between them $\langle f|Z_P|i\rangle$. The early calculations of Kogut *et al.* (1976) were done 'by hand'. Hamer *et al.* (1986) carried the series to order x^8, using a recoupling scheme for electric fields at the vertices of the lattice proposed by Robson and Webber (1982), and a table of $6j$ symbols and $3j$ phases for the SU(3) group calculated by Butler (1981) and Bickerstaff *et al.* (1982). This is a rather clumsy and elaborate approach, and we shall not go into further details here. The series for the ground-state energy is listed in Table 10.2.

In the U(1) case, a Dlog Padé analysis of the series for the 'specific heat' $C = -\frac{2}{N}\lambda^2\partial^2\omega_0/\partial\lambda^2$ indicates a singularity at $x_c = 0.79(5)$, with an index $\alpha = 0.9(3)$; but a Monte Carlo study (Hamer and Aydin, 1991) finds that the phase transition

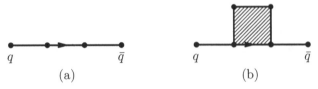

q $\bar{q}$ q $\bar{q}$

(a)   (b)

Fig. 10.10. (a) A string of electric flux along an axis of the lattice connecting a static quark–antiquark pair; (b) A cluster consisting of a single plaquette rooted to the axial string.

occurs somewhat earlier, at $x_c = 0.675(25)$, which would imply that the transition is probably first order, in agreement with the Euclidean analyses. In the SU(3) case, there is no evidence of a singularity at finite x, arguing that the system remains in a confining phase at all couplings.

10.2.3 *The string tension*

In the Hamiltonian formulation, a measure of the string tension can be obtained very directly as the energy per unit length of a 'string' of electric flux connecting a hypothetical quark-antiquark pair lying along one axis of the lattice, as illustrated in Figure 10.10.

In the strong-coupling limit $x \to 0$, the string tension is simply the excitation energy on a single link

$$K = \begin{cases} 1, & U(1) \\ 4/3, & SU(3) \end{cases} \tag{10.70}$$

and both models are manifestly confining, with a finite string tension. At higher orders the string begins to fluctuate, and the string tension can be expressed as a sum of contributions from linked clusters which are 'line-rooted' or attached to the original straight string (Irving and Hamer, 1984c). By these means, series for the string tension have been calculated to order x^8 for the U(1) model (Irving and Hamer, 1984a) and x^7 for the SU(3) model in (3+1) dimensions (Hamer *et al.*, 1986) – see Table 10.2. As in the Euclidean case, however, a 'roughening' transition occurs for this quantity, which prevents analytic continuation of the series beyond that point. Attempts were made to circumvent this by calculating an 'exact linked cluster expansion' (ELCE), where the contribution of each cluster is calculated *exactly*, rather than as a power series in x, and an extrapolation to the bulk limit is made at each value of x. For further discussion of this technique, see the references above.

Figure 10.11 shows some illustrative results for the SU(3) model. The ELCE estimates of the axial string tension are accurate at strong coupling (small $\sqrt{\lambda}$),

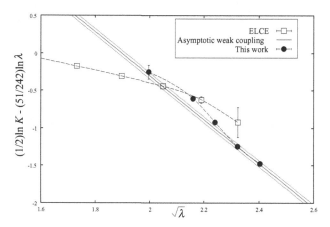

Fig. 10.11. Numerical results for the string tension as a function of $\sqrt{\lambda}$ (where $\lambda = 3x$) for the SU(3) Yang–Mills theory in (3+1)D. The straight lines show the expected scaling behaviour, the filled circles are path-integral Monte Carlo estimates (Byrnes *et al.*, 2004), and the open squares are ELCE estimates of the axial string tension (Irving and Hamer, 1984b).

but become uncertain in the more interesting crossover region; whereas the Monte Carlo estimates show much better convergence, and better scaling behaviour, at intermediate to weak couplings.

10.2.4 Glueball masses

In the pure Yang–Mills theory, the lowest energy 'glueball' excited state of H_0 is the single-plaquette excitation as in Figure 10.9 with energy gap

$$\omega_1 - \omega_0 = \begin{cases} 4, & U(1) \\ 16/3, & SU(3) \end{cases} \tag{10.71}$$

For the U(1) case, strong-coupling series have been carried to order x^{14} for the scalar and axial-vector masses, M_S and M_A respectively (Hamer *et al.*, 1994a). Figure 10.12 shows approximants to the mass gaps as functions of x, cut off at the critical value $x_c = 0.675(25)$ estimated from the Monte Carlo study of Hamer and Aydin (1991). On the face of it, it seems clear that the mass gaps remain finite at x_c, and the transition is first order.

The first strong-coupling series expansions for the SU(3) glueball states were again carried out by Kogut *et al.* (1976). The series were extended to order x^7 by Hamer (1989) using the 'overlapping cluster' technique, but nothing further has been done since then. Table 10.2 lists the series for the energy gaps corresponding to the $J^P = 0^{++}$, 1^{+-} and 2^{++} glueball states. Figure 10.13 shows differential

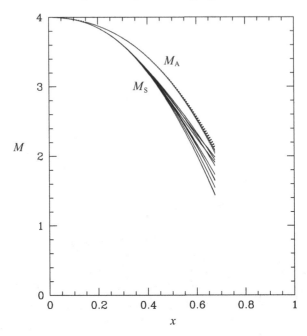

Fig. 10.12. Graph of the symmetric mass gap M_S and the antisymmetric mass gap M_A against x for the pure U(1) lattice gauge theory in (3+1)D. From Hamer *et al.* (1994a).

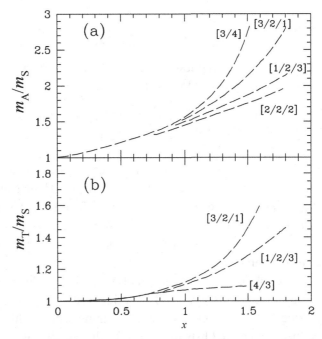

Fig. 10.13. Series estimates of the mass ratios (a) m_A/m_S, and (b) m_T/m_S as functions of x. Padé and Shafer approximants are shown. From Hamer (1989).

approximant estimates of the mass ratios m_A/m_S and m_T/m_S of the axial vector and tensor glueball masses to the scalar mass, as functions of x. Note from Table 10.2 that the series for the scalar and tensor masses are almost identical up to order x^3, and consequently m_T/m_S stays close to 1 for small x values. At one stage, Berg *et al.* (1986) suggested on the basis of a Monte Carlo study that m_T/m_S might even be less than 1, a counter-intuitive result. The series calculations show conclusively, however, that this is not the case. All the coefficients in the expansion for the ratios m_A/m_S and m_T/m_S are positive or zero to the order calculated, and the series approximants remain strictly greater than one for $x > 0$.

The series approximants diverge at large x, and so unfortunately it is difficult to obtain any reliable estimate of the mass ratios in the continuum limit (i.e. as $x \to \infty$).

10.3 Models with dynamical fermions

As regards the full theory of lattice QCD including dynamical quarks, rather little has yet been done by means of strong coupling expansions. In the Euclidean formulation with Wilson fermions, the so-called 'hopping parameter expansion' is used to study the effects of dynamical fermions on physical observables at large quark masses. It can be combined with the strong coupling expansion to give a physical picture of how hadrons propagate on a lattice (Rothe, 1997). This approach has been used by Hoek and Smit (1986) to compute strong-coupling predictions for the hadron masses including $O(1/g^2)$ corrections, but has not been carried any further. The approach is technically complicated, and we shall not pursue it. In the Hamiltonian formulation, some calculations of the hadron spectrum have been done using both strong-coupling expansions and the t-expansion, as outlined in later sections. But first, let us illustrate the effects of fermions in a simpler case, the Schwinger model in (1+1) dimensions.

10.3.1 The lattice Schwinger model

The Schwinger model is quantum electrodynamics in two space–time dimensions. It is a fascinating model in its own right, and shows many of the same phenomena as QCD, including confinement, chiral symmetry breaking with a U(1) axial anomaly, and a topological θ vacuum (Coleman *et al.*, 1975). It is also perhaps the simplest non-trivial gauge theory, and has been a standard test-bed for new lattice techniques.

In the massless case, the theory has been solved by Schwinger (1962), and is equivalent to a theory of free, massive bosons, with mass

$$\frac{M_1}{g} = \frac{1}{\pi} = 0.564\cdots \tag{10.72}$$

where g is the electromagnetic coupling constant, which in (1+1) dimensions has the dimensions of mass. Choosing the time-like axial gauge,

$$A_0 = 0 \tag{10.73}$$

the Hamiltonian is found to be

$$H = \int dx \left(-i\bar{\psi}\gamma^1(\partial_1 + igA_1)\psi + m\bar{\psi}\psi + \frac{1}{2}E^2 \right) \tag{10.74}$$

where the electric field E has only one component

$$E = F^{10} = -\dot{A}^1 \tag{10.75}$$

The remaining gauge field is not an independent degree of freedom, but can be eliminated if desired, using the constraint provided by Gauss' law

$$\partial_1 E = -\partial_1 \dot{A}^1 = g\bar{\psi}\gamma_0\psi \tag{10.76}$$

The model can conveniently be formulated on a Hamiltonian lattice with staggered fermions, as shown by Banks *et al.* (1976). Let the lattice spacing be a, and label sites of the one-dimensional spatial lattice with an integer n. Define a single-component fermion field $\phi(n)$ at each site n, obeying anti-commutation relations

$$\{\phi^\dagger(n), \phi(m)\} = \delta_{mn}, \qquad \{\phi(n), \phi(m)\} = 0 \tag{10.77}$$

The gauge field is defined on links $(n, n+1)$ connecting each pair of sites by

$$U(n, n+1) = e^{i\theta(n)} = e^{-iagA^1(n)} \tag{10.78}$$

Then the lattice Hamiltonian equivalent to equation (10.74) is

$$H = -\frac{i}{2a} \sum_{n=1}^{N} [\phi^\dagger(n)e^{i\theta(n)}\phi(n+1) - \text{h.c.}]$$

$$+m \sum_{n=1}^{N} (-1)^n \phi^\dagger(n)\phi(n) + \frac{g^2 a}{2} \sum_{n=1}^{N} L^2(n), \tag{10.79}$$

where the number of lattice sites N is even, and the correspondence between lattice and continuum fields is

$$\frac{1}{ag}\theta(n) \to -A^1(x), \tag{10.80a}$$

$$gL(n) \to E(x), \tag{10.80b}$$

$$\frac{\phi(n)}{\sqrt{a}} \to \begin{cases} \psi_{\text{upper}}(x), & n \text{ even} \\ \psi_{\text{lower}}(x), & n \text{ odd} \end{cases} \tag{10.80c}$$

The γ matrices are represented by

$$\gamma^0 = \begin{pmatrix} 1 & 0 \\ 0 & -1 \end{pmatrix}, \quad \gamma^1 = \begin{pmatrix} 0 & 1 \\ -1 & 0 \end{pmatrix} \tag{10.81}$$

and the lattice gauge variables in this 'compact' formulation obey commutation relations

$$[\theta(n), L(m)] = i\delta_{nm} \tag{10.82}$$

so that $L(N)$ has integer eigenvalues $L(n) = 0, \pm1, \pm2, \cdots$. The staggering of the fermion field neatly avoids the fermion doubling problem in this case.

Define the dimensionless Hamiltonian

$$W = \frac{2}{ag^2}H = W_0 + xV \tag{10.83}$$

where

$$W_0 = \sum_n L^2(n) + \mu \sum_n (-1)^n \phi^\dagger(n)\phi(n), \tag{10.84a}$$

$$V = -i \sum_n [\phi^\dagger(n)e^{i\theta(n)}\phi(n) - \text{h.c.}] \tag{10.84b}$$

and

$$\mu = \frac{2m}{g^2 a}, \quad x = \frac{1}{g^2 a^2} \tag{10.85}$$

Treat W_0 as the unperturbed Hamiltonian and V as a perturbation, then a perturbation series in x can be developed for the eigenvalues of the Hamiltonian, as discussed by Banks *et al.* (1976). In the strong-coupling limit, the unperturbed ground state $|0\rangle$ has

$$L(n) = 0, \quad \phi^\dagger(n)\phi(n) = \frac{1}{2}[1 - (-1)^n], \quad \text{all } n. \tag{10.86}$$

A 'string' state of unit flux along the x-axis has

$$L(n) = 1, \quad \text{all } n \tag{10.87}$$

with a finite string tension or energy per unit length $K = 1$ in lattice units, showing manifest confinement. Finally, the two lowest fermion–antifermion bound states in this limit are

$$|\psi_{1,2}\rangle = \frac{1}{\sqrt{N}} \sum_{n=1}^{N} [\phi^\dagger(n)e^{i\theta(n)}\phi(n) \pm \text{h.c.}]|0\rangle \tag{10.88}$$

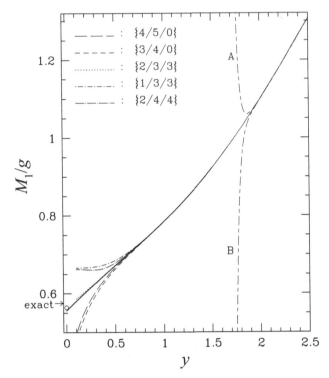

Fig. 10.14. The vector mass gap M_1/g plotted as a function of $y = 1/\sqrt{x}$ for the massless Schwinger model. Lines A,B: partial series sums. Dashed lines: representative series approximants. Solid points: finite lattice estimates. Solid line: a polynomial fit to the series approximants over a 'window' $0.7 \le y \le 1.5$. Open circle: the exact result at $y = 0$. From Hamer *et al.* (1997).

where the $\pm$ sign corresponds to the 'vector'('scalar') state, respectively, with equal mass gaps of $(1 + 2\mu)$. Strong-coupling series have been calculated up to order x^{28} for the energy gaps corresponding to these two fermion–antifermion states in the massless Schwinger model (Hamer *et al.*, 1997).

We now face the problem of extrapolating these series from the strong-coupling limit $x = 0$ (or $y = 1/\sqrt{x} = \infty$) to the continuum limit $a \to 0$, i.e. $x = 1/(g^2 a^2) \to \infty$, or $y = 0$. The results for the massless Schwinger model ($m = 0$) are illustrated in Figure 10.14 for the vector state. Simply summing the terms in the series gives results such as curves A and B – there is no convergence below $y \simeq 2$. Using the standard Padé approximants and integrated differential approximants, as shown by the dashed curves, much better convergence is obtained but still the series approximants splay away, and fail to converge below $y \simeq 0.5$. Higher-order approximants usually do converge better, but only for a little deeper into the weak-coupling regime.

Now a weak-coupling analysis shows that the energy gaps have an asymptotic expansion

$$\frac{M_1}{g} = \sum_{n=0}^{\infty} c_n y^n + O(e^{-b/y}) \tag{10.89}$$

The essential singularity $e^{-b\sqrt{x}}$ in the weak-coupling limit is due to the presence of multiple vacua in the model, and is characteristic of lattice gauge models. The presence of this complicated singularity structure in the continuum limit explains why the series approximants fail to converge in this region.

Quite a good result can be obtained by 'matching' the series approximants to a polynomial in y over the range $0.8 \le y \le 1.5$. The polynomial in y, shown by a solid line in the figure, passes very close to the numerical results obtained by an alternate finite lattice method, and gives an estimate $M_1/g = 0.56(2)$, compared to the exact result $0.564\cdots$. For the scalar state, however, this procedure is less successful (see Figure 10.15). The series approximants fail to converge below $y \le 1$, and furthermore there is a 'bump' in the energy function around $y = 1$, so that any polynomial matching procedure is highly uncertain, giving

$$\frac{M_2}{g} = 1.25(15) \tag{10.90}$$

compared to the exact result $1.128\cdots$. So despite the relatively long series available, there remains considerable uncertainty in the continuum limit estimates. Better methods of analytic continuation are needed here.

10.3.2 Hamiltonian strong-coupling expansions in $(3+1)D$

In the Hamiltonian formulation, some strong-coupling calculations of the hadron spectrum were performed 'by hand' by the COTY collaboration (Banks *et al.*, 1977b). The model considered was QCD with two dynamical quarks in a staggered lattice formulation. The pure gauge Hamiltonian is given by the standard expression

$$H = \frac{g^2}{2a} \left[\sum_l E_l^2 + x \sum_P (6 - \mathrm{Tr} U_P - \mathrm{Tr} U_P^\dagger) \right] \tag{10.91}$$

where $x = 2/g^4$, and

$$[E_l^a, U_{l'}] = t^a U_l \delta_{ll'} \tag{10.92}$$

For fermions, the Kogut–Susskind staggered lattice formulation is used, with a single fermion degree of freedom per site

$$\{ \chi_i^\dagger(\mathbf{n}), \chi_j(\mathbf{m}) \} = \delta_{\mathbf{n},\mathbf{m}} \delta_{ij} \tag{10.93}$$

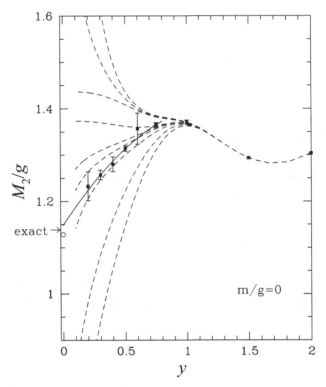

Fig. 10.15. The scalar mass gap M_2/g for the massless Schwinger model. The solid points are finite lattice estimates. Notation is otherwise as in Figure 10.14, except that here the solid line is a second-order polynomial fit to the finite lattice data. From Hamer *et al.* (1997).

where i, j are colour indices. The fermion part of the Hamiltonian is

$$H_F = \frac{1}{2a} \sum_{\mathbf{n},\mu} \eta_\mu(\mathbf{n})[\chi^\dagger(\mathbf{n})U(\mathbf{n}, \mu)\chi(\mathbf{n}+\mu) + \text{h.c.}] \qquad (10.94)$$

where

$$\eta_x(\mathbf{n}) = (-1)^z, \qquad \eta_y(\mathbf{n}) = (-1)^x, \qquad \eta_z(\mathbf{n}) = (-1)^y \qquad (10.95)$$

and μ runs over the three space directions. The eight fermion degrees of freedom on the corners of the unit cube together represent two flavours of Dirac fermion in the continuum. Thus the lattice Hamiltonian describes QCD with two massless quarks (u and d). We shall not give the detailed correspondence between the lattice and continuum quark fields. The continuous chiral symmetry of the continuum theory has been replaced by a discrete chiral symmetry in the lattice model, corresponding to lattice translation by a single link. This symmetry ensures that if the bare lattice Hamiltonian is written for massless quarks, they will remain massless order by order in perturbation theory. A quark mass term would violate the discrete chiral symmetry.

The dimensionless Hamiltonian is now written as

$$W = \frac{2a}{g^2} H = W_0 + y W_q - 2y^2 W_m \tag{10.96}$$

with

$$W_0 = \sum_l \mathbf{E}_l^2 \tag{10.97a}$$

$$W_q = \sum_{\mathbf{n}, \mu} [\chi^\dagger(\mathbf{n}) U(\mathbf{n}, \mu) \chi(\mathbf{n} + \hat{\mu}) \eta_\mu(\mathbf{n}) + \text{h.c.}] \tag{10.97b}$$

$$W_m = \sum_P (\text{Tr} U_P + \text{Tr} U_P^\dagger) \tag{10.97c}$$

and $y = 1/g^2 = \sqrt{x/2}$. Perturbation expansions in y are then calculated for the eigenvalues of the system.

In the strong-coupling limit $y = 0$, the ground state has $\{\mathbf{E}_l^2 = 0, \text{ all } l\}$ as usual. The different fermion states are all degenerate in this limit, however, and degenerate perturbation theory must be applied to find the preferred fermion configuration. It turns out that the fermionic ground state is doubly degenerate, with the quarks at each site forming a colour singlet. If $|0\rangle$ represents the fermion vacuum state at each site, obeying

$$\chi_i |0\rangle = 0, \quad \text{all } i, \tag{10.98}$$

then the ground state fermionic configuration is either

$$\begin{cases} |0\rangle, & \mathbf{n} \text{ even} \\ \frac{1}{6} \epsilon_{ijk} \chi_i^\dagger \chi_j^\dagger \chi_k^\dagger |0\rangle, & \mathbf{n} \text{ odd} \end{cases} \tag{10.99}$$

or vice versa, where $\mathbf{n}$ even (odd) corresponds to $(x + y + z)$ even (odd), respectively. This twofold degeneracy is connected with the chiral symmetry. By choosing to occupy one of these two states, the system will spontaneously break the discrete chiral symmetry.

A complication occurs because the energy of a 'nucleon' excitation consisting of three quarks in a colour singlet combination at a single site is zero in the strong-coupling limit. To remove this unwanted degeneracy, Banks *et al.* (1977b) added an extra 'irrelevant' interaction term to the Hamiltonian, which lifts the degeneracy in the strong-coupling limit, but vanishes in the continuum limit $g^2 \to 0$. We will not need to consider the specific form of this interaction.

Strong coupling series in y were calculated up to order y^4 for meson states consisting in the strong-coupling limit of quark–antiquark pairs connected by a single link of colour flux, with quantum numbers of the $\pi, \rho, \omega, \sigma, B, f$ and A_1 mesons, and for the nucleon state corresponding to a quark triplet excited at a single site. Padé approximants were then calculated for the mass ratios, and extrapolated directly to the continuum limit $y \to \infty$, with the nucleon taken as the reference state. The resulting estimates in the continuum limit were

$$m_\rho/m_N = 0.822$$
$$m_\omega/m_N = 0.824$$
$$m_\pi/m_N = 0.820$$
$$m_\sigma/m_N = 1.01 \qquad\qquad (10.100)$$
$$m_B/m_N = 0.95$$
$$m_f/m_N = 1.17$$
$$m_{A_1}/m_N = 1.00$$

for one particular value of the irrelevant coupling parameter. The results were found to be almost independent of this parameter.

There are several points which may be noted here. Since the series were short, only [1,1] Padé approximants could be calculated, and there was no possibility of checking the convergence of the approximants at different couplings. Most of the ratios are in the ballpark of the physical values, however, with the glaring exception of the pion. The pion remains almost degenerate with the ρ and ω, and shows no sign of dropping towards zero mass.

In principle, a massless pion is expected to emerge in the continuum limit in this model. As one approaches the continuum limit, which is a critical point of the lattice model, the discrete lattice symmetries are expected to be 'promoted' back to their continuous counterparts in the continuum theory. The asymptotic freedom of the theory assures us that the continuous symmetries are restored when $a \to 0$ since they are present in the long wavelength behaviour of the lattice free fields. Then according to the Goldstone theorem, spontaneous breaking of the chiral symmetry should result in a massless pion triplet in the continuum.

The failure of the strong-coupling series to follow this prediction is a major problem for this approach. It might be explained by the lack of spin–spin forces in the low-order calculations. Magnetic field effects due to loops of flux are not important up to this order, but may become important at higher orders. Alternatively, it might be that the pion triplet arises from different basis states, such as the quark–antiquark pairs connected by three links of flux on the unit cube at strong coupling, which would imply that some level crossing takes place. Higher-order calculations are required to explore these possibilities further.

Other properties of the hadrons can also be calculated by means of strong-coupling series. Banks *et al.* (1977b) also calculated the axial charge of the nucleon g_A, and from the [1/1] Padé approximant obtained $g_A \to 1.81$ as $x \to \infty$, which is an improvement on the zeroth order value $g_A = 3$, but not very close to the experimental value $g_A = 1.24(2)$. Jones *et al.* (1978) also calculated some decay constants.

10.4 The *t*-expansion

The *t*-expansion method is an alternative to the strong-coupling expansion method for quantum Hamiltonians, first proposed by Horn and Weinstein (1984). We follow the outline of Schreiber (1993). The method relies on the idea of projecting out the ground state by evolving the system in imaginary time: if $|\psi_0\rangle$ is an arbitrary trial state having a finite overlap with the ground state, then the one-parameter family of states

$$|\psi_t\rangle = \frac{1}{\sqrt{\langle \psi_0 | e^{-tH} | \psi_0 \rangle}} e^{-tH/2} |\psi_0\rangle \tag{10.101}$$

contracts onto the ground state of H as $t \to \infty$. Usually, the strong-coupling ground state is taken as the trial state. It follows that the energy function

$$E(t) = \frac{\langle \psi_0 | H e^{-tH} | \psi_0 \rangle}{\langle \psi_0 | e^{-tH} | \psi_0 \rangle} \tag{10.102}$$

approaches the ground-state energy in the same limit. The function $E(t)$ can be expanded as a Taylor series in t:

$$E(t) = \sum_{n=0}^{\infty} \frac{(-t)^n}{n!} \langle H^{n+1} \rangle_c, \tag{10.103}$$

where the connected matrix elements $\langle H^{n+1} \rangle_c$ are defined recursively by

$$\langle H^{n+1} \rangle_c = \langle \psi_0 | H^{n+1} | \psi_0 \rangle - \sum_{p=0}^{n-1} \binom{n}{p} \langle H^{p+1} \rangle_c \langle \psi_0 | H^{n-p} | \psi_0 \rangle \tag{10.104}$$

The connected matrix elements in the Taylor expansion of $E(t)$ are all proportional to the volume, and this allows them to be computed by means of either a connected diagram expansion or a linked cluster expansion. For a given cluster g, one computes the function $E^g(t)$, then obtains the 'cumulant' energy for cluster g by subtracting sub-graph contributions

$$\epsilon^g(t) = E^g(t) - \sum_{g'} c(g'/g) \epsilon^g(t) \tag{10.105}$$

and then the ground-state energy is

$$E(t)/N = \sum_g c(g)\epsilon^g(t). \tag{10.106}$$

The calculation of the connected moments of the Hamiltonian $\langle H^n \rangle_0^c$ forms the heart of the problem. It is quite straightforward for the U(1) model, but less so for the SU(3) model. A discussion of this problem has been given by Van den Doel and Roskies (1986), who used a 'loop diagram' approach. It involves a considerable amount of combinatorial complexity, but can in principle be carried to high orders: they computed the connected moments up to $\langle H^{10} \rangle_c$. We do not have space to go into the details of this approach here.

Equation (10.102) is an expansion of the function $E(t)$ as a power series in the auxiliary variable t. The standard method of evaluating this function in the limit $t \to \infty$ is the method of 'D-Padé approximants'. The derivative $dE(t)/dt$ is approximated by $[L,M]$ Padé approximants (with $M \geq L + 2$), which are integrated out to infinity.

The t-expansion has one important advantage over the strong-coupling series expansion. For the axial string tension, the strong-coupling expansion cannot be continued beyond the roughening transition, and provides no estimate of the weak-coupling behaviour, whereas the t-expansion estimates can be continued quite happily beyond the transition.

The lowest lying glueball states, except for the 0^{++} state, are the ground states in their corresponding sectors, and their energies can be evaluated as follows. Since we always work with particle states that carry zero momentum, we can use translation invariance to write the t-expansion for the mass of any particular excitation in an explicit volume-independent expansion. Represent the trial wave function for the excitation

$$|\phi\rangle = \frac{1}{\sqrt{N}} \sum_i |\phi_i\rangle \tag{10.107}$$

where the index i runs over 'unit cells' of the lattice. Use translation invariance to single out the unit cell at the origin ($i = 0$) and express

$$\langle \phi | H^n | \phi \rangle = \sum_i \langle \phi_i | H^n | \phi_0 \rangle \tag{10.108}$$

Hence we may define

$$\langle H^{n+1} \rangle_\phi^c = \sum_i \langle \phi_i | H^{n+1} | \phi_0 \rangle - \sum_{p=0}^{n-1} \binom{n}{p} \langle H^{p+1} \rangle_\phi^c \sum_i \langle \phi_i | H^{n-p} | \phi_0 \rangle, \tag{10.109}$$

which serves the same purpose as (10.104) for the vacuum. The matrix elements

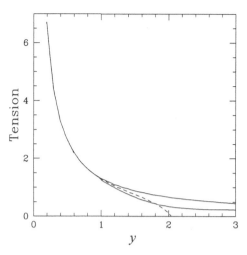

Fig. 10.16. Estimates from the *t*-expansion of the SU(3) string tension as a function of $y = \sqrt{2x}$. The solid lines are *t*-expansion results, the dashed line is the series expansion. From Van den Doel and Horn (1986).

can be computed in terms of diagrams or clusters which are 'rooted to' the unit cell at $i = 0$, as in Chapter 4. Note that the vacuum energy must be subtracted before carrying out this procedure.

To illustrate the results of this approach, Figure 10.16 shows values for the string tension as calculated from the *t*-expansion (Van den Doel and Horn, 1986). Whereas the strong-coupling expansion curve crosses zero beyond the roughening transition, the *t*-expansion results continue to behave in a plausible fashion. As it turns out, however, the predicted asymptotic slope is nothing like that shown by the Monte Carlo data in Figure 10.11.

Figure 10.17 shows estimates of $(M/\sqrt{K})$ for the scalar (S), tensor (T) and axial vector (A) glueball states (Horn and Lana, 1991). From these data they concluded $m_S \simeq 1.3\,\text{GeV}$, $1.4 < m_A/m_S < 1.8$, and $1.06 < m_T/m_S < 1.3$, although the mass ratios do not show good scaling behaviour at weak coupling.

10.4.1 Results for dynamical fermions

The *t*-expansion has been applied by Horn and Schreiber (1993) to the same Hamiltonian model of QCD with two dynamical quarks on a staggered lattice as was treated by the COTY collaboration. Since the *t*-expansion is not an expansion in x, there is no need in this case to add an artificial term to lift the degeneracy of the nucleon states in the strong-coupling limit. The *t*-expansion has been carried up to order H^7 (or in some cases H^9) for states corresponding to the π, ρ and ω mesons and the nucleon. Figures 10.18 and 10.19 show examples of the results.

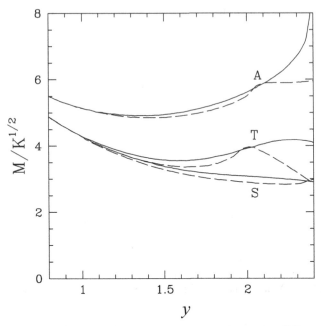

Fig. 10.17. Estimates from the t-expansion of the ratios $M/\sqrt{K}$ for the scalar (S), tensor (T) and axial vector (A) glueball states as functions of $y = \sqrt{2x}$. From Horn and Lana (1991).

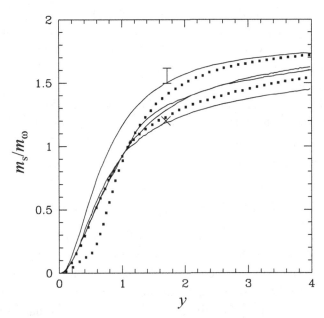

Fig. 10.18. D-Padé approximants to the mass ratio between the scalar glueball and the ω meson obtained from the t-expansion. From Horn and Schreiber (1993).

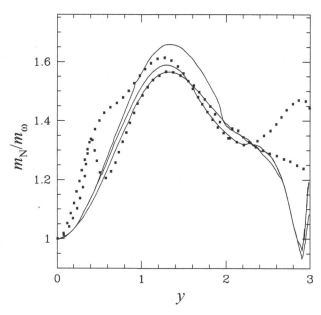

Fig. 10.19. D-Padé approximants to the mass ratio between the nucleon and the ω meson obtained from the *t*-expansion. From Horn and Schreiber (1993).

The mass ratio N/ω rises from zero and appears to reach a fairly stable value around 1.2–1.5, which is consistent with experiment. The mass ratio σ/ω is less stable, but an asymptotic value 1.55–1.45 is estimated from the peak value. The series for the meson states are the *same* up to order H^6, and so the predicted differences in their masses are small. This means that the same major problem has emerged as found by the COTY collaboration: there is no sign of a massless pion triplet emerging in the continuum limit.

Further work was done by Schreiber (1993, 1994) using the *t*-expansion to calculate the masses of excited baryon states, and of 'heavy–light' quark–antiquark meson states, as well as the chiral condensate $\langle \bar{\psi} \psi \rangle$. The mass splittings between the heavy–light states turned out to be too small compared with experiment. The three-quark baryon states were generally found to become heavier than the quark–antiquark meson states as the continuum limit is approached, as expected.

10.5 Conclusions

We have outlined the work that has been done up to date in developing strong-coupling series and *t*-expansions for lattice gauge theories, and in particular for lattice QCD. It is clear that very little has been done in recent years. These approaches have fallen into disfavour by comparison with the remarkable results that have been achieved using path integral Monte Carlo methods.

There are a number of reasons for this. Firstly, the discovery of the roughening transition meant that strong-coupling series for the axial string tension could not be continued to the continuum limit. This does not affect other quantities, however. Secondly, the low-order preliminary calculations of Banks *et al.* (1977b) for full dynamical QCD revealed a major problem in that no massless pion triplet seemed to emerge in the continuum limit, as expected from Goldstone's theorem. The higher-order calculations in the t-expansion seem to indicate that spin–spin forces coming in at higher orders are not the answer to the problem. This remains an outstanding question.

The most serious problem of all, however, is that one cannot see convincing evidence of convergence or scaling behaviour in the weak-coupling region from these calculations. This is evident in the figures we have shown both for strong-coupling expansions and the t-expansion. The example of the Schwinger model shows how the presence of essential singularities in the continuum limit appears to ruin the convergence of the standard Padé or integrated differential approximants.

There is an urgent need, therefore, to find better means of analytic continuation for the series for these models. Some attempts have been made in this direction: for instance, Jones *et al.* (1979) tried a matrix Padé approximant method, and Byrnes *et al.* (2003) applied Feynman–Kleinert approximants (Kleinert, 1995) to the problem, while Stubbins (1988) has explored alternate means of continuation for the t-expansion, but no great improvement was obtained. This remains an open problem.

A great deal remains to be done. For QCD with dynamical fermions, only hand calculations of the strong-coupling series for meson and baryon states have been done. The calculations are technically complicated, but there is no doubt the series could be pushed considerably further by automated techniques such as linked cluster expansions. An economic method of handling the SU(3) symmetry factors is needed: it may well be that the 'loop diagram' approach is more efficient than the Robson–Webber recoupling scheme here.

There are other problems where a series expansion approach might be of great utility. There has been considerable discussion recently of the properties of QCD at finite baryon density, in connection with astrophysical problems. The use of Monte Carlo methods is hampered here by the presence of complex fermion determinants, so that a series approach could be very useful. A discussion of the equation of state at strong coupling has recently been given by Umino (2002) and Fang and Luo (2004).

11

Additional topics

11.1 Disordered systems

The physics of systems with a large degree of random structural disorder is a large and fascinating field (see e.g. Ziman, 1979). We will describe, in this section, three major areas of work where series expansion methods have been successfully used. The discussion will be quite brief but should suffice to give the reader a flavour of the field and a guide to possible future work. We will restrict the discussion, at the outset, in two important ways. Firstly, we will consider only lattice systems, leaving aside the important areas of amorphous and 'glassy' systems. Secondly, we will only consider the case of 'quenched' disorder, which is frozen into place when the system is created. Strictly speaking, such systems are not in complete thermodynamic equilibrium. However in a large system with short range interactions, all configurations are effectively sampled: the system is said to be 'self averaging'. In model calculations, such as those discussed here, an average over different disordered configurations has to be taken.

11.1.1 Percolation

Let us consider a square lattice containing two kinds of atoms or sites arranged randomly. We will call them 'white' and 'black'. They might, for example, be non-magnetic and magnetic atoms in an alloy such as CuMn. Let us denote by p the fraction of black sites. As p is increased larger clusters of connected black sites will be expected and, at a critical probability or 'percolation threshold' p_c, a black cluster of macroscopic (infinite) size will occur for the first time. It is usual to define a 'percolation probability' $P(p)$ as the probability that a randomly chosen black site will be in an infinite cluster. This example is technically called 'site percolation', and the percolation threshold for the square lattice is known to high accuracy (but not exactly!) as $p_c \simeq 0.592746....$ A good introduction to the whole field is Stauffer and Aharony (1992).

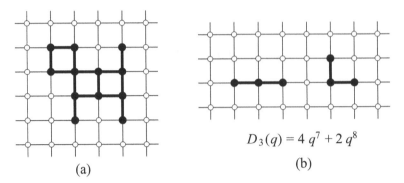

$$D_3(q) = 4\,q^7 + 2\,q^8$$

(a)   (b)

Fig. 11.1. (a) A 'lattice animal' on the SQ lattice; (b) the two possible three-site clusters on the SQ lattice and the corresponding perimeter polynomial.

The most comprehensive series studies of percolation were carried out in the 1970s and 1980s by Sykes and coworkers at King's College, London (Sykes and Glen, 1976; Sykes and Wilkinson, 1986). We briefly outline their approach. In the 'low density' limit ($p \ll 1$) the black sites will form small clusters of various size and shapes – these are sometimes referred to as 'lattice animals'. Figure 11.1(a) shows an example of a cluster with 12 sites and 15 'perimeter sites'.

If we denote by g_{st} the total number of cluster configurations of size s and perimeter t (divided by the number of lattice sites N) then the average number of s-clusters will be

$$n_s = \sum_t g_{st} p^s (1 - p)^t = p^s D_s(q) \tag{11.1}$$

where $q = 1 - p$, and the $D_s(q)$ are known as 'perimeter polynomials'. Figure 11.1(b) shows the possible three-site clusters on the square lattice and the polynomial D_3. Small clusters can be enumerated directly by computer. Sykes and Glen (1976) give the complete perimeter polynomials through D_{17} for the SQ lattice.

A quantity of particular interest is the 'mean cluster size' $S(p)$. This can be obtained as follows. The quantity $w_s = s n_s / p$ is the probability that an arbitrarily chosen black site belongs to an s-cluster. The mean cluster size is then

$$S(p) = \sum_s w_s s = \frac{1}{p} \sum_s s^2 n_s = \sum_s s^2 p^{s-1} D_s(q) \tag{11.2}$$

From this expression we obtain a series in p for $S(p)$. For the SQ lattice it is known to 18 terms

$$S = 1 + 4p + 12p^2 + 24p^3 + 52p^4 + 108p^5 + \cdots + 929368p^{18} + \cdots \tag{11.3}$$

$S(p)$ is assumed to diverge at the percolation threshold, as

$$S(p) \sim (p_c - p)^{-\gamma} \quad p \to p_c- \tag{11.4}$$

and the series (11.3) can be analysed by standard methods to estimate $p_c \simeq 0.593$ and $\gamma \simeq 2.4$. It is also possible to obtain series in the high-density region, in powers of $q = 1 - p$, for various quantities including the percolation probability $P(p)$, which vanishes at p_c according to

$$P(p) \sim (p - p_c)^{\beta} \quad p \to p_c+ \tag{11.5}$$

with $\beta \simeq 0.138$. The reader is referred to the literature for further details, as well as for work on 'bond percolation', and for results on other two and three-dimensional lattices.

Before leaving this topic we mention a computationally simpler problem, 'directed percolation', where it is possible to use methods far more efficient than direct enumeration. In this case the bonds have a direction associated with them and a percolation process, starting from some origin site, can only progress in the allowed directions. Much longer series can be obtained for this class of problem (Jensen, 1999) and consequently the critical parameters can be determined to much greater precision.

11.1.2 Random bond Ising models

Here we consider a spin model on a regular lattice where the exchange interactions, assumed, for simplicity, to extend only to nearest neighbours, can have a strength which is random, and determined by some probability distribution $P(J)$. Almost all of the existing work has been for Ising systems and a common choice for the probability is

$$P(J_{ij}) = p\delta(J_{ij} - J) + (1 - p)\delta(J_{ij} + J) \tag{11.6}$$

i.e. we have a fraction p of ferromagnetic bonds of strength J and a fraction $(1 - p)$ of antiferromagnetic bonds of strength $-J$. The earliest comprehensive series work on such problems was done at King's College, London, particularly by Rapaport. We refer to the work of Fisch (1991) for later results and for references to earlier work.

We show here how to derive high-temperature series for the free energy and susceptibility of the random bond Ising model, following the lines of Chapter 2 for the pure case. We could use the first principles approach, keeping track of the bond variables $v_{ij} = \tanh \beta J_{ij}$, and finally averaging over the bond distribution. However it is simpler to start from the multigraph formulation discussed in Section 2.5.1. The first point to note is that different bond factors v_{ij} are statistically independent.

Table 11.1. *Data for the high-temperature series for*
$\frac{1}{N} \ln \mathcal{Z}$ *for the random bond Ising model on the SC lattice.*

Graph	Weight	LC	μ-factor	Contribution
	1	3	μ^4	$3\mu^4 v^4$
	1	22	μ^6	$22\mu^6 v^6$
	1	207	μ^8	$207\mu^8 v^6$
	-1	18	μ^6	$-18\mu^6 v^8$
	$-\frac{1}{2}$	3	1	$-\frac{3}{2} v^8$
	1	2412	μ^{10}	$2412\mu^{10} v^{10}$
	-2	12	μ^{10}	$-24\mu^{10} v^{10}$
	-1	324	μ^8	$-324\mu^8 v^{10}$
	-1	48	μ^6	$-48\mu^6 v^{10}$
	-1	36	μ^4	$-36\mu^4 v^{10}$

Secondly we have, for the $\pm J$ distribution (11.6)

$$\overline{(v_{ij})^m} = \begin{cases} v^m; & m \text{ even} \\ \mu v^m; & m \text{ odd} \end{cases} \tag{11.7}$$

where $v = \tanh \beta J$ and $\mu = 2p - 1$.

Table 11.1 shows the data needed for the zero-field free energy series at high temperatures, to order v^{10}. The lattice constants are for the SC lattice.

Table 11.2. *Data for the high-temperature susceptibility series for the random bond Ising model on the SC lattice.*

Order	Graph	Weight	LC	μ-factor	Contribution
1		1	3	μ	$3\mu v$
2		1	15	μ^2	$15\mu^2 v^2$
3		1	75	μ^3	$75\mu^3 v^3$
4		1	363	μ^4	$363\mu^4 v^4$
5		1	1767	μ^5	$1767\mu^5 v^5$
		-1	12	μ^3	$-12\mu^3 v^5$
6		1	8463	μ^6	$8463\mu^6 v^6$
		-1	96	μ^4	$-96\mu^4 v^6$
		-1	12	μ^2	$-12\mu^2 v^6$
7		1	40695	μ^7	$40695\mu^7 v^7$
		-2	18	μ^7	$-36\mu^7 v^7$
		-1	132	μ^5	$-132\mu^5 v^7$
		-1	480	μ^5	$-480\mu^5 v^7$
		-1	192	μ^5	$-192\mu^5 v^7$
		-1	96	μ^3	$-96\mu^3 v^7$
		-1	12	μ	$-12\mu v^7$

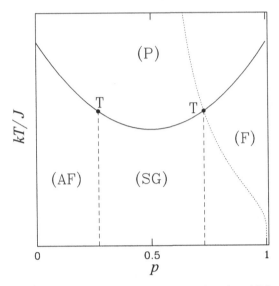

Fig. 11.2. Schematic phase diagram for the $\pm J$ random bond Ising model on the simple cubic lattice. The Nishimori line is shown dotted. The points T are tricritical points.

The resulting series is

$$
-\beta f = \frac{1}{N} \ln \mathcal{Z}
$$
$$
= \ln 2 + 3 \ln \cosh K + (3\mu^2)v^4 + (22\mu^6)v^6 + (207\mu^8 - 18\mu^6 - \tfrac{3}{2})v^8
$$
$$
+(2388\mu^{10} - 324\mu^8 - 48\mu^6 - 36\mu^4)v^{10} + \cdots \tag{11.8}
$$

For $p = 1$ (pure ferromagnet) or $p = 0$ (pure antiferromagnet), the known series are reproduced.

Table 11.2 shows the data needed to obtain the series for the uniform susceptibility to order v^7, for the SC lattice.

The resulting series is

$$
\chi = 1 + (6\mu)v + (30\mu^2)v^2 + (150\mu^3)v^3 + (726\mu^4)v^4
$$
$$
+(353\mu^5 - 24\mu^3)v^5 + (16926\mu^6 - 192\mu^4 - 24\mu^2)v^6
$$
$$
+(81318\mu^7 - 1608\mu^5 - 192\mu^3 - 24\mu)v^7 + \cdots \tag{11.9}
$$

Fisch (1991) has evaluated this series up to order v^{15}.

Analysis of this series yields the paramagnetic to ferromagnetic critical line, from the pure ferromagnet limit ($p = 1$) to the tricritical point $(p^* \simeq 0.77, (kT/J)^* \simeq 1.69)$ where ferromagnetic (F), paramagnetic (P) and spin glass (SG) phases coexist.

This is shown schematically in Figure 11.2. For the Ising case this phase diagram is completely symmetric about $p = \frac{1}{2}$, and so there is another tricritical point terminating the antiferromagnetic (AF) to paramagnetic critical line.

The spin-glass phase will be discussed in the following sub-section. The spin glass/ferromagnetic and spin glass/antiferromagnetic phase boundaries are inaccessible to high-temperature series, but are believed to be vertical. The special line $v = 2p - 1$ (or $e^{-2\beta J} = (1 - p)/p$) is known as the Nishimori line (Nishimori, 1986). This is shown as a dotted line in Figure 11.2 (there is a second line which is the reflected image with respect to $p = \frac{1}{2}$). The internal energy can be obtained exactly along this line and shows no singular behaviour on crossing the phase boundary. The tricritical point is believed to lie on the Nishimori line.

11.1.3 Ising spin glasses

A 'spin glass' is a system in which the atomic magnetic moments are frozen into a random state. Such phases have been observed in many materials where disorder of some kind is present. The simplest theoretical model, first proposed by Edwards and Anderson (1975) is a random bond Ising model with equal probability of $+J$ and $-J$ bonds: the model discussed in the previous sub-section, for $p = \frac{1}{2}$. Even this simple model is quite subtle, and the field remains very active. The unfamiliar reader may wish to consult the review by Binder and Young (1986).

Although, as can be seen from Figure 11.2, the spin-glass phase extends over a range of p values, we will describe here series expansion work precisely at $p = \frac{1}{2}$. The main quantity of interest is the 'spin-glass susceptibility' defined by

$$\chi_{\text{SG}} = \frac{1}{N^2} \overline{\sum_{ij} \langle \sigma_i \sigma_j \rangle^2} \tag{11.10}$$

where the bar is an average over configurations, as before. This is also known as the 'Edwards–Anderson susceptibility'. It is the quantity which is expected to diverge at the spin-glass transition, if a transition exists. Many authors have studied the spin-glass problem via high-temperature series. The most comprehensive results for the $\pm J$ distribution are due to Singh and Chakravarty (1986), who obtained series for the SQ, SC and 4D cubic lattices to order 19, 17, 15 respectively in powers of $w = \tanh^2 \beta J$. Analysis of the resulting series confirmed a finite temperature spin-glass transition in three and four dimensions, but not in two dimensions. This is consistent with Monte Carlo studies, and with our current understanding of Ising spin glasses.

Singh and Chakravarty used a 'star graph expansion' for χ_{SG}^{-1} (see Section 2.5.2). While this may be the most effective method for obtaining long series, we will use a simpler approach, based on connected multi-graph expansions. Let us first

Table 11.3. *Low-order graphs and graph weights for the correlator expansion.*

Order	Graph	Weight	Order	Graph	Weight
1	•—•	1	6		−1
2	⋀	1	7		−1
3	⋀⋁	1	8		−1
4	⋀⋀	1	9		1
5	⋀⋁⋀	1			
	▢	−1			

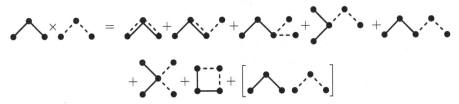

Fig. 11.3. An example showing multiplication of graphs.

consider the graphical expansion for the correlator $\langle \sigma_i \sigma_j \rangle$. A list of graphs and graph weights, at low orders, for the SC lattice, is shown in Table 11.3. The list is complete to order 5. The only graphs shown are those that will contribute to our final expansion to order w^5.

Consider now the process of calculating $\langle \sigma_i \sigma_j \rangle^2$. We need to multiply the set of graphs with itself. Figure 11.3 shown an example.

Thus $\langle \sigma_i \sigma_j \rangle^2$ will consists of all possible overlaps of all the graphs in the list, as well as disconnected graphs. Let us now consider the configuration average. From the previous result (11.7) (with $\mu = 0$) we see that the only graphs which survive are those in which all bonds have an even multiplicity. In the example shown in Figure 11.3 only the first overlap survives.

This observation allows us to construct a list of connected multigraphs for the spin-glass susceptibility. Each bond contributes, as usual, a factor $v = \tanh \beta J$, but

Table 11.4. *Graph data to derive the spin-glass*
susceptibility χ_{SG} *to order* w^5 *for the simple cubic lattice.*

Order	Graph	Weight	LC	Contribution
1		1	3	$3w$
2		1	15	$15w^2$
3		1	75	$75w^3$
4		1	363	$363w^4$
		−24	3	$-72w^4$
5		1	1767	$1767w^5$
		3	12	$36w^5$
		−12	48	$-576w^5$

since each bond has even multiplicity the natural expansion variable is $w = v^2$.
Table 11.4 gives all the graphs, their weights and SC lattice constants to order w^5.

Combining the data in Table 11.4, and multiplying by 2 from the summation in
(11.10), gives

$$\chi_{SG} = 1 + 6w + 30w^2 + 150w^3 + 582w^4 + 2454w^5 + \cdots \tag{11.11}$$

which reproduces the leading terms of the known series.

11.1.4 Further topics

In this section we have described three applications of series expansion methods
to systems with structural disorder. The discussion has, of necessity, been brief
and the reader will need to consult the published literature for details. These topics
by no means cover all of the areas of interest. One obvious extension is to more
general distributions $P(J)$ of bond randomness. Such studies have been reported by
Ditzian and Kadanoff (1979), and by Rajan and Riseborough (1983). Substantial
extension of these series is desirable. Another class of problems involves random

fields rather than bonds. Shapir and Aharony (1982) have obtained short series for this problem but, again, further work is possible. The whole field of random systems with classical vector spins is largely unstudied by series methods. An exception is the random anisotropy model, for which Fisch and Harris (1990) have derived 14th-order series for 'cubic' lattices in dimension 2–5. Finally, for quantum systems, again little series work exists. An interesting recent study of a 'quantum precolation model', whuch is of relevance to metal–insulator transitions, has been reported by Daboul *et al.* (2000).

11.2 Other series expansion methods

The earlier chapters have described, in some detail, the main methods of deriving series expansions for lattice models. There are, however, other approaches that have been used, in some cases very successfully, and for completeness we will give brief outlines of these.

11.2.1 The 'finite lattice' method of Enting

This method originated in the 1970s (de Neef and Enting, 1977) and has been subsequently developed and refined by Enting and coworkers (Enting, 1978; Adler and Enting, 1984; Guttmann and Jensen, 1994). The method is similar in spirit to the Ising star-graph expansion described in Section 2.5.2, and can be regarded as an extension of it. The method is applicable, in principle, to any classical lattice statistical problem, and has proved to be particularly powerful on the square lattice, where it has yielded by far the longest low-temperature series for Ising models with $S \geq 1$ (to u^{113} for $S = 1$, and comparable lengths for $1 < S < 3$). It has also been used with considerable success for the Potts model and enumeration of self-avoiding polygons.

The method is based on the following results (we confine ourselves here to the Ising model on the square lattice):

$$\frac{1}{N} \ln \mathcal{Z} = \sum_{\{r\}} a_r \ln \mathcal{Z}_r \qquad (11.12)$$

where the set $\{r\}$ denotes all $n \times m$ rectangular arrays of sites and bonds, and a_r are numerical coefficients, which must be determined. This will yield a high-temperature series correct to v^r, where $v = \tanh K$, if all rectangles with perimeters $p \leq r$ are included.

Let us illustrate this to low order, where the calculations can be done by hand. Table 11.5 shows the graphs with $p \leq 10$, and their corresponding $\ln \mathcal{Z}$. We omit the prefactors involving multiples of $\ln 2$ and $\ln \cosh K$.

Table 11.5. *Rectangle subgraphs on the square lattice with perimeter*
$p \leq 10$, *and their* $\ln \mathcal{Z}$ *expansions.*

Graph	$\ln \mathcal{Z}$ (to order v^{10})
1	$\ln(1 + v^4) = v^4 - \frac{1}{2}v^8 + \cdots$
2	$\ln(1 + 2v^4 + v^6) = 2v^4 + v^6 - 2v^8 - 2v^{10} + \cdots$
3	$\ln(1 + 3v^4 + 2v^6 + 2v^8) = 3v^4 + 2v^6 - \frac{5}{2}v^8 - 6v^{10} + \cdots$
4	$\ln(1 + 4v^4 + 4v^6 + 7v^8) = 4v^4 + 4v^6 - v^8 - 16v^{10} + \cdots$
5	$\ln(1 + 4v^4 + 3v^6 + 5v^8 + 3v^{10}) = 4v^4 + 3v^6 - 3v^8 - 9v^{10} + \cdots$
6	$\ln(1 + 6v^4 + 7v^6 + 20v^8 + 21v^{10} + 9v^{12}) = 6v^4 + 7v^6$ $+2v^8 - 21v^{10} + \cdots$

The easiest way to implement (11.12) is to adopt the sub-graph subtraction idea, and to define 'reduced partition functions'

$$
\begin{aligned}
\phi_1 &= \ln \mathcal{Z}_1 = v^4 - \tfrac{1}{2}v^8 + \cdots \\
\phi_2 &= \ln \mathcal{Z}_2 - 2\phi_1 = v^6 - v^8 - 2v^{10} + \cdots \\
\phi_3 &= \ln \mathcal{Z}_3 - 2\phi_2 - 3\phi_1 = v^8 - 2v^{10} + \cdots \\
\phi_4 &= \ln \mathcal{Z}_4 - 4\phi_2 - 4\phi_1 = 5v^8 - 8v^{10} + \cdots \\
\phi_5 &= \ln \mathcal{Z}_5 - 2\phi_3 - 3\phi_2 - 4\phi_1 = v^{10} + \cdots \\
\phi_6 &= \ln \mathcal{Z}_6 - 2\phi_4 - 2\phi_3 - 7\phi_2 - 6\phi_1 = 13v^{10} + \cdots \quad\quad (11.13)
\end{aligned}
$$

and then evaluate the bulk partition function from

$$
\begin{aligned}
\tfrac{1}{N} \ln \mathcal{Z} &= \phi_1 + 2\phi_2 + 2\phi_3 + \phi_4 + 2\phi_5 + 2\phi_6 + \cdots \\
&= v^4 + 2v^6 + \tfrac{9}{2}v^8 + 12v^{10} + \cdots \quad\quad (11.14)
\end{aligned}
$$

where the factors of 2 in the first line allow for the two possible orientations of these clusters. The reader can confirm that (11.14) is correct to order v^{10}.

The essence of this method is to replace combinatorial complexity (exponentially increasing number of graphs and lattice constants) with algebraic complexity (the need to compute partition functions of rectangle graphs). This can be done efficiently, in two dimensions, using transfer matrix methods.

Table 11.6. *Spin correlators for an Ising plaquette.*

$$C_{12} = \langle \sigma_1 \sigma_2 \rangle = \frac{v + v^3}{1 + v^4}$$

$$C_{14} = \langle \sigma_1 \sigma_4 \rangle = \frac{2v^2}{1 + v^4}, \qquad v = \tanh K$$

$$C_{1234} = \langle \sigma_1 \sigma_2 \sigma_3 \sigma_4 \rangle = \frac{2v^2}{1 + v^4}$$

In most work it has proven more effective to work with the partition functions directly, using the product

$$\mathcal{Z} = \prod_{\{r\}} \mathcal{Z}_r^{a_r} \qquad (11.15)$$

Explicit values for the a_r, for different problems, are given in the literature. We refer the reader to a recent review (Enting, 1996) for further details of this method.

Although this method appears to have its greatest advantage in two dimensions, it has also been applied to three-dimensional systems (Guttmann and Enting, 1993). Similar methods have been developed by Bhanot and coworkers (Bhanot *et al.* 1994). Another algebraic technique for series expansions, more powerful than the finite lattice method but with, so far, a smaller range of applications is based on corner transfer matrices (Baxter and Enting, 1979).

11.2.2 The Schwinger–Dyson equation method

The Schwinger–Dyson (SD) equations are a set of linear relations between the correlation functions of a field theory (see e.g. Zinn-Justin, 1996). The iterative solution of the SD equations to yield high-temperature expansions for a lattice model was first achieved by Butera *et al.* (1986), for the plane-rotator model on the square lattice. They also showed how Ising models could be treated in a similar way.

To illustrate the idea, in the simplest possible way, we consider an Ising model on a cluster of four sites. This is shown in Table 11.6, together with expressions for the independent spin correlators, which can easily be calculated directly.

Now let us calculate the same correlators via a less direct route. From the definition we have

$$C_{12} = \langle \sigma_1 \sigma_2 \rangle = \frac{1}{\mathcal{Z}_4} \sum_{\sigma_1, \sigma_2, \sigma_3, \sigma_4} \sigma_1 \sigma_2 e^{K(\sigma_1 \sigma_2 + \sigma_1 \sigma_3 + \sigma_2 \sigma_4 + \sigma_3 \sigma_4)} \qquad (11.16)$$

If we replace the dummy summation variable σ_1 by $-\sigma_1$, this yields

$$C_{12} = \langle \sigma_1 \sigma_2 \rangle = \frac{1}{Z_4} \sum_{\sigma_1, \sigma_2, \sigma_3, \sigma_4} (-\sigma_1 \sigma_2) e^{-2K(\sigma_1 \sigma_2 + \sigma_1 \sigma_3)} e^{-\beta H}$$

$$= \langle -\sigma_1 \sigma_2 e^{-2K(\sigma_1 \sigma_2 + \sigma_1 \sigma_3)} \rangle$$

$$= a^2 w - a^2(1 + w^2) C_{12} + a^2 w C_{14} \tag{11.17}$$

with $a = \cosh 2K$, $w = \tanh 2K$. In exactly the same way we can obtain two further equations

$$C_{14} = \langle \sigma_1 \sigma_4 \rangle = 2a^2 w C_{12} - a^2 C_{14} - a^2 w^2 C_{1234} \tag{11.18}$$

$$C_{1234} = \langle \sigma_1 \sigma_2 \sigma_3 \sigma_4 \rangle = 2a^2 w C_{12} - a^2 w^2 C_{14} - a^2 C_{1234} \tag{11.19}$$

solving these equations gives

$$C_{12} = \frac{w}{2 - w^2}, \qquad C_{14} = C_{1234} = \frac{w^2}{2 - w^2} \tag{11.20}$$

which, in terms of $v = \tanh K$, are identical to the expressions in Table 11.6.

This is, of course, an unnecessarily lengthy way to solve this simple case. The power of the method comes at higher order! Let us generalize the previous argument. Consider a general correlator (still for the Ising model on the square lattice)

$$C_N(\mathbf{r}) = \frac{1}{Z} \sum_{\{\sigma\}} (-\sigma_{r_1} \sigma_{r_2} \cdots \sigma_{r_N}) e^{-2K\sigma_{r_1} \sum_\delta \sigma_{r_1 + \delta}} e^{-\beta H}$$

$$= \langle -\sigma_{r_1} \sigma_{r_2} \cdots \sigma_{r_N} e^{-2K\sigma_{r_1} \sum_\delta \sigma_{r_1 + \delta}} \rangle \tag{11.21}$$

where δ runs over all nearest neighbours of site r_1. Expanding the exponential, and a small amount of algebra, gives

$$C_N(\mathbf{r}) = \frac{a^4}{1 + a^4} \langle -\sigma_{r_1} \sigma_{r_2} \cdots \sigma_{r_N} \left[1 - \prod_\sigma (1 - w\sigma_{r_1} \sigma_{r_1 + \delta}) \right] \rangle \tag{11.22}$$

This is a linear relation between the correlators of the initial configuration (which we can refer to as a 'state') and a set of new states. Applying the SD equation to each of these will generate new states. This is repeated until the desired order is obtained.

Butera *et al.* (1987) describe a FORTRAN program for the SD expansion for the plane–rotator model on the triangular lattice. The Milan group used this method to high order, through K^{20} for the $N = 2$ model on the square lattice (Butera and Comi, 1993). However in their more recent work they have used the renormalized free graph expansion.

11.2.3 The continuous unitary transformation method

An alternative to the linked cluster method for deriving series expansions at $T = 0$ is the continuous unitary transformation (CUT) method, developed and used extensively by the Cologne group (see e.g. Knetter and Uhrig, 2000). The basic idea, due to Wegner (1994), is to introduce a generalized Hamiltonian $H(\ell)$, parametrized by a continuous variable ℓ, which evolves according to the 'flow equation'

$$\frac{dH(\ell)}{d\ell} = [\eta(\ell), H(\ell)] \tag{11.23}$$

where $\eta(\ell)$ is an anti-Hermitian operator $(\eta^\dagger(\ell) = -\eta(\ell))$, termed the *generator*. It is easy to show that (11.23) is equivalent to a continuous unitary transformation

$$H(\ell) = U(\ell)H(0)U^\dagger(\ell) \tag{11.24}$$

with $U(\ell)$ satisfying the evolution equation

$$\frac{dU(\ell)}{d\ell} = \eta(\ell)U(\ell) \tag{11.25}$$

The initial Hamiltonian $H(0)$ is assumed to have the usual form $H = H_0 + \lambda V$, and the transformation (11.24) aims to produce an effective Hamiltonian at $\ell = \infty$, with the usual block structure

$$\widetilde{H} = H(\infty) = \begin{pmatrix} \blacksquare & 0 & 0 \\ 0 & \blacksquare & 0 \\ 0 & 0 & \blacksquare \end{pmatrix} \tag{11.26}$$

where each block corresponds to a different set of degenerate eigenfunctions of H_0.

Knetter and Uhrig showed how to implement this procedure for a special type of Hamiltonian which can be written in the form

$$H = H_0 + \lambda \sum_{n=-N}^{N} T_n \tag{11.27}$$

where it is assumed that the unperturbed Hamiltonian H_0 has a set of equally spaced energy levels, which may be degenerate, and which is bounded from below. The perturbation is a sum of $2N + 1$ operators T_n, which act as raising or lowering operators, by a step n, for the ladder of energy levels of H_0, and N is a (small) positive integer. Specifically the T_n satisfy the commutation relation

$$[H_0, T_n] = nT_n \tag{11.28}$$

These conditions are not as restrictive as they may initially appear. In particular, any fully dimerized spin-$\frac{1}{2}$ system will have a singlet ground state for H_0 and successive equidistant excited levels with $1,2,3,\cdots$ triplets. The operator T_n will then serve to excite or de-excite triplons (we will give a specific example below). Another example is the Hubbard model, discussed in Section 8.2 where the energy levels of $H_0 = U \sum_i n_{i\uparrow} n_{i\downarrow}$ form a discrete ladder with $0,1,2,\cdots$ doubly occupied sites. On the other hand, the decomposition of H used for Ising expansions, discussed in Section 5.1.1, does not have this form.

To show how this procedure is implemented, we start by writing the Hamiltonian $H(\ell)$ in the general form

$$H(\ell) = H_0 + \sum_{k=1}^{\infty} \lambda^k \sum_{\{m_i\}} F^{(k)}(\ell; m_1, m_2, \cdots, m_k) T_{m_1} T_{m_2} \cdots T_{m_k} \qquad (11.29)$$

or, in a shorter notation,

$$H(\ell) = H_0 + \sum_{k=1}^{\infty} \lambda^k \sum_{|\mathbf{m}|=k} F^{(k)}(\ell; \mathbf{m}) T^{(k)}(\mathbf{m}) \qquad (11.30)$$

where $\mathbf{m}$ is a vector $\mathbf{m} = (m_1, m_2, m_3, \cdots, m_k)$, with each element taking values $-N, -N+1, \cdots, N-1, N$, and F is a weight function, to be determined. This form simply reflects the outcome of applying the perturbation k-times, giving a product of k T_n operators. The choice of generator $\eta(\ell)$ is not unique. Knetter and Uhrig (2000) propose the following optimum generator

$$\eta(\ell) = \sum_{k=1}^{\infty} \lambda^k \sum_{|\mathbf{m}|=k} \text{sgn}(M(\mathbf{m})) F^{(k)}(\ell; \mathbf{m}) T^{(k)}(\mathbf{m}) \qquad (11.31)$$

where $M(\mathbf{m}) = \sum_{i=1}^{k} m_i$, and sgn is the sign function, with $\text{sgn}(0) = 0$. Inserting the $H(\ell)$ and $\eta(\ell)$ into (11.23), and comparing the coefficients of $T^{(k)}(\mathbf{m})$, one obtains the following recurrence relations

$$\frac{d F^{(k)}(\ell; \mathbf{m})}{d\ell} = -|M(\mathbf{m})| F^{(k)}(\ell; \mathbf{m}) + \sum_{k_1=1}^{k-1} \sum_{\substack{|\mathbf{m}_1|=k_1 \\ \{\mathbf{m}_1, \mathbf{m}_2\}=\mathbf{m}}} [\text{sgn}(M(\mathbf{m}_1))$$

$$-\text{sgn}(M(\mathbf{m}_2))] F^{(k_1)}(\ell; \mathbf{m}_1) F^{(k-k_1)}(\ell; \mathbf{m}_2) \qquad (11.32)$$

Here the summation condition $\{\mathbf{m}_1, \mathbf{m}_2\} = \mathbf{m}$ means that one sums over all possible nontrivial breakups of $\mathbf{m}$. The above differential equations can be solved recursively for a given value of N and the following initial conditions

$$F^{(k)}(0; \mathbf{m}) = 1 \quad \text{for} \quad |\mathbf{m}| = k = 1$$
$$F^{(k)}(0; \mathbf{m}) = 0 \quad \text{for} \quad |\mathbf{m}| = k > 1 \qquad (11.33)$$

We refer the reader to the literature for further details.

Our ultimate object is the effective Hamiltonian $\tilde{H}$ at $\ell = \infty$. In this limit only terms with $M(\mathbf{m}) = 0$ will remain, reflecting the block structure (11.26). Knetter and Uhrig (2000) have computed $\tilde{H}$ to order λ^{15}, for the case $N = 1$, and to order λ^{10} for $N = 2$. In the former case, the general result to third order is

$$\tilde{H} = H_0 + \lambda T_0 + \lambda^2(T_1 T_{-1} - T_{-1} T_1) + \frac{1}{2}\lambda^3(2T_1 T_0 T_{-1} + 2T_{-1} T_0 T_1$$
$$- T_0 T_1 T_{-1} - T_0 T_{-1} T_1 - T_1 T_{-1} T_0 - T_{-1} T_1 T_0) + \cdots \quad (11.34)$$

This result is very reminiscent of the derivation of an effective Hamiltonian for the Hubbard model, discussed in Section 8.2.1 (see also MacDonald *et al.*, 1988). Indeed it is rather straightforward to derive (11.34) directly from the initial Hamiltonian

$$H = H_0 + \lambda(T_{-1} + T_0 + T_1) \quad (11.35)$$

via a sequence of unitary transformations which successively eliminate terms with $M(\mathbf{m}) \neq 0$. Stein (1997) has in fact used the CUT method to generate an effective Hamiltonian for the Hubbard model, and has compared the CUT method with the more traditional one.

There are two ways to apply the general formalism above to derive series expansions for a specific model. One approach is to use the CUT to derive an effective Hamiltonian for each cluster, provided the necessary properties are satisfied, and then use the standard linked cluster formalism, with subgraph subtractions, to obtain the bulk series. For an excited state which is in the same sector as the ground state, the CUT method turns out to be more efficient than the multi-block methods discussed in Appendix 5. The second approach, which has been used in most applications of the CUT method, is to carry out the calculation on a large enough finite lattice. The advantage of this second approach is that it avoids the need for generation of clusters and multiple calculations. However in two or three dimensions this would be computationally very demanding, as it would require quite large finite lattices.

Here we demonstrate the finite lattice approach, taking the transverse field Ising chain in the disordered phase as an example. The Hamiltonian is written in the form

$$H = \sum_i (1 - \sigma_i^z)/4 + \lambda \sum_i (\sigma_i^+ \sigma_{i+1}^+ + \sigma_i^+ \sigma_{i+1}^- + \sigma_i^- \sigma_{i+1}^+ + \sigma_i^- \sigma_{i+1}^-) \quad (11.36)$$

where $\lambda = -4J/\Gamma$. This can be written in the form (11.27) immediately, with

$$T_0 = \sum_i (\sigma_i^+ \sigma_{i+1}^- + \sigma_i^- \sigma_{i+1}^+)$$

$$T_1 = \sum_i \sigma_i^- \sigma_{i+1}^-$$

$$T_{-1} = \sum_i \sigma_i^+ \sigma_{i+1}^+ \quad (11.37)$$

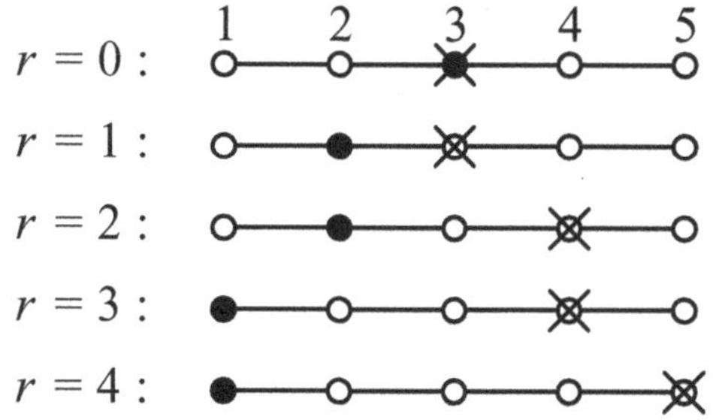

Fig. 11.4. Illustration of finite cluster calculation of transition amplitudes. The filled circles and crosses represent the initial and final positions of the excitations.

The ground-state energy is now given by

$$E_0 = \langle 0|\tilde{H}|0\rangle \tag{11.38}$$

where $|0\rangle$ is the unperturbed ground state, i.e. the state with all spins up. Since $T_{-1}|0\rangle = 0$ and $T_0|0\rangle = 0$, many of the terms vanish, and we obtain, to order λ^4,

$$E_0 = -\lambda^2\langle 0|T_{-1}T_1|0\rangle + \lambda^4\Big(\langle 0|T_{-1}T_1T_{-1}T_1|0\rangle$$
$$-\langle 0|T_{-1}T_0T_0T_1|0\rangle - \frac{1}{2}\langle 0|T_{-1}T_{-1}T_1T_1|0\rangle\Big) + \cdots \tag{11.39}$$

To obtain the ground-state energy correct to order λ^n requires, in this case, a rung of $n + 1$ sites with *periodic* boundary conditions, this ensures that there are no end sites, which would not contribute the same amount of energy as the middle sites. The result, which the reader may wish to verify, is

$$E_0 = -5\lambda^2 - 5\lambda^4 + \cdots \tag{11.40}$$

The energy per site agrees with the result obtained previously (Section 4.3).

To obtain the one-particle excitation spectrum one needs to compute the transition amplitudes $\Delta(\mathbf{r})$ in the single particle subspace

$$\Delta(\mathbf{r}) = \langle i|\tilde{H}|j\rangle - E_0\delta_{ij}, \quad \mathbf{r} \equiv \mathbf{r}_i - \mathbf{r}_j \tag{11.41}$$

where $|i\rangle$ is the unperturbed one-particle state with excitation located at site i. The calculation can be carried out on a finite cluster with *open* boundary conditions. An open chain of $n + 1$ sites is sufficient to obtain the series correct to order λ^n, provided one chooses the proper initial and final positions of the excitation to avoid boundary effects. To demonstrate this, we take a cluster of five sites as an example. Figure 11.4 shows the initial and final positions of the excitation required to get the correct series to order 4. Note that the energy E_0, which is subtracted off in (11.41), is the ground-state energy of the open chain of five sites, which is found to be

$$E_0 = -4\lambda^2 - 2\lambda^4 + \cdots \tag{11.42}$$

We find, to order λ^4,

$$\langle 3|\widetilde{H}|3\rangle = \tfrac{1}{2} - 2\lambda^2 + 0\lambda^4 + \cdots$$
$$\langle 2|\widetilde{H}|3\rangle = \lambda - 2\lambda^3 + \cdots$$
$$\langle 2|\widetilde{H}|4\rangle = -\lambda^2 + 4\lambda^4 + \cdots$$
$$\langle 1|\widetilde{H}|4\rangle = 2\lambda^3 + \cdots$$
$$\langle 1|\widetilde{H}|5\rangle = -5\lambda^4 + \cdots \tag{11.43}$$

which can then be combined to yield the one-particle excitation energy

$$\epsilon(k) = \tfrac{1}{2} + 2\lambda^2 + 2\lambda^4 + (2\lambda - 4\lambda^3)\cos(k) + (-2\lambda^2 + 8\lambda^4)\cos(2k)$$
$$+ 4\lambda^3 \cos(3k) - \lambda^4 \cos(4k) + O(\lambda^5) \tag{11.44}$$

which agrees with the result obtained earlier (Section 4.5.3).

The CUT technique can also be used to compute the energies and structure factors for two-particle excitations (Knetter *et al.*, 2003) but we will not elaborate further on this.

Appendix 1: some graph theory ideas

This appendix provides a brief introduction to topics in the mathematical field of Graph Theory, which are pertinent to the subject of series expansions. For further details the reader is referred to Domb (1974) and to Chartrand (1977).

We start with some definitions.

(i) A *graph* is a collection of points (vertices) and lines (bonds) (see Figure A1.1).

(ii) A *connected graph* is one in which there is a path between any pair of points. A *disconnected graph* is one which is not connected. The number of components of a disconnected graph can be 2, 3, ...

(iii) An *articulation point (articulation vertex)* is a vertex, the removal of which, with all of its incident lines, breaks the connectivity of the graph.

(iv) The *order* (degree) of a vertex is the number of lines incident on the vertex. Note that if a vertex is of order 1, then the vertex to which it is joined is an articulation vertex.

(v) A *star graph* is a connected graph with no articulation points.

(vi) A *tree graph* is a connected graph with at least one vertex of order 1 (Note that this differs from more usual definitions, but is most convenient for our purposes).

(vii) A *simple graph* is one in which there is at most one line joining any pair of vertices. A *multi-graph* is one in which there is more than one line between at least one pair of vertices.

(viii) A *labelled graph* is one in which the n vertices are labelled. An *unlabelled graph* is one which is not labelled.

(ix) A *rooted graph* (one-rooted, two-rooted, . . .) is one in which one or more vertices is considered as fixed, in some sense.

(x) A *polygon* is a closed loop with all vertices of order 2.

(xi) A *chain* (or 'self-avoiding walk', SAW) is a tree graph with no vertex of order greater than 2.

These definitions are illustrated with simple examples in Figure A1.1.

It is important to note that a graph is defined only by the topology/connectivity between points, and not by a particular geometric picture.

Table A1.1. *Number of simple connected graphs $N(l, p)$ with l lines, and p points.*

l	$p = 2$	3	4	5	6	7	8	9	10
1	1	0	0	0	0	0	0	0	0
2	0	1	0	0	0	0	0	0	0
3	0	1	2	0	0	0	0	0	0
4	0	0	2	3	0	0	0	0	0
5	0	0	1	5	6	0	0	0	0
6	0	0	1	5	13	11	0	0	0
7	0	0	0	4	19	33	23	0	0
8	0	0	0	2	22	67	89	47	0
9	0	0	0	1	20	107	236	240	106
10	0	0	0	1	14	132	486	797	657
11	0	0	0	0	9	138	814	2075	2678
12	0	0	0	0	5	126	1169	4495	8548

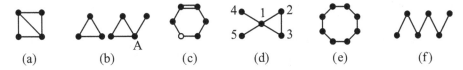

(a)	(b)	(c)	(d)	(e)	(f)

Fig. A1.1. (a) A connected unlabelled star; (b) a disconnected graph with an articulation point (A); (c) a one-rooted multi-graph; (d) a labelled tree graph; (e) a polygon; (f) a chain.

The number of graphs proliferates rapidly with increasing number of points and/or lines. There are mathematical results for the numbers of graphs of various classes (Domb, 1974), but it is more important for our purposes is to be able to generate them systematically, as discussed in Section 2.2. However, for interest we provide a table giving $N(l, p)$, the number of simple connected graphs with l lines and p points (Table A1.1). Worthwhile exercises for the reader are:

(i) construct by hand all graphs with, say, $l \leq 7$, $p \leq 6$; and
(ii) use the computer program gen.f supplied (www.cambridge.org/9780521842426), with input nvmax=7, nbmax=6 to generate these graphs. Try to go further, and note how fast the program is. However, beware of rapidly increasing output data.

An important concept is the *symmetry number* of an unlabelled graph. This is the number of permutations of the vertices which leave the graph invariant. Graph generation and counting programs, of necessity, deal with labelled graphs and so, for instance, the lattice constant of a bare graph is obtained by dividing the number of embeddings of the labelled graph by the symmetry number. Figure A1.2 shows some unlabelled simple graphs and their symmetry numbers.

Table A1.2. *Strong and weak subgraphs, and corresponding embedding constants, of graph (a) of Figure A1.1*

Graph:								
Subgraph:							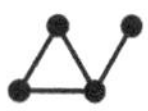	
Strong:	4	5	2	2	0	0	0	0
Weak:	4	5	8	2	2	6	4	1

2	8	2	5! = 120	8	2 × 6 × 6 = 72

Fig. A1.2. Examples of unlabelled graphs and symmetry numbers.

Care needs to be taken in some of the computational procedures when dealing with graphs of high symmetry. For example, if we are generating all graphs with, say, up to 12 edges and 13 vertices, then one of the graphs generated (the analogue of the fourth graph in Figure A1.2) will have symmetry number 12! Unless the program is suitably modified it will go through all of these 12! permutations in checking for canonical labelling. The way to avoid this potential problem is to separate the vertices into equivalence classes, order each class, and only consider permutations which preserve the order within classes.

A *sub-graph G'* of a connected graph G is a subset of the vertices and bonds of G which forms a smaller connected graph. In this context we need to distinguish between *weak sub-graphs*, which may omit some bonds joining the vertices of the subgraph in G, and *strong sub-graphs*, where such omission is not allowed. The corresponding *embedding constant* gives the number of ways in which a subgraph G' can be formed from G. This is illustrated in Table A1.2.

The concept of sub-graphs and embedding can be extended to the case where G is an infinite lattice $\mathcal{L}$ of a particular type (strictly a finite lattice of N sites with periodic boundary conditions). The list of possible sub-graphs is then the list of graphs embeddable in $\mathcal{L}$, and the embedding constants become polynomials in N. Connected graphs clearly have embedding constants proportional to N, while the embedding constants of disconnected graphs contain higher powers. In either case the coefficient of the term linear in N is termed the strong/weak *lattice constant* of the graph for the particular lattice $\mathcal{L}$. Examples are given in Chapter 2.

Appendix 2: the 'pegs in holes' algorithm

Many of the computational aspects of deriving series expansions are of a combinatorial nature and efficient algorithms can reduce the computational effort by orders of magnitude! Many of the computer programs included in this book have, at their heart, a fundamental enumeration problem: that of putting a set of 'objects' into a set of 'places', in all possible ways, subject to some possible constraints.

To be less abstract, suppose we have a puzzle with M 'pegs' which have to be inserted into N 'holes' in a board, with $M \le N$. Both pegs and holes are numbered, and we assume a constraint of no more than one peg per hole. Every possible placement constitutes a 'success' or 'count'. For example, if $M = 2$ and $N = 4$ the successes are (obviously)

| 1 2 | | | | 1 | 2 | | | 1 | | 2 | | 1 2 | | 1 | | 2 | | | 1 2 |
| 2 1 | | | | 2 | 1 | | | 2 | | 1 | | 2 1 | | 2 | | 1 | | | 2 1 |

To write a computer code for this procedure we need the following elements (at least):

- a counter k to designate the pegs;
- an array $s(k)$ to designate the location of each peg k;
- an array occ to keep track of the occupancy (0 or 1) of each hole.

A section of (Fortran 77) code which does this might read

```
---------------------------------------------------------------
      DO 10 i=1,m
10    s(i)=0
      DO 12 i=1,n
12    occ(i)=0
      count=0
      k=1
```

```
20    sk=s(k)+1                     !try next location of kth peg
      s(k)=sk                       !place peg
      IF(sk.GT.n) GOTO 22           !check if last hole reached
      IF(occ(sk).EQ.1) GOTO 20      !check if hole already full
      occ(sk)=1                     !update occ array
      k=k+1                         !go to next peg
      IF(k.LE.m) GOTO 20            !check if all pegs placed
      count=count+1
      WRITE(6,78) count,s(1:m),occ(1:n)
78    FORMAT(i3,2x,<m>i3,2x,<n>i3)
22    CONTINUE
      s(k)=0                        !remove kth peg
      k=k-1                         !backstep to previous peg
      IF(k.EQ.0) GOTO 30            !loop complete - exit
      occ(s(k))=0
      GOTO 20
30    CONTINUE
```

A nice example to illustrate this principle in use is to compute the symmetry number of a graph. In this case the pegs are the n vertices of the graph and the holes are the n vertices of an identical graph. We seek to place the vertices of the graph onto the vertices of an identical graph in all possible ways, subject to the constraint that the connectivity (or equivalently, the adjacency matrix) is unchanged. The number of successes is the symmetry number of the graph. A computer program *symm.f* can be accessed at www.cambridge-org/978052184242 and the reader is encouraged to experiment with this.

Appendix 3: free graph expansion technicalities

This Appendix provides technical details needed to implement the free graph expansion, as described in Chapter 3.

The bare vertex functions (BVF's) $M_n^0(\alpha\beta\cdots\nu)$ were defined in Eqs (3.9)–(3.17), via

$$M_n^0(\alpha\beta\cdots\nu) = \frac{\partial^n}{\partial h_\alpha \partial h_\beta \cdots \partial h_\nu} M_0^0(h) \tag{A3.1}$$

where we have suppressed the site index i. We are interested here only in their zero-field values. M_0^0 is an even function of h, and so can be expanded as a series

$$M_0^0(h) = \sum_{n=0}^{\infty} a_{2n} \left(h_x^2 + h_y^2 + h_z^2 \right)^n \tag{A3.2}$$

where the coefficients a_s depend on N, and can be obtained from (3.10)–(3.13). Of course, (A3.2) is for $N = 3$; simpler expressions exist for $N = 2$ and $N = 1$. Expanding (A3.2) in a trinomial expansion gives

$$M_0^0(h) = \sum_{n=0}^{\infty} a_{2n} \sum_{\substack{r,s,t=0 \\ r+s+t=n}}^{n} \frac{n!}{r!\,s!\,t!} h_x^{2r} h_y^{2s} h_z^{2t} \tag{A3.3}$$

From these results it is clear that in zero-field only the BVFs of even order are non-zero and, furthermore, each distinct Cartesian index must occur an even number of times. Otherwise the order of indices is immaterial (this is not the case in quantum systems where non-commuting operators occur!). We can now obtain the BVFs explicitly, and these are given, up to $n = 10$ in Table A3.1.

The free graph expansion also requires a knowledge of the free lattice constants (FLCs) of unrooted, one-rooted, and two-rooted graphs. We mention three different approaches.

Table A3.1. *Bare vertex functions in zero field for* $N = 1, 2, 3$. *Each Cartesian index represents a pair, e.g.* $(xy) \equiv (xxyy)$, *and only inequivalent cases are shown.*

Order		$N = 1$	$N = 2$	$N = 3$
2	$M_2^0(x)$	1	1/2	1/3
4	$M_4^0(xx)$	-2	$-3/8$	$-2/15$
	$M_4^0(xy)$	$-$	$-1/8$	$-2/45$
6	$M_6^0(xxx)$	16	5/4	16/63
	$M_6^0(xxy)$	$-$	1/4	16/315
	$M_6^0(xyz)$	$-$	$-$	16/945
8	$M_8^0(xxxx)$	-272	$-1155/128$	$-16/15$
	$M_8^0(xxxy)$	$-$	$-165/128$	$-16/105$
	$M_8^0(xxyy)$	$-$	$-99/128$	$-16/175$
	$M_8^0(xxyz)$	$-$	$-$	$-16/525$
10	$M_{10}^0(xxxxx)$	7936	3591/32	256/33
	$M_{10}^0(xxxxy)$	$-$	399/32	256/297
	$M_{10}^0(xxxyy)$	$-$	171/32	256/693
	$M_{10}^0(xxxyz)$	$-$	$-$	256/2079
	$M_{10}^0(xxyyz)$	$-$	$-$	256/3465

(1) Direct counting via computer. The program *count.f* which was used to obtain high-temperature or weak lattice constants can be easily modified (by removing the single occupancy constraint) to count free graphs. For unrooted and one-rooted graphs the procedure is identical. For two-rooted graphs one needs to fix the two roots. However for the susceptibility all possible separations of the root points are summed over and so again the same program can be used.

(2) It is possible to express the free lattice constant of a graph in terms of weak lattice constants of itself and its 'collapse chain' (Wortis 1974). This does not seem particularly useful, and it is mentioned only for completeness.

(3) Algebraic procedures, based on the idea that free embeddings are essentially random walks. We give some examples:

- n-link chain: FLC $= q^n$ ($q =$ coordination number);

- articulated graph: FLC $=$ FLC(1) $\times$ FLC(2);

- polygons: an n-sided polygon is an n-step random walk which returns to the origin. Consider the case of the square lattice. The generating function for n-step random walks from the origin is

$$g_n(x, y) = (x + y + x^{-1} + y^{-1})^n = \sum_{r,s=-n}^{n} c_{r,s} x^r y^s \qquad \text{(A3.4)}$$

where $c_{r,s}$ is number of n-step walks from $(0,0)$ to (r, s). Thus $\text{FLC}(P_n)$ is given by the coefficient of the constant term in the expansion of $(x + y + x^{-1} + y^{-1})^n$ which is $\sum_{p=0}^{n/2} n!/([p!(n/2 - p)!]^2)$.

For other lattices the generating functions will be more complicated, but the idea is the same.

Appendix 4: matrix perturbation theory

An essential part of the derivation of long perturbation series for ground-state properties, and excitations, of quantum lattice models is an efficient computational procedure for determining the corresponding series for finite clusters. We outline the basic procedure here – it is incorporated into our computer programs *tim1.f*, *tim2.f*, which compute series for the transverse field Ising model, and which can be accessed at www.cambridge.org/9780521842426.

We write the Hamiltonian as $H = H_0 + \lambda V$, as usual, and seek to expand quantities in power of λ. We denote the ground state of H_0 by $|0\rangle$, with energy e_0, and assume, for the moment, that this state is non-degenerate. The Hilbert space for any finite cluster is of finite dimension and hence H_0, V and other operators can be represented by square symmetric matrices. It is efficient to start from the state $|0\rangle$ and, by operating on this state recursively with V, to generate other eigenstates of H_0, $|i\rangle$, with energies e_i. To a given order, λ^p, only those states in the first $p/2$ generations will contribute, and these will be the only states needed – they may be substantially fewer than the complete unperturbed basis!

Let us first consider the ground state problem. We denote the ground state energy and wavefunction of the complete Hamiltonian by E and $|\Psi\rangle$ and expand these in powers of λ in the usual way

$$E = \sum_{r=0}^{\infty} E_r \lambda^r \tag{A4.1}$$

$$|\Psi\rangle = \sum_{r=0}^{\infty} \lambda^r |\Psi_r\rangle \tag{A4.2}$$

with $E_0 = e_0$ and $|\Psi_0\rangle = |0\rangle$. Substitution into the Schrödinger equation $H|\Psi\rangle = E|\Psi\rangle$ then leads to

$$(H_0 - E_0)|\Psi_r\rangle = -V|\Psi_{r-1}\rangle + \sum_{s=0}^{r-1} E_{r-s}|\Psi_s\rangle \tag{A4.3}$$

291

Taking the scalar product on the left with $\langle k|$, first with $k = 0$ and then with $k \neq 0$, and denoting $\langle k|\Psi_r\rangle = c_{k,r}$, yields

$$E_r = \sum_j V_{0j} c_{j,r-1} \tag{A4.4}$$

and

$$c_{k,r} = \frac{1}{e_k - e_0}\left[-\sum_j V_{kj} c_{j,r-1} + \sum_{s=0}^{r-1} E_{r-s} c_{k,s}\right], \quad k \neq 0 \tag{A4.5}$$

where $V_{kj} \equiv \langle k|V|j\rangle$ are the matrix elements of V in the unperturbed basis, which can be computed and stored at the same time as the basis states are generated. Note that the matrix V is usually very sparse and only the positions and values of the non-zero elements need to be stored. The above equations do not determine the coefficients $c_{0,r}$ $(r > 0)$. These can be set to zero, meaning that the wave function corrections $|\Psi_r\rangle$, at all orders, are chosen to be orthogonal to the unperturbed ground state wavefunction.

This formalism is easily generalized if the Hamiltonian includes, in addition, a field term, so that

$$H = H_0 + \lambda V + hX \tag{A4.6}$$

where X is an operator (in the simplest case just the magnetization). Keeping terms to order h^2 suffices to allow computation of the expectation value $\langle X\rangle$ and of the generalized susceptibility $\chi = -\lim_{h\to 0}(\partial^2 E/\partial h^2)$. We simply write the r-th order energy and wavefunction as

$$E_r = E_r^{(0)} + hE_r^{(1)} + h^2 E_r^{(2)} + \cdots \tag{A4.7a}$$
$$|\Psi_r\rangle = |\Psi_r^{(0)}\rangle + h|\Psi_r^{(1)}\rangle + h^2|\Psi_r^{(2)}\rangle + \cdots \tag{A4.7b}$$

and introduce coefficients

$$c_{k,r}^{(i)} \equiv \langle k|\Psi_r^{(i)}\rangle \quad i = 0, 1, 2 \tag{A4.8}$$

The Schrödinger equation then yields three equations

$$(H_0 - E_0)|\Psi_r^{(0)}\rangle = -V|\Psi_{r-1}^{(0)}\rangle + \sum_{s=0}^{r-1} E_{r-s}^{(0)}|\Psi_s^{(0)}\rangle$$

$$(H_0 - E_0)|\Psi_r^{(1)}\rangle = -X|\Psi_r^{(0)}\rangle - V|\Psi_{r-1}^{(1)}\rangle + \sum_{s=0}^{r-1} E_{r-s}^{(0)}|\Psi_s^{(1)}\rangle$$

$$+ \sum_{s=0}^{r} E_{r-s}^{(1)}|\Psi_s^{(0)}\rangle$$

$$(H_0 - E_0)|\Psi_r^{(2)}\rangle = -X|\Psi_r^{(1)}\rangle - V|\Psi_{r-1}^{(2)}\rangle + \sum_{s=0}^{r-1} E_{r-s}^{(0)}|\Psi_s^{(2)}\rangle$$

$$+ \sum_{s=0}^{r} \left[E_{r-s}^{(1)}|\Psi_s^{(1)}\rangle + E_{r-s}^{(2)}|\Psi_s^{(0)}\rangle \right] \tag{A4.9}$$

the first of which is just (A4.3) again. These are solved as before, yielding the results

$$E_r^{(0)} = \sum_j V_{0j} c_{j,r-1}^{(0)}$$

$$E_r^{(1)} = \sum_j \left(X_{0j} c_{j,r-1}^{(0)} + V_{0j} c_{j,r-1}^{(1)} \right)$$

$$E_r^{(2)} = \sum_j \left(X_{0j} c_{j,r-1}^{(1)} + V_{0j} c_{j,r-1}^{(2)} \right) \tag{A4.10}$$

$$c_{k,r}^{(0)} = \frac{1}{e_k - e_0} \left[-\sum_j V_{kj} c_{j,r-1}^{(0)} + \sum_{s=1}^{r-1} E_{r-s}^{(0)} c_{k,s}^{(0)} \right]$$

$$c_{k,r}^{(1)} = \frac{1}{e_k - e_0} \left[-\sum_j \left(X_{kj} c_{j,r}^{(0)} + V_{kj} c_{j,r-1}^{(1)} \right) \right.$$

$$\left. + \sum_{s=1}^{r-1} \left(E_{r-s}^{(0)} c_{k,s}^{(1)} + \sum_{s=1}^{r} E_{r-s}^{(1)} c_{k,s}^{(0)} \right) \right]$$

$$c_{k,r}^{(2)} = \frac{1}{e_k - e_0} \left[-\sum_j \left(X_{kj} c_{j,r}^{(1)} + V_{kj} c_{j,r-1}^{(2)} \right) \right.$$

$$\left. + \sum_{s=1}^{r-1} E_{r-s}^{(0)} c_{k,s}^{(2)} + \sum_{s=1}^{r} \left(E_{r-s}^{(1)} c_{k,s}^{(1)} + E_{r-s}^{(2)} c_{k,s}^{(0)} \right) \right] \tag{A4.11}$$

These equations are solved recursively.

Appendix 5: matrix block diagonalization

This appendix provides technical details of the perturbative block diagonalization procedures used to obtain an effective Hamiltonian matrix for one-particle states, as discussed initially in Section 4.5, or for two-particle states, as discussed in Section 6.4.

The Hamiltonian of the cluster is written as $H = H_0 + \lambda V$ and a basis set is formed by N eigenstates of H_0, with the first n corresponding to the degenerate unperturbed excited states of interest: to be specific we consider one-particle states. As in the previous Appendix we denote the basis states by $|i\rangle$ and their unperturbed energies by e_i.

We now construct an orthogonal matrix O, such that the transformed Hamiltonian

$$\widetilde{H} = O^{-1} H O \tag{A5.1}$$

has the block diagonalized form

$$\widetilde{H} = \begin{pmatrix} h_{\text{eff}} & 0 & 0 \\ 0 & \blacksquare & 0 \\ 0 & 0 & \blacksquare \end{pmatrix} \tag{A5.2}$$

This transformation can be constructed order-by-order in the perturbation parameter λ by expanding the matrices

$$\widetilde{H} = \widetilde{H}^{(0)} + \lambda \widetilde{H}^{(1)} + \lambda^2 \widetilde{H}^{(2)} + \cdots \tag{A5.3}$$

$$O = O^{(0)} + \lambda O^{(1)} + \lambda^2 O^{(2)} + \cdots \tag{A5.4}$$

with $\widetilde{H}^{(0)} = H_0$, and $O^{(0)} = I$, the identity matrix. Rewriting (A5.1) in the form $O\widetilde{H} = HO$, substituting the expansions (A5.4) and collecting terms, gives

$$\widetilde{H}^{(r)} + O^{(r)} H_0 - H_0 O^{(r)} = V O^{(r-1)} - \sum_{s=1}^{r-1} O^{(r-s)} \widetilde{H}^{(s)} \tag{A5.5}$$

Taking matrix elements $A_{ij} \equiv \langle i|A|j \rangle$ gives

$$\widetilde{H}_{ij}^{(r)} + (e_j - e_i)O_{i,j}^{(r)} = \left(VO^{(r-1)} - \sum_{s=1}^{r-1} O^{(r-s)}\widetilde{H}^{(s)} \right)_{ij}$$
$$= X_{ij}^{(r)} \qquad (A5.6)$$

where for conciseness we denote the right-hand side of (A5.6) by a matrix $X^{(r)}$. Note that $X^{(r)}$ is determined fully by the O and $\widetilde{H}$ matrices at lower order. Thus (A5.6) yields immediately

$$\widetilde{H}_{ij}^{(r)} = X_{ij}^{(r)}, \qquad e_i = e_j \qquad (A5.7)$$

which gives the rth-order elements of the upper left block h_{eff}, and

$$O_{i,j}^{(r)} = \frac{1}{e_j - e_i} X_{ij}^{(r)}, \qquad e_i \neq e_j. \qquad (A5.8)$$

This latter result determines completely the off-diagonal blocks of $O^{(r)}$.

The diagonal blocks of $O^{(r)}$ are not yet uniquely determined, and their form depends on the method chosen. In the two-block approach, discussed in Section 4.5, one requires the diagonal blocks to be symmetric. Then the expansion (A5.4), when substituted into the orthogonality relation $O^T O = I$, yields

$$O^{(r)} + [O^T]^{(r)} = -\sum_{s=1}^{r-1} O^{(s)}[O^T]^{(r-s)} \qquad (A5.9)$$

which can now be used to determine the diagonal blocks of $O^{(r)}$. From (A5.6) one can see that only the matrix elements $O_{ij}^{(r)}$ with $j \leq n$ are needed to evaluate h_{eff}, i.e. we only need the left hand $N \times n$ segment of each $O^{(r)}$ matrix. This makes the TBOT method computationally efficient.

The procedure described above is incorporated in the computer program *tim3.f*, which is supplied (www.cambridge.org/9780521842426). This program computes the effective Hamiltonian and one-particle transition amplitudes for the transverse field Ising model in the disordered phase, discussed in Section 4.5.3. The reader can easily adapt it to other cases.

It is worth noting that the formalism discussed here can also be used for the ground state sector. In that case $n = 1$ and h_{eff} is simply the perturbed ground state energy. It is also worth noting that if the diagonal blocks of the transformation matrix $O^{(r)}$ ($r > 0$) are simply set to zero, one obtains the similarity transformation introduced by Gelfand (1996), which is no longer orthogonal. For reasons discussed in Section 4.5, we do not use this approach for excited states. However the method described in Appendix 4, and used for the ground-state sector is, in fact, a similarity transformation.

When the TBOT method fails it is necessary to use the more computationally demanding multiblock method (MBOT) approach, which requires the computation and storage of the full $N \times N$ $O^{(r)}$ matrices. A systematic procedure (Zheng *et al.*, 2001b) is as follows. We write $O = e^S$, where S is a real antisymmetric matrix, with the block structure

$$
S = \begin{pmatrix}
0 & -s_1^T & -s_2^T \\
\hline
s_1 & 0 & -s_3^T \\
\hline
s_2 & s_3 & 0
\end{pmatrix}
$$

i.e. the diagonal blocks of S are assumed to be zero. We also introduce an auxiliary matrix R

$$
R = O - S = e^S - S \tag{A5.10}
$$

and expand

$$
S = \lambda S^{(1)} + \lambda^2 S^{(2)} + \cdots
$$
$$
R = I + \lambda^2 R^{(2)} + \lambda^3 R^{(3)} + \cdots \tag{A5.11}
$$

Then the $R^{(r)}$ matrix is fully determined by lower order S matrices, via

$$
R^{(r)} = \frac{1}{2!} \sum_{k,l=1}^{r-1} S^{(k)} S^{(l)} \delta_{k+l,r} + \frac{1}{3!} \sum_{k,m,l=1}^{r-2} S^{(k)} S^{(l)} S^{(m)} \delta_{k+l+m,r} + \cdots \tag{A5.12}
$$

The procedure then is as follows:

- determine the off-diagonal blocks of $O^{(r)}$ from (A5.8);
- determine the full matrix $R^{(r)}$ from (A5.12);
- determine the diagonal blocks of $O^{(r)}$ from $O^{(r)} = R^{(r)} + S^{(r)}$, remembering that the diagonal blocks of $S^{(r)}$ are zero;
- determine the full $S^{(r)}$ matrix from $S^{(r)} = O^{(r)} - R^{(r)}$;
- repeat successively at higher orders.

The whole procedure is easily computerized.

Appendix 6: the moment–cumulant expansion

For our purposes, the *moment* $\langle\ \rangle$ of a set of lattice variables $\{\alpha \cdots \zeta\}$ is the average value of that combination of variables, with respect to some partition function or path integral. The *cumulant* $[\ ]$ of those variables is then defined by

$$\langle \alpha \cdots \zeta \rangle = \sum_P [\alpha \cdots \beta][\gamma \cdots \zeta] \tag{A6.1}$$

where the sum goes over all possible partitions P of the set of variables, or symbolically

$$\langle\ \rangle = \sum_{n=1}^{\infty} \frac{1}{n!} [\ ]^n \tag{A6.2}$$

For example,

$$\langle \alpha \rangle = [\alpha]$$
$$\langle \alpha\beta \rangle = [\alpha\beta] + [\alpha][\beta]$$
$$\langle \alpha\beta\gamma \rangle = [\alpha\beta\gamma] + [\alpha\beta][\gamma] + [\beta\gamma][\alpha]$$
$$+ [\gamma\alpha][\beta] + [\alpha][\beta][\gamma] \tag{A6.3}$$

Equation A6.2 can be inverted to give symbolically

$$[\ ] = \sum_{n=1}^{\infty} (-1)^{n-1} \frac{1}{n} \langle\ \rangle^n, \tag{A6.4}$$

for example

$$[\alpha] = \langle \alpha \rangle$$
$$[\alpha\beta] = \langle \alpha\beta \rangle - \langle \alpha \rangle \langle \beta \rangle$$
$$[\alpha\beta\gamma] = \langle \alpha\beta\gamma \rangle - \langle \alpha\beta \rangle \langle \gamma \rangle - \langle \beta\gamma \rangle \langle \alpha \rangle$$
$$- \langle \gamma\alpha \rangle \langle \beta \rangle + 2 \langle \alpha \rangle \langle \beta \rangle \langle \gamma \rangle \tag{A6.5}$$

Let F be the generating function for the moments:

$$F(\{\phi_\alpha\}) = \sum_{n=1}^{\infty} \sum_{\alpha_1, \cdots \alpha_n} \frac{1}{n!} \langle \alpha_1 \cdots \alpha_n \rangle \phi_{\alpha_1} \cdots \phi_{\alpha_n} \qquad (A6.6)$$

where the ϕ_α are complex variables indexed by the elements α; and let f be the corresponding generating function for the cumulants. Then the central theorem of the moment-cumulant expansion is

$$1 + F = \exp(f) \qquad (A6.7)$$

– see Baker (1990).

An important property is that if any moment *factorizes*

$$\langle \alpha_1 \cdots \alpha_n \beta_1 \cdots \beta_n \rangle = \langle \alpha_1 \cdots \alpha_n \rangle \langle \beta_1 \cdots \beta_n \rangle \qquad (A6.8)$$

then the corresponding cumulant *vanishes*. In lattice parlance, only *connected graphs* contribute to the cumulants.

In quantum field theory, this formalism can be used to show that the generating functional of connected Green's functions equals the logarithm of the generating functional of all Green functions. In statistical mechanics, it can be used to show that the free energy, i.e. the logarithm of the partition function, can be expressed in terms of connected graphs alone.

Appendix 7: integral equation approach to the two-particle Schrödinger equation

In Section 6.4.2, we showed how to use the finite lattice approach to solve the two-particle Schrödinger equation

$$h_{\text{eff}}|\Psi\rangle = E|\Psi\rangle \tag{A7.1}$$

Here we consider an alternative approach. We express the Schrödinger equation as an integral equation in momentum space, for various cases, as follows.

Consider an unsymmetrized state of non-identical particles, types a and b. Then there are $N(N-1)$ possible states on an N-site lattice, labelled by positions $|\mathbf{i}, \mathbf{j}\rangle$, where $\mathbf{i}, \mathbf{j}$ refer to the positions of particles a and b, respectively. We have assumed here that two particles may not reside at the same position. Then the irreducible two-particle matrix element is:

$$\Delta_2^{ab}(\mathbf{i}, \mathbf{j}; \mathbf{m}, \mathbf{n}) = E_2^{ab}(\mathbf{i}, \mathbf{j}; \mathbf{m}, \mathbf{n}) - E_0\delta_{\mathbf{i},\mathbf{m}}\delta_{\mathbf{j},\mathbf{n}} - \Delta_1^a(\mathbf{i}, \mathbf{m})\delta_{\mathbf{j},\mathbf{n}} - \Delta_1^b(\mathbf{j}, \mathbf{n})\delta_{\mathbf{i},\mathbf{m}}$$

$$\tag{A7.2}$$

Let the two-particle eigenstate be:

$$|\psi\rangle = \sum_{\mathbf{i},\mathbf{j}} f_{\mathbf{ij}}|\mathbf{i}, \mathbf{j}\rangle \quad (i \neq j) \tag{A7.3}$$

substitute in the Schrödinger equation

$$\sum_{\mathbf{m},\mathbf{n}} E_2(\mathbf{i}, \mathbf{j}; \mathbf{m}, \mathbf{n})f_{\mathbf{mn}} = Ef_{\mathbf{ij}} \tag{A7.4}$$

one obtains

$$(E - E_0)f_{\mathbf{ij}} - \sum_{\mathbf{m}\neq\mathbf{j}} \Delta_1^a(\mathbf{i}, \mathbf{m})f_{\mathbf{mj}} - \sum_{\mathbf{m}\neq\mathbf{i}} \Delta_1^b(\mathbf{j}, \mathbf{m})f_{\mathbf{im}}$$

$$= \sum_{\mathbf{m},\mathbf{n}} \Delta_2^{ab}(\mathbf{i}, \mathbf{j}; \mathbf{m}, \mathbf{n})f_{\mathbf{mn}} \quad (i \neq j) \tag{A7.5}$$

299

Completing the sums on the left-hand side, one obtains:

$$(E - E_0)f_{ij} - \sum_{m}[\Delta_1^a(\mathbf{i}, \mathbf{m})f_{mj} + \Delta_1^b(\mathbf{j}, \mathbf{m})f_{im}] =$$

$$\sum_{m,n}\Delta_2^{a,b}(\mathbf{i}, \mathbf{j}; \mathbf{m}, \mathbf{n})f_{mn} - \Delta_1^a(\mathbf{i}, \mathbf{j})f_{jj} - \Delta_1^b(\mathbf{j}, \mathbf{i})f_{ii} \quad (\mathbf{i} \neq \mathbf{j}) \quad (A7.6)$$

We assume the above equation holds for the case $\mathbf{i} = \mathbf{j}$, where the fictitious amplitudes f_{ii} are introduced to simplify the calculations.

Next, perform a Fourier transformation,

$$f(\mathbf{k}, \mathbf{q}) = \frac{1}{N} \sum_{\mathbf{i}, \mathbf{j}} e^{i(\mathbf{k}_1 \cdot \mathbf{i} + \mathbf{k}_2 \cdot \mathbf{j})} f_{ij} \quad (A7.7)$$

where $\mathbf{k}, \mathbf{q}$ are the centre-of-mass and relative momenta:

$$\mathbf{k} = (\mathbf{k}_1 + \mathbf{k}_2) \quad \mathbf{q} = \frac{1}{2}(\mathbf{k}_1 - \mathbf{k}_2) \quad (A7.8)$$

For example for the Δ_1^a term

$$\frac{1}{N} \sum_{\mathbf{m}, \mathbf{i}, \mathbf{j}} \Delta_1^a(\mathbf{i}, \mathbf{m}) e^{i(\mathbf{k}_1 \cdot \mathbf{i} + \mathbf{k}_2 \cdot \mathbf{j})} f_{mj}$$

$$= \frac{1}{N} \sum_{\mathbf{m}, \delta, \mathbf{j}} \Delta_1^a(\delta) e^{i(\mathbf{k}_1 \cdot (\mathbf{m} - \delta) + \mathbf{k}_2 \cdot \mathbf{j})} f_{mj}, \quad \delta \equiv \mathbf{m} - \mathbf{i}$$

$$= \sum_{\delta} \Delta_1^a(\delta) e^{-i\mathbf{k}_1 \cdot \delta} f(\mathbf{k}, \mathbf{q})$$

$$= \sum_{\delta} \Delta_1^a(\delta) e^{-i(\mathbf{k}/2 + \mathbf{q}) \cdot \delta} f(\mathbf{k}, \mathbf{q})$$

$$= \sum_{\delta} \Delta_1^a(\delta) \cos((\mathbf{k}/2 + \mathbf{q}) \cdot \delta) \quad (A7.9)$$

where we have assumed inversion symmetry for the one-particle dispersion:

$$\Delta_1^{a,b}(\delta) = \Delta_1^{a,b}(-\delta) \quad (A7.10)$$

Then, equation (A7.6) leads to

$$\left[E - E_0 - \sum_{\delta}[\Delta_1^a(\delta) \cos(\mathbf{k} \cdot \delta/2 + \mathbf{q} \cdot \delta) + \Delta_1^b(\delta) \cos(\mathbf{k} \cdot \delta/2 - \mathbf{q} \cdot \delta)] \right]$$

$$f(\mathbf{k}, \mathbf{q}) = \frac{1}{N} \sum_{\mathbf{q}'} f(\mathbf{k}, \mathbf{q}') \left[\sum_{\mathbf{r}, \delta_1, \delta_2} \Delta_2^{ab}(\mathbf{r}, \delta_1, \delta_2) e^{i(\mathbf{k} \cdot \mathbf{r} + \mathbf{q} \cdot \delta_1 - \mathbf{q}' \cdot \delta_2)} \right.$$

$$\left. - \sum_{\delta}[\Delta_1^a(\delta) \cos(\mathbf{k} \cdot \delta/2 + \mathbf{q} \cdot \delta) + \Delta_1^b(\delta) \cos(\mathbf{k} \cdot \delta/2 - \mathbf{q} \cdot \delta)] \right] \quad (A7.11)$$

Finally, look for solutions with definite *exchange symmetry*.

Symmetric states

$$f_{ij} = +f_{ji} \tag{A7.12}$$

therefore

$$f(\mathbf{k}, \mathbf{q}) = +f(\mathbf{k}, -\mathbf{q}). \tag{A7.13}$$

'Averaging' over $f(\mathbf{k}, \pm\mathbf{q})$ (i.e. taking $\frac{1}{2}[f(\mathbf{k}, \mathbf{q}) + f(\mathbf{k}, -\mathbf{q})]$), we get:

$$\{E - E_0 - \sum_{\delta}[\Delta_1^a(\delta) + \Delta_1^b(\delta)]\cos(\mathbf{k} \cdot \delta/2)\cos(\mathbf{q} \cdot \delta)\} f(\mathbf{k}, \mathbf{q}) =$$

$$\frac{1}{N}\sum_{\mathbf{q}'} f(\mathbf{k}, \mathbf{q}')[\sum_{\mathbf{r},\delta_1,\delta_2} \frac{1}{2}\Delta_2(\mathbf{r}, \delta_1, \delta_2)e^{i\mathbf{k}\cdot\mathbf{r}}\cos(\mathbf{q} \cdot \delta_1)\cos(\mathbf{q}' \cdot \delta_2)$$

$$- \sum_{\delta}[\Delta_1^a(\delta) + \Delta_1^b(\delta)]\cos(\mathbf{k} \cdot \delta/2)\cos(\mathbf{q} \cdot \delta)] \tag{A7.14}$$

where Δ_2 is the irreducible two-particle matrix element for the symmetrized two-particle state, which is related to Δ_2^{ab} by

$$\Delta_2(\mathbf{i}, \mathbf{j}; \mathbf{m}, \mathbf{n}) \equiv \frac{1}{2}(\langle \mathbf{ij}| + \langle \mathbf{ji}|)\Delta_2^{ab}(|\mathbf{mn}\rangle + |\mathbf{nm}\rangle)$$

$$= \frac{1}{2}[\Delta_2^{ab}(\mathbf{i}, \mathbf{j}; \mathbf{m}, \mathbf{n}) + \Delta_2^{ab}(\mathbf{m}, \mathbf{n}; \mathbf{i}, \mathbf{j}) + \Delta_2^{ab}(\mathbf{j}, \mathbf{i}; \mathbf{m}, \mathbf{n}) + \Delta_2^{ab}(\mathbf{i}, \mathbf{j}; \mathbf{n}, \mathbf{m})] \tag{A7.15}$$

Antisymmetric states

$$f_{ij} = -f_{ji} \tag{A7.16}$$

therefore

$$f(\mathbf{k}, \mathbf{q}) = -f(\mathbf{k}, -\mathbf{q}). \tag{A7.17}$$

'Averaging' over $f(\mathbf{k}, \pm\mathbf{q})$ (i.e. taking $\frac{1}{2}[f(\mathbf{k}, \mathbf{q}) - f(\mathbf{k}, -\mathbf{q})]$), we get:

$$\{E - E_0 - \sum_{\delta}[\Delta_1^a(\delta) + \Delta_1^b(\delta)]\cos(\mathbf{k} \cdot \delta/2)\cos(\mathbf{q} \cdot \delta)\} f(\mathbf{k}, \mathbf{q}) =$$

$$\frac{1}{N}\sum_{\mathbf{q}'} f(\mathbf{k}, \mathbf{q}')\sum_{\mathbf{r},\delta_1,\delta_2} \frac{1}{2}\Delta_2(\mathbf{r}, \delta_1, \delta_2)e^{i\mathbf{k}\cdot\mathbf{r}}\sin(\mathbf{q} \cdot \delta_1)\sin(\mathbf{q}' \cdot \delta_2)$$

$$\tag{A7.18}$$

where Δ_2 is the irreducible two-particle matrix element for the antisymmetrized two-particle state, which is related to Δ_2^{ab} by

$$\Delta_2(\mathbf{i}, \mathbf{j}; \mathbf{m}, \mathbf{n}) \equiv \frac{1}{2}((\langle \mathbf{ij}| - \langle \mathbf{ji}|)\Delta_2^{ab}(|\mathbf{mn}\rangle - |\mathbf{nm}\rangle))$$

$$= \frac{1}{2}[\Delta_2^{ab}(\mathbf{i}, \mathbf{j}; \mathbf{m}, \mathbf{n}) + \Delta_2^{ab}(\mathbf{m}, \mathbf{n}; \mathbf{i}, \mathbf{j}) - \Delta_2^{ab}(\mathbf{j}, \mathbf{i}; \mathbf{m}, \mathbf{n}) - \Delta_2^{ab}(\mathbf{i}, \mathbf{j}; \mathbf{n}, \mathbf{m})]$$

$$(A7.19)$$

Identical particles

If the particles a and b are identical, the solution is the same as for symmetric states except the labels a and b must now be dropped, and to avoid double counting it turns out that the Δ_2 term must be multiplied by an extra factor of $1/2$:

$$[E - E_0 - 2\sum_{\delta} \Delta_1(\delta)\cos(\mathbf{k} \cdot \delta/2)\cos(\mathbf{q} \cdot \delta)]f(\mathbf{k}, \mathbf{q}) =$$

$$\frac{1}{N}\sum_{\mathbf{q}'} f(\mathbf{k}, \mathbf{q}')\Bigg[\frac{1}{2}\sum_{\mathbf{r}, \delta_1, \delta_2} \Delta_2(\mathbf{r}, \delta_1, \delta_2)\cos(\mathbf{k} \cdot \mathbf{r})\cos(\mathbf{q} \cdot \delta_1)\cos(\mathbf{q}' \cdot \delta_2)$$

$$-2\sum_{\delta} \Delta_1(\delta)\cos(\mathbf{k} \cdot \delta/2)\cos(\mathbf{q} \cdot \delta)\Bigg] \qquad (A7.20)$$

The above integral equations can be solved, for a given value of $\mathbf{k}$, using standard discretization techniques. For example for the one-dimensional case, instead of using continuous momentum $\mathbf{q}$, one can use N discretized and equally spaced values of momentum $q_i \equiv 2\pi(i - 1)/N$ $(i = 1, 2, \cdots, N)$, so that instead of solving the complicated integral equation, one only needs to compute the eigenvalue and eigenvector of an $N \times N$ matrix for the discretized system. For example for two identical particles, the matrix needing to be diagonalized is

$$H_{i,j} = \frac{1}{2N}\sum_{\mathbf{r}, \delta_1, \delta_2} \Delta_2(\mathbf{r}, \delta_1, \delta_2)\cos(\mathbf{k} \cdot \mathbf{r})\cos(\mathbf{q}_i \cdot \delta_1)\cos(\mathbf{q}_j \cdot \delta_2)$$

$$+(2\delta_{ij} - \frac{2}{N})\sum_{\delta} \Delta_1(\delta)\cos(\mathbf{k} \cdot \delta/2)\cos(\mathbf{q}_i \cdot \delta) \qquad (A7.21)$$

Notice that this matrix is non-symmetric due to the unphysical f_{ii} term we have introduced in Eq. (A7.6), but even so the eigenvalues obtained from this matrix are real. The solutions we obtain also include an unphysical one with eigenvalue equal to 0, due to the unphysical f_{ii} term.

The results obtained from the calculation with discretized momenta will converge to those with continuous momentum as $N \to \infty$. Actually for those bound states with finite coherence length, the calculation will normally be well converged for quite small values of N, but for unbound states we have an infinite coherence length, so one may need to do finite N extrapolations to get results at $N = \infty$.

As in the finite lattice approach, there are two methods to compute the eigenvalues of the matrix for the discretized system. Obviously one can get numerical results for the eigenvalues, for a given value of coupling λ and momentum $\mathbf{k}$, via standard numerical techniques where we just perform a naive sum for the series in Δ_1 and Δ_2. A better technique is to compute the series in λ for the eigenvalues through degenerate perturbation theory.

Return to our example of the transverse Ising chain. If we take the series Δ_1 and Δ_2 up to order μ^4, and take two-particle total momentum $k = 0$, and $\mu = 0.1$, the number of discretized momenta is $N = 4$, and one would get the following 4×4 matrix

$$
H = \begin{pmatrix}
2.705534375 & -0.900265625 & -0.905003125 & -0.900265625 \\
-1.005253125 & 3.014978125 & -1.004753125 & -1.004971875 \\
-1.105003125 & -1.099765625 & 3.304534375 & -1.099765625 \\
-1.005253125 & -1.004971875 & -1.004753125 & 3.014978125
\end{pmatrix}
$$

$$(\text{A7.22})$$

The eigenvalues are 0, 3.7279019, 4.01995, 4.292173, where the first one is unphysical.

Appendix 8: correspondences between field theory and statistical mechanics

The correspondence between field theory and statistical mechanics rests on the mathematical equivalence between the Euclidean path integral in field theory

$$Z_{QFT} = \int D\phi e^{-S_E/\hbar} \qquad (A8.1)$$

and the partition function in classical statistical mechanics

$$Z_{SM} = \text{Tr}\{e^{-\beta H_{SM}}\}. \qquad (A8.2)$$

Both quantities involve a sum over all configurations of the system, each weighted by an exponential weighting factor.

The path integral in field theory can initially be defined as the trace of the time development operator in the Schrödinger representation between initial time t and final time t':

$$Z_{QFT} = \text{Tr}\{e^{-iH_{QFT}(t'-t)}\}. \qquad (A8.3)$$

so that in Euclidean space-time $t \to -i\tau$

$$Z_{QFT} \to \text{Tr}\{e^{-H_{QFT}T}\} = \sum_i e^{-E_i T}. \qquad (A8.4)$$

where T is the size of the system in Euclidean time, and $\{E_i\}$ are the energy eigenvalues of the quantum Hamiltonian. Therefore

$$Z_{QFT} \sim e^{-E_0 T}, \quad T \to \infty \qquad (A8.5)$$

i.e. the ground-state, with minimum energy E_0, dominates the sum as T gets large.

A Greens function

$$\langle Q(t)Q(0)\rangle_0 = \frac{\langle 0|Q(t)e^{-iHt}Q(0)|0\rangle}{\langle 0|e^{-iHt}|0\rangle} = \frac{1}{Z_{QFT}}\int D\phi\, Q(t)Q(0)e^{iS/\hbar}, \qquad (A8.6)$$

Table A8.1. *Corresponding quantities in statistical mechanics (d space dimensions) and quantum field theory (one time, (d − 1) space dimensions)*

Statistical mechanics	Quantum field theory
Partition function	Path integral
Inverse temperature, $\beta = 1/kT$	Inverse coupling, (e.g.) $1/g^2$
Classical Hamiltonian, H_{SM}	Euclidean action, S_E
− Log(Transfer matrix), − ln(V)	Quantum Hamiltonian, H_{QFT}
Equilibrium state	Ground state
Correlation functions	Propagators
Free energy density	Ground state energy density
Inverse correlation length, $1/\xi$	Energy gap, $E_n - E_0$
Soliton	Instanton

after inserting a complete set of intermediate states at times t and 0, can be expressed in Euclidean time τ as

$$\langle Q(\tau)Q(0)\rangle_0 = \sum_i c_i e^{-(E_i - E_0)\tau},$$

$$\sim c_n e^{-(E_n - E_0)\tau}, \quad \tau \to \infty \tag{A8.7}$$

where n is the lowest energy state created by the operator Q from the vacuum (we assume for simplicity there is no overlap with the vacuum itself). Thus the exponential decay of the Greens function is controlled by the energy gap in the corresponding sector of states.

In statistical mechanics, the lattice partition function is often written in terms of the 'transfer matrix' V:

$$Z_{SM} = \text{Tr}\{e^{-\beta H_{SM}}\} = \text{Tr}\{V^N\} \tag{A8.8}$$

where V is a matrix connecting configurations of the system on successive 'layers' in a chosen direction (the 'time' direction), and N is the number of layers. For N large,

$$Z_{SM} \sim \lambda_M^N, \quad N \to \infty \tag{A8.9}$$

where λ_M is the maximum eigenvalue of the transfer matrix.

The decay of a correlation function in statistical mechanics is parametrized in terms of the correlation length ξ:

$$\langle Q(0)Q(r)\rangle \sim e^{-r/\xi}, \quad r \to \infty \tag{A8.10}$$

Using the equivalences of equation (A8.1) to (A8.2), (A8.4) to (A8.8), and (A8.7) to (A8.10), we can now draw up a table of corresponding quantities in the two formalisms, between a quantum field theory in one time and $(d-1)$ space dimensions, and a classical statistical mechanics system in d space dimensions (Table A8.1). Constant factors have been ignored.

An often-quoted example is the two-dimensional classical Ising model, whose field theory equivalent is the transverse Ising model in one space and one time dimension (Suzuki, 1971; Fradkin and Susskind, 1978).

Appendix 9: computer programs

Various computer programs have been referred to in the text and used to obtain some of the results presented in the preceding chapters. These programs can be accessed at www.cambridge.org/9780521184242. This Appendix lists the various programs, together with a paragraph or two about each one.

The programs are written in fairly old-fashioned FORTRAN, and have been tested and used in our work over the years. They are relatively efficient and, as a result, not always as transparent as they might be. We attempt to point out particularly subtle sections when these occur. However we do not claim the ultimate in efficiency or complete freedom from 'bugs'. Readers are invited to let us know of any.

A few comments on programming style are worth making. Some may prefer a single program of tens of thousands of lines which essentially does everything and covers all possible cases. In our view this is generally both inefficient and is difficult to adapt to particular problems. Our preference is for maximum modularity where a problem is broken down into different parts which are performed sequentially. This does lead to greater overheads, requiring storage of intermediate results, but has major advantages in efficiency, flexibility, and transparency.

The programs dealing with graph generation and counting are largely based on programs written by Dr C. J. (Chuck) Elliott at the University of Alberta in the early 1970s and we gratefully acknowledge this. Others have been developed within our group. A few have been developed specifically for this book.

The following gives a list of all files that are available, with a number, name, filename and description

(1) **Summary** (readme.txt)
   A text file which summarizes what is available (essentially the same information as here). Each program file contains the following:
   • a program listing, with comment statements at various points;

- sample input files;
- sample output files, or parts thereof; and
- some general notes.

The user will need to edit each file appropriately to create a compilable FORTRAN program $\cdots$ .f and to provide their own input files in the correct format.

(2) **Graph symmetry number** (*symm*)

A simple program to read in graphs from a file and compute their symmetry numbers. The input graphs are described in pair format. The symmetry number is computed by a subroutine symm which uses the 'pegs in holes' algorithm (Appendix 2).

(3) **Graph generator** (*gen*)

This program generates graphs without constraints, up to a maximum number of bonds and vertices, as described in Section 2.2. Subroutine *symred* checks for canonical labelling. The output is a list of graphs and a distribution table (cf. Table A1.1).

(4) **Graph sorter** (*sort*)

A simple program to take the output graph list from gen.f (or elsewhere) and sort in order of increasing number of bonds and number of vertices. For large lists ($> 10^5$), this program is inefficient and it is better to use successive filtering.

(5) **Graph counter** (*count*)

This program obtains weak, strong, or free lattice constants of connected graphs on an arbitrary lattice.

(6) **Disconnected graph generator** (*disc*)

A program to generate disconnected graphs up to a specific number of bonds, vertices, and components from an input list of connected graphs.

(7) **Graph generator with constraints** (*gensc*)

A modified version of the graph generating program, with constraints for the simple cubic lattice (see Section 2.3).

(8) **Tree graph counter** (*treecnt*)

A counting program for tree graphs using the efficient algebraic procedure discussed in Section 2.2. This version is for high-temperature (weak) lattice constants.

(9) **Disconnected graph counter** (*sepcnt*)

A counting program for disconnected graphs, to obtain high-temperature lattice constants. An algebraic procedure, based on embedding into graphs with fewer vertices and components is used.

(10) *G*-**graph generator for free graph expansion** (*ggrafs*)

A program to generate *G* graphs for the free graph expansion method, as described in Section 3.3. The program 'decorates' bonds of input bare graphs with multiplicities and identifies all inequivalent root points.

(11) χ**-graph generator for free graph expansion** (*xgrafs*)

A program to generate two-rooted irreducible graphs for pair correlators/susceptibility for the Free Graph Expansion Method, as discussed in Section 3.3. The procedure is similar to that used in the *ggrafs* program (10).

(12) **Free graph expansion program** (*free2*)

Main program to implement the free graph expansion method for the plane rotator ($N = 2$) model, as described in Section 3.3.

(13) **Computation of subgraph data** (*subgr*)

A program to obtain subgraphs and their embedding factors, for use in linked cluster perturbation expansions. The program given is for strong embeddings. The necessary modification for weak embeddings is straightforward.

(14) **Cluster perturbation theory** (*tim1*)

Program to carry out matrix perturbation theory for finite clusters, following the procedure of Appendix 4. This particular program computes energy, magnetization and susceptibility series for single clusters, for the transverse field Ising model in the ordered phase, as described in Section 4.3. It can be used as a model for programs for other Hamiltonians.

(15) **Bulk perturbation series** (*tim2*)

A generalization of the cluster perturbation theory program (*tim1*), to compute bare and reduced series for a list of clusters, and to compute series for the bulk lattice. This particular program is for the transverse field Ising model in the ordered phase. It can easily be adapted for other systems.

(16) **One-particle excitation series** (*tim3*)

A program to compute the effective Hamiltonian and transition amplitudes for one-particle excitations for individual clusters. The two-block orthogonal transformation (TBOT) method is used, as described in Appendix 5. This particular program is for the transverse field Ising chain in the disordered phase. It can easily be modified for other systems, and generalized to include subtraction of sub-cluster terms and generation of bulk lattice excitation energy series.

(17) **Thermodynamic perturbation theory** (*tpert*)

A program to compute a finite-T perturbation expansion for the partition function and free energy of individual clusters, for the alternating Heisenberg chain, as described in Section 7.5.1 (dimer expansion).

We welcome feedback from readers on any the the computer codes.

Readers may also be interested in the following web sites:

http://www.ms.unimelb.edu.au/~tonyg/: This provides codes for series analysis, written by A.J. Guttmann (see Guttmann (1989) for listings). These codes can be freely downloaded.

http://brahms.th.physik.uni-bonn.de/ClusterExpansion/DOC/-Intro.html: This site provides information from the University of Bonn group, on various aspects of cluster expansions. Program listings are not provided but can, as we understand, be obtained from the authors.

http://www.uni-saarland.de/fak7/ulrig: This site provides information from the University of Köln Saarland group, on various aspects of series expansions via the continuous unitary transformation method.

http://www.phys.unsw.edu.au/ ~zwh/: This site provides additional information on linked cluster series expansions from our group.

References

Abramowitz, M. and Stegun, I. A. (1965). *Handbook of Mathematical Functions: with Formulas, Graphs, and Mathematical Tables*. (New York: Dover Publications).

Aeppli, G., Hayden, S. and Perring, T. (1997). Seeing the spins in solids. *Physics World,* **10**(12), 33–7.

Affleck, I. (1998). *Dynamical Properties of Unconventional Magnetic Systems*. NATO Advanced Study Institute, Series E, Vol. 349. (Dordrecht: Kluwer Academic).

Aoki, S., Boyd, G., Burkhalter, R., *et al.* (2000). Quenched light hadron spectrum. *Physical Review Letters,* **84**(2), 238–41.

Baeriswyl, D., Campbell, D., Carmelo, J., Guinea, F. and Louis, E. (1995). *The Hubbard Model: Its Physics and its Mathematical Physics*. (New York: Plenum Press).

Baker, G. A. (1961). Application of Padé approximant method to investigation of some magnetic properties of Ising model. *Physical Review,* **124**(3), 768–74.

Baker, G. A. (1990). *Quantitative Theory of Critical Phenomena*. (London, New York: Academic Press).

Baker, G. A. and Johnson, J. D. (1985). Universality among scalar spin systems. *Physical Review Letters,* **54**(22), 2461–1.

Baker, G. A., Zheng, W. H. and Oitmaa, J. (1994). Series study of the continuous-spin Ising model. *Journal of Physics A – Mathematical and General,* **27**(10), 3403–5.

Baker, S. J., Bishop, R. F. and Davidson, N. J. (1996). Coupled cluster analysis of the U(1) lattice gauge model using a correlated mean-field reference state. *Physical Review D,* **53**(5), 2610–18.

Balian, R., Drouffe, J. M. and Itzykson, C. (1975). Gauge fields on a lattice. III: strong-coupling expansions and transition points. *Physical Review D,* **11**(8), 2104–19; *Physical Review* D, **19**, 2514.

Banks, T., Myerson, R. and Kogut, J. (1977a). Phase transitions in abelian lattice gauge theories. *Nuclear Physics B,* **129**(3), 493–510.

Banks, T., Raby, S., Susskind, L., *et al.* (1977b). Strong-coupling calculations of hadron spectrum of quantum chromodynamics. *Physical Review D,* **15**(4), 1111–27.

Banks, T., Susskind, L. and Kogut, J. (1976). Strong-coupling calculations of lattice gauge theories – (1+1)-dimensional exercises. *Physical Review D,* **13**(4), 1043–53.

Bares, P. A., Blatter, G. and Ogata, M. (1991). Exact solution of the $t - J$ model in one dimension at $2t = \pm J$ – ground-state and excitation spectrum. *Physical Review B,* **44**(1), 130–54.

Barnes, T., Riera, J. and Tennant, D. A. (1999). $S = \frac{1}{2}$ alternating chain using multiprecision methods. *Physical Review B,* **59**(17), 11384–97.

Bartkowiak, M., Henderson, J. A., Oitmaa, J. and de Brito, P. E. (1995). High-temperature series expansion for the extended Hubbard model. *Physical Review B*, **51**(20), 14077–84.

Batista, C. D., Gubernatis, J. E., Bonca, J. and Lin, H. Q. (2004). Intermediate coupling theory of electronic ferroelectricity. *Physical Review Letters*, **92**(18), 187601.

Baxter, R. J. and Enting, I. G. (1979). Series expansions from corner transfer matrices – square lattice Ising model. *Journal of Statistical Physics*, **21**(2), 103–23.

Berezinski, V. (1971). Destruction of long-range order in one-dimensional and 2-dimensional systems having a continuous symmetry group 1 – classical systems. *Soviet Physics Jetp–USSR*, **32**(3), 493.

Berg, B. A., Billoire, A. and Vohwinkel, C. (1986). 0^{++}-2^{++} glueball mass ratio in non-abelian gauge-theories. *Physical Review Letters*, **57**(4), 400–3.

Bernu, B. and Misguich, G. (2001). Specific heat and high-temperature series of lattice models: interpolation scheme and examples on quantum spin systems in one and two dimensions. *Physical Review B*, **6313**(13), 134409.

Betts, D. D. (1974). *Phase Transitions and Critical Phenomena*. Vol. 3. (New York: Academic Press).

Bickerstaff, R. P., Butler, P. H., Butts, M. B., Haase, R. W. and Reid, M. F. (1982). 3jm and 6j tables for some bases of SU(6) and SU(3). *Journal of Physics A – Mathematical and General*, **15**(4), 1087–1117.

Binder, K. and Young, A. P. (1986). Spin-glasses – experimental facts, theoretical concepts, and open questions. *Reviews of Modern Physics*, **58**(4), 801–976.

Boos, H. E. and Korepin, V. E. (2001). Quantum spin chains and Riemann zeta function with odd arguments. *Journal of Physics A – Mathematical and General*, **34**(26), 5311–16.

Bugg, D. V., Peardon, M. and Zou, B. S. (2000). The glueball spectrum. *Physics Letters B*, **486**(1–2), 49–53.

Bühler, A., Elstner, N. and Uhrig, G. S. (2000). High temperature expansion for frustrated and unfrustrated S = 1/2 spin chains. *European Physical Journal B*, **16**(3), 475–86.

Buonaura, M. C. and Sorella, S. (1998). Numerical study of the two-dimensional Heisenberg model using a Green function Monte Carlo technique with a fixed number of walkers. *Physical Review B*, **57**(18), 11446–56.

Butera, P., Cabassi, R., Comi, M. and Marchesini, G. (1987). High-temperature expansion via Schwinger–Dyson equations – the planar rotator model on a triangular lattice. *Computer Physics Communications*, **44**(1–2), 143–56.

Butera, P., Comi, M. and Marchesini, G. (1986). New algorithm for high-temperature series – the planar rotator model. *Physical Review B*, **33**(7), 4725–33.

Butera, P. and Comi, M. (1993). Quantitative study of the Kosterlitz–Thouless phase-transition in a system of 2-dimensional plane rotators (XY model) – high-temperature expansions to order β^{20}. *Physical Review B*, **47**(18), 11969–79.

Butera, P. and Comi, M. (1994). High-temperature study of the Kosterlitz–Thouless phase-transition in the XY model on the triangular lattice. *Physical Review B*, **50**(5), 3052–7.

Butera, P. and Comi, M. (1997). N-vector spin models on the simple-cubic and the body-centered-cubic lattices: A study of the critical behavior of the susceptibility and of the correlation length by high-temperature series extended to order β^{21}. *Physical Review B*, **56**(13), 8212–40.

Butera, P. and Comi, M. (2000). Extension to order β^{23} of the high-temperature expansions for the spin-1/2 Ising model on simple cubic and body-centered cubic lattices. *Physical Review B*, **62**(22), 14837–43.

Butera, P. and Comi, M. (2002). An on-line library of extended high-temperature expansions of basic observables for the spin-S Ising models on two- and three-dimensional lattices. *Journal of Statistical Physics*, **109**(1–2), 311–15.

Butler, P. H. (1981). *Point Group Symmetry Applications: Methods and Tables*. (New York: Plenum Press).

Byrnes, T. M. R., Hamer, C. J., Zheng, W. H. and Morrison, S. (2003). Application of Feynman–Kleinert approximants to the massive Schwinger model on a lattice. *Physical Review D*, **68**(1), 016002.

Byrnes, T. M. R., Loan, M., Hamer, C. J., *et al.* (2004). Hamiltonian limit of (3+1)-dimensional SU(3) lattice gauge theory on anisotropic lattices. *Physical Review D*, **69**(7), 074509.

Byrnes, T. M. R., Sriganesh, P., Bursill, R. J. and Hamer, C. J. (2002). Density matrix renormalization group approach to the massive Schwinger model. *Physical Review D*, **66**(1), 013002.

Campostrini, M., Pelissetto, A., Rossi, P. and Vicari, E. (1996). Strong-coupling analysis of two-dimensional O(N) sigma models with $N \leq 2$ on square, triangular, and honeycomb lattices. *Physical Review B*, **54**(10), 7301–17.

Campostrini, M., Pelissetto, A., Rossi, P., and Vicari, E. (1999). Improved high-temperature expansion and critical equation of state of three-dimensional Ising-like systems. *Physical Review E*, **60**(4), 3526–63.

Campostrini, M., Pelissetto, A., Rossi, P. and Vicari, E. (2002). 25th-order high-temperature expansion results for three-dimensional Ising-like systems on the simple-cubic lattice. *Physical Review E*, **65**(6), 066127.

Carmelo, J. and Baeriswyl, D. (1988). Solution of the one-dimensional Hubbard model for arbitrary electron-density and large U. *Physical Review B*, **37**(13), 7541–8.

Chakrabarti, B. K., Dutta, A. and Sen, P. (1996). *Quantum Ising Phases and Transitions in Transverse Ising Models*. (New York: Springer).

Chandra, P., Coleman, P., and Larkin, A. I. (1990). Ising transition in frustrated Heisenberg models. *Physical Review Letters*, **64**(1), 88–91.

Chartrand, G. (1977). *Graphs as Mathematical Models*. (Boston: Prindle, Weber & Schmidt).

Christensen, N. B., McMorrow, D. F., Ronnow, H. M., Harrison, A., Perring, T. G. and Coldea, R. (2004). Deviations from linear spin wave theory in the 2D, $S = 1/2$ Heisenberg antiferromagnet CFTD. *Journal of Magnetism and Magnetic Materials*, **272–76**, 896–7.

Coldea, R., Hayden, S. M., Aeppli, G., *et al.* (2001). Spin waves and electronic interactions in La_2CuO_4. *Physical Review Letters*, **86**(23), 5377–80.

Coleman, S., Jackiw, R. and Susskind, L. (1975). Charge shielding and quark confinement in the massive Schwinger model. *Annals of Physics*, **93**(1–2), 267–75.

Creutz, M. (1979). Confinement and the critical dimensionality of space–time. *Physical Review Letters*, **43**(8), 553–6.

Creutz, M. (1983). *Quarks, Gluons, and Lattices*. Cambridge Monographs on Mathematical Physics. (Cambridge: Cambridge University Press).

Creutz, M. and Moriarty, K. J. M. (1982). Numerical studies of Wilson loops in SU(3) gauge theory in 4 dimensions. *Physical Review D*, **26**(8), 2166–8.

Czycholl, G. (1999). Influence of hybridization on the properties of the spinless Falicov–Kimball model. *Physical Review B*, **59**(4), 2642–8.

Daboul, D., Chang, I. and Aharony, A. (2000). Series expansion study of quantum percolation on the square lattice. *European Physical Journal B*, **16**(2), 303–16.

Dagotto, E. and Rice, T. M. (1996). Surprises on the way from one- to two-dimensional quantum magnets: The ladder materials. *Science*, **271**(5249), 618–23.

Davies, C. T. H., Follana, E., Gray, A., *et al.* (2004). High-precision lattice QCD confronts experiment. *Physical Review Letters*, **92**(2), 022001.

de Neef, T. and Enting, I. G. (1977). Series expansions from finite lattice method. *Journal of Physics A – Mathematical and General* **10**(5), 801–5.

Ditzian, R. V. and Kadanoff, L. P. (1979). High-temperature expansion methods for Ising systems with quenched impurities. *Physical Review B*, **19**(9), 4631–45.

Domb, C. (1970). Self-avoiding walks and Ising and Heisenberg models. *Journal of Physics C: Solid State Physics*, **3**(2), 256.

Domb, C. (1974). *Phase Transitions and Critical Phenomena*. Vol. 3. (New York: Academic Press).

Drouffe, J. M. and Zuber, J. B. (1981). Roughening transition in lattice gauge-theories in arbitrary dimension. II. The groups Z_3, U(1), SU(2), SU(3). *Nuclear Physics B*, **180**(2), 264–74.

Drouffe, J. M. and Zuber, J. B. (1983). Strong coupling and mean field methods in lattice gauge theories. *Physics Reports – Review Section of Physics Letters*, **102**(1–2), 1–119.

Edwards, R. G., Heller, U. M. and Klassen, T. R. (1998). Accurate scale determinations for the Wilson gauge action. *Nuclear Physics B*, **517**(1–3), 377–92.

Edwards, S. F. and Anderson, P. W. (1975). Theory of spin glasses. *Journal of Physics F – Metal Physics*, **5**(5), 965–74.

Elitzur, S. (1975). Impossibility of spontaneously breaking local symmetries. *Physical Review D*, **12**(12), 3978–82.

Elstner, N. and Singh, R. R. P. (1998a). Field-dependent thermodynamics and quantum critical phenomena in the dimerized spin system $Cu_2(C_5H_{12}N_2)_2Cl_4$. *Physical Review B*, **58**(17), 11484–7.

Elstner, N. and Singh, R. R. P. (1998b). Strong-coupling expansions at finite temperatures: application to quantum disordered and quantum critical phases. *Physical Review B*, **57**(13), 7740–8.

Enting, I. G. (1996). Series expansions from the finite lattice method. *Nuclear Physics B – (Proc. Suppl.)*, **47**, 180–7.

Falcioni, M., Marinari, E., Paciello, M. L., Parisi, G., and Taglienti, B. (1981). Phase-transition analysis in Z_2 and U(1) lattice gauge-theories. *Physics Letters B*, **105**(1), 51–4.

Fang, Y. Z. and Luo, X. Q. (2004). Hamiltonian lattice quantum chromodynamics at finite density with Wilson fermions. *Physical Review D*, **69**(11), 114501.

Ferer, M. and Velgakis, M. J. (1983). High-temperature critical behavior of two-dimensional planar models – a series investigation. *Physical Review B*, **27**(1), 314–25.

Fisch, R. (1991). High-temperature series for the $\pm J$ random-bond Ising-model. *Physical Review B*, **44**(2), 652–7.

Fisch, R. and Harris, A. B. (1990). High-temperature series for random-anisotropy magnets. *Physical Review B*, **41**(16), 11305–13.

Fradkin, E. and Susskind, L. (1978). Order and disorder in gauge systems and magnets. *Physical Review D*, **17**(10), 2637–58.

Gelfand, M. P. (1996). Series expansions for excited states of quantum lattice models. *Solid State Communications*, **98**(1), 11–14.

Gelfand, M. P. and Singh, R. R. P. (2000). High-order convergent expansions for quantum many particle systems. *Advances in Physics*, **49**(1), 93–140.

Gottlieb, S., Liu, W., Toussaint, D., Renken, R. L. and Sugar, R. L. (1987). Hybrid-molecular-dynamics algorithms for the numerical simulation of quantum chromodynamics. *Physical Review D*, **35**(8), 2531–42.

Gross, D. J. and Wilczek, F. (1973). Ultraviolet behavior of non-abelian gauge theories. *Physical Review Letters*, **30**(26), 1343–6.

Guida, R. and Zinn-Justin, J. (1998). Critical exponents of the N-vector model. *Journal of Physics A – Mathematical and General*, **31**(40), 8103–21.

Guo, S. H., Chen, Q. Z., Fang, X. Y., Liu, J. M., Luo, X. Q. and Zheng, W. H. (1996). Truncated eigenvalue equation and long wavelength behavior of lattice gauge theory. *Nuclear Physics B (Proc. Suppl.)*, **47**, 827–830.

Guth, A. H. (1980). Existence proof of a non-confining phase in 4-dimensional U(1) lattice gauge-theory. *Physical Review D*, **21**(8), 2291–2307.

Guttmann, A. J. (1989). *Phase Transitions and Critical Phenomena*, Vol. 13. (New York: Academic Press).

Hamer, C. J. (1989). Hamiltonian strong coupling expansions for glueball masses in SU(3). *Physics Letters B*, **224**(3), 339–42.

Hamer, C. J. and Aydin, M. (1991). Lattice U(1) gauge-model in 3+1 dimensions. *Physical Review D*, **43**(12), 4080–7.

Hamer, C. J. and Irving, A. C. (1984). Methods in Hamiltonian lattice field-theory. I. Connected diagram expansions. *Nuclear Physics B*, **230**(3), 336–60.

Hamer, C. J. Irving, A. C., and Preece, T. E. (1986). Cluster expansion approach to non-Abelian lattice gauge-theory in (3+1)D. II. SU(3). *Nuclear Physics B*, **270**(4), 553–74.

Hamer, C. J., Zheng, W. H. and Arndt, P. (1992). Third-order spin-wave theory for the Heisenberg antiferromagnet. *Physical Review B*, **46**(10), 6276–92.

Hamer, C. J., Oitmaa, J. and Zheng, W. H. (1994a). Strong-coupling series for abelian lattice gauge models in 3+1 dimensions. *Physical Review D*, **49**(1), 535–42.

Hamer, C. J., Zheng, W. H. and Oitmaa, J. (1994b). Spin-wave stiffness of the Heisenberg antiferromagnet at zero-temperature. *Physical Review B*, **50**(10), 6877–88.

Hamer, C. J., Zheng, W. H. and Oitmaa, J. (1997). Series expansions for the massive Schwinger model in Hamiltonian lattice theory. *Physical Review D*, **56**(1), 55–67.

Hamer, C. J., Weihong, Z. and Oitmaa, J. (1998). One- and two-hole states in the two-dimensional t-J model via series expansions. *Physical Review B*, **58**(23), 15508–19.

Hamer, C. J., Zheng, W. H. and Singh, R. R. P. (2003). Dynamical structure factor for the alternating Heisenberg chain: A linked cluster calculation. *Physical Review B*, **68**(21), 214408.

Hamermesh, M. (1962). *Group Theory and its Application to Physical Problems*. Addison-Wesley Series in Physics. (Reading, MA: Addison-Wesley).

Hart, A. and Teper, M. (2002). Glueball spectrum in O(a)-improved lattice QCD. *Physical Review D*, **65**(3), 034502.

Hasenbusch, M., Pinn, K. and Vinti, S. (1999). Critical exponents of the three-dimensional Ising universality class from finite-size scaling with standard and improved actions. *Physical Review B*, **59**(17), 11471–83.

Hasenfratz, A. and Hasenfratz, P. (1980). The connection between the lambda-parameters of lattice and continuum QCD. *Physics Letters B*, **93**(1–2), 165–9.

He, H. X., Hamer, C. J. and Oitmaa, J. (1990). High-temperature series expansions for the (2+1)-dimensional Ising model. *Journal of Physics A – Mathematical and General*, **23**(10), 1775–87.

Henderson, J. A., Oitmaa, J. and Ashley, M. C. B. (1992). High-temperature expansion for the single-band Hubbard model. *Physical Review B*, **46**(10), 6328–37.

Hirsch, J. E. (1985). Attractive interaction and pairing in fermion systems with strong on-site repulsion. *Physical Review Letters*, **54**(12), 1317–20.

Hoek, J. and Smit, J. (1986). On the $1/g^2$ corrections to hadron masses. *Nuclear Physics B*, **263**(1), 129–54.

Hohenberg, P. (1967). Existence of long-range order in 1 and 2 dimensions. *Physical Review*, **158**(2), 383–6.

Hollenberg, L. C. L. (1994). 1st-order analytic diagonalization of lattice QCD. *Physical Review D*, **50**(11), 6917–20.

Horn, D. and Lana, G. (1991). t-expansion calculation of the SU(3) axial and tensor glueballs. *Physical Review D*, **44**(9), 2864–8.

Horn, D. and Schreiber, D. (1993). t-expansion of lowest hadron masses. *Physical Review D*, **47**(5), 2081–8.

Horn, D. and Weinstein, M. (1984). The t-expansion – a nonperturbative analytic tool for Hamiltonian systems. *Physical Review D*, **30**(6), 1256–70.

Huse, D. A. (1988). Ground-state staggered magnetization of two-dimensional quantum Heisenberg antiferromagnets. *Physical Review B*, **37**(4), 2380–2.

Irving, A. C. and Hamer, C. J. (1984a). Abelian lattice gauge theories in 3+1 dimensions – the linked cluster approach. *Nuclear Physics B*, **240**(3), 362–76.

Irving, A. C. and Hamer, C. J. (1984b). Linked cluster expansions for U(1) lattice gauge theory in 2+1 and 3+1 dimensions. *Nuclear Physics B*, **235**(3), 358–82.

Irving, A. C. and Hamer, C. J. (1984c). Methods in Hamiltonian lattice field-theory. II. linked cluster expansions. *Nuclear Physics B*, **230**(3), 361–84.

Itzykson, C. and Zuber, J. B. (1980). *Quantum Field Theory*. (New York: McGraw-Hill).

Jaime, M., Correa, V. F., Harrison, N., *et al.* (2004). Magnetic-field-induced condensation of triplons in han purple pigment BaCuSi$_2$O$_6$. *Physical Review Letters*, **93**(8), 087203.

Jensen, I. (1999). Low-density series expansions for directed percolation: I. a new efficient algorithm with applications to the square lattice. *Journal of Physics A – Mathematical and General*, **32**(28), 5233–49.

Jensen, I. and Guttmann, A. J. (1996). Extrapolation procedure for low-temperature series for the square lattice spin-1 Ising model. *Journal of Physics A – Mathematical and General*, **29**(14), 3817–36.

Jensen, I., Guttmann, A. J. and Enting, I. G. (1996). Low-temperature series expansions for the square lattice Ising model with spin $S > 1$. *Journal of Physics A – Mathematical and General*, **29**(14), 3805–15.

Jones, D. R. T., Kenway, R. D., Kogut, J. B. and Sinclair, D. K. (1979). Lattice gauge-theory calculations using an improved strong-coupling expansion and matrix Pade approximants. *Nuclear Physics B*, **158**(1), 102–22.

Jones, D. R. T., Scharbach, P. N., Sinclair, D. K. and Kogut, J. (1978). Lattice gauge theory calculation of f_π and f_ρ. *Physical Review D*, **17**(7), 1871–5.

Kageyama, H., Yoshimura, K., Stern, R., *et al.* (1999). Exact dimer ground state and quantized magnetization plateaus in the two-dimemsional spin system SrCu$_2$(BO$_3$)$_2$. *Physical Review Letters*, **82**(15), 3168–71.

Karsten, L. H. and Smit, J. (1981). Lattice fermions – species doubling, chiral invariance and the triangle anomaly. *Nuclear Physics B*, **183**(1–2), 103–40.

Kato, G., Shiroishi, M., Takahashi, M. and Sakai, K. (2004). Third-neighbour and other four-point correlation functions of spin-1/2 XXZ chain. *Journal of Physics A – Mathematical and General*, **37**(19), 5097–123.

Kennedy, A. D. (1989). Status of lattice gauge theory calculations. *Computer Physics Communications*, **57**(1–3), 57–67.

Kim, Y. J., Aharony, A., Birgeneau, R. J., *et al.* (1999). Ordering due to quantum fluctuations in $Sr_2Cu_3O_4Cl_2$. *Physical Review Letters*, **83**(4), 852–5.

Kim, Y. J., Birgeneau, R. J., Chou, F. C., *et al.* (2001). Neutron scattering study of $Sr_2Cu_3O_4Cl_2$. *Physical Review B*, **64**(2), 024435.

Klaus, B. and Roiesnel, C. (1998). High-statistics finite size scaling analysis of U(l) lattice gauge theory with a Wilson action. *Physical Review D*, **58**(11), 114509.

Kleinert, H. (1995). Variational interpolation algorithm between weak-coupling and strong-coupling expansions – application to the polaron. *Physics Letters A*, **207**(3–4), 133–9.

Knetter, C., Schmidt, K. P. and Uhrig, G. S. (2003). High order perturbation theory for spectral densities of multi-particle excitations: $S = 1/2$ two-leg Heisenberg ladder. *European Physical Journal B*, **36**(4), 525–44.

Knetter, C. and Uhrig, G. S. (2000). Perturbation theory by flow equations: dimerized and frustrated $S = 1/2$ chain. *European Physical Journal B*, **13**(2), 209–25.

Kogut, J. B., Pearson, R. B., Shigemitsu, J. and Sinclair, D. K. (1980). Z_n and N-state Potts lattice gauge-theories – phase-diagrams, 1st-order transitions, β-functions, and $1/N$ expansions. *Physical Review D*, **22**(10), 2447–64.

Kogut, J., Sinclair, D. K. and Susskind, L. (1976). Quantitative approach to low-energy quantum chromodynamics. *Nuclear Physics B*, **114**(2), 199–236.

Kogut, J. and Susskind, L. (1975). Hamiltonian formulation of Wilson's lattice gauge theories. *Physical Review D*, **11**(2), 395–408.

Kosterlitz, J. M. and Thouless, D. J. (1973). Ordering, metastability and phase-transitions in 2 dimensional systems. *Journal of Physics C – Solid State Physics*, **6**(7), 1181–203.

Kotov, V. N. and Sushkov, O. P. (2004). Stability of the spiral phase in the two-dimensional extended t–J model. *Physical Review B*, **70**(19), 195105.

Kotov, V. N., Oitmaa, J., Sushkov, O. and Weihong, Z. (2000). Spontaneous dimer order, excitation spectrum and quantum-phase transitions in the J_1–J_2 Heisenberg model. *Philosophical Magazine B – Physics of Condensed Matter Statistical Mechanics Electronic Optical and Magnetic Properties*, **80**(8), 1483–98.

Kubo, K. and Tada, M. (1983). The high-temperature series for the single-band Hubbard-model in the strong correlation limit. I. *Progress of Theoretical Physics*, **69**(5), 1345–57.

Kubo, K. and Tada, M. (1984). The high-temperature series for the single-band Hubbard-model in the strong correlation limit. II. *Progress of Theoretical Physics*, **71**(3), 479–86.

Laughlin, R. B. (1998). A critique of two metals. *Advances in Physics*, **47**(6), 943–58.

Lieb, E. H. and Wu, F. Y. (1968). Absence of Mott transition in an exact solution of short-range 1-band model in 1 dimension. *Physical Review Letters*, **20**(25), 1445.

Lin, H. Q., Flynn, J. S. and Betts, D. D. (2001). Exact diagonalization and quantum Monte Carlo study of the spin-1/2 XXZ model on the square lattice. *Physical Review B*, **64**(21), 214411.

Lüscher, M. and Weisz, P. (1988). Application of the linked cluster-expansion to the n-component ϕ^4 theory. *Nuclear Physics B*, **300**(3), 325–59.

MacDonald, A. H., Girvin, S. M. and Yoshioka, D. (1988). t/U expansion for the Hubbard-model. *Physical Review B*, **37**(16), 9753–6.

MacDonald, A. H., Girvin, S. M. and Yoshioka, D. (1990). t/U expansion for the Hubbard-model – reply. *Physical Review B*, **41**(4), 2565–8.

Marland, L. G. (1981). Series expansions for the zero-temperature transverse Ising model. *Journal of Physics A – Mathematical and General*, **14**(8), 2047–57.

Marshall, W. and Lovesey, S. W. (1971). *Theory of Thermal Neutron Scattering: the Use of Neutrons for the Investigation of Condensed Matter*. The International Series of Monographs on Physics. (Oxford: Clarendon Press).

Martinez, G. and Horsch, P. (1991). Spin polarons in the $t-J$ model. *Physical Review B*, **44**(1), 317–31.

McKenzie, S. (1975). High-temperature reduced susceptibility of Ising-model. *Journal of Physics A – Mathematical and General*, **8**(10), L102–L105.

McKenzie, S. (1982). *Phase Transitions*. Nato Advanced Study Institute Series, Vol. 72. (New York: Plenum).

Melzi, R., Carretta, P., Lascialfari, A., *et al.* (2000). $Li_2VO(Si, Ge)O_4$, a prototype of a two-dimensional frustrated quantum Heisenberg antiferromagnet. *Physical Review Letters*, **85**(6), 1318–21.

Mermin, N. D. and Wagner, H. (1966). Absence of ferromagnetism or antiferromagnetism in 1- or 2-dimensional isotropic Heisenberg models. *Physical Review Letters*, **17**(22), 1133–6.

Misguich, G., Bernu, B. and Pierre, L. (2003). Determination of the exchange energies in Li_2VOSiO_4 from a high-temperature series analysis of the square-lattice J_1-J_2 Heisenberg model. *Physical Review B*, **68**(11), 113409.

Montvay, I. and Münster, G. (1994). *Quantum Fields on a Lattice*. Cambridge Monographs on Mathematical Physics. (Cambridge: Cambridge University Press).

Morningstar, C. J. and Peardon, M. (1997). Efficient glueball simulations on anisotropic lattices. *Physical Review D*, **56**(7), 4043–61.

Münster, G. (1981a). High-temperature expansions for the free-energy of vortices and the string tension in lattice gauge theories. *Nuclear Physics B*, **180**(1), 23–60.

Münster, G. (1981b). Strong coupling expansions for the mass gap in lattice gauge theories. *Nuclear Physics B*, **190**(2), 439–53; (E) *ibid* B205, 648 (1982); *ibid* B256, 67 (1985).

Negele, J. and Orland, H. (1988). *Quantum Many-Particle Systems*. (Reading, MA: Addison-Wesley Publishing Co.).

Nickel, B. (1982). *The Problem of Confluent Singularities in Phase Transitions*. 1980 Cargese Lectures. (New York: Plenum Press).

Nickel, B. G. and Rehr, J. J. (1990). High-temperature series for scalar-field lattice models – generation and analysis. *Journal of Statistical Physics*, **61**(1–2), 1–50.

Nielsen, H. B. and Ninomiya, M. (1981). A no-go theorem for regularizing chiral fermions. *Physics Letters B*, **105**(2–3), 219–223.

Nishimori, H. (1986). Exact results on the Ising spin-glass in finite dimensions. *Progress of Theoretical Physics*, **76**(1), 305–6.

Obokata, T., Ono, I. and Oguchi, T. (1967). Pade approximation to ferromagnet with anisotropic exchange interaction. *Journal of the Physical Society of Japan*, **23**(3), 516.

Ogata, M., Luchini, M. U., Sorella, S. and Assaad, F. F. (1991). Phase-diagram of the one-dimensional $t - J$ model. *Physical Review Letters*, **66**(18), 2388–91.

Oitmaa, J. (1981). High-temperature multigraph expansions for general Ising systems. *Canadian Journal of Physics*, **59**(1), 15–21.

Oitmaa, J. and Bornilla, E. (1996). High-temperature-series study of the spin-1/2 Heisenberg ferromagnet. *Physical Review B*, **53**(21), 14228–35.

Oitmaa, J., Hamer, C. J., and Weihong, Z. (1991). Low-temperature series expansions for the (2+1)-dimensional Ising-model. *Journal of Physics A – Mathematical and General*, **24**(12), 2863–7.

Oitmaa, J., Hamer, C. J., and Zheng, W. H. (1994). Heisenberg antiferromagnet and the XY model at $T = 0$ in 3-dimensions. *Physical Review B*, **50**(6), 3877–93.

Oitmaa, J., Singh, R. R. P. and Zheng, W. H. (1996). Quantum spin ladders at $T = 0$ and at high temperatures studied by series expansions. *Physical Review B*, **54**(2), 1009–18.

Oitmaa, J. and Zheng, W. H. (1996). Series expansion for the J_1–J_2 Heisenberg antiferromagnet on a square lattice. *Physical Review B*, **54**(5), 3022–3025.

Oitmaa, J. and Zheng, W. H. (2003). Finite-temperature strong-coupling expansions for the Kondo lattice model. *Physical Review B*, **67**(21), 214407.

Oitmaa, J. and Zheng, W. H. (2004a). Curie and Néel temperatures of quantum magnets. *Journal of Physics C – Solid State Physics* **16**, 8653–8660.

Oitmaa, J. and Zheng, W. H. (2004b). Phase diagram of the BCC $S = \frac{1}{2}$ Heisenberg antiferromagnet with first and second neighbor exchange. *Physical Review B*, **69**(6), 064416.

Osterwalder, K. and Seiler, E. (1978). Gauge field theories on a lattice. *Annals of Physics*, **110**(2), 440–71.

Paiva, T., Scalettar, T., Zheng, W., Singh, R. R. P. and Oitmaa, J. (2005). Ground state and finite temperature signatures of quantum phase transitions in the half-filled Hubbard model on a honeycomb lattice. *cond-mat/0406535*.

Pan, K. K. (2000). Neel temperature of quantum quasi-two-dimensional Heisenberg antiferromagnets. *Physics Letters A*, **271**(4), 291–5.

Pan, K. K. and Wang, Y. L. (1991). Magnetic susceptibility of the strongly correlated Hubbard model. *Physical Review B*, **43**(4), 3706–9.

Pan, K. K. and Wang, Y. L. (1997). Magnetic phase diagram of the half-filled three-dimensional Hubbard model. *Physical Review B*, **55**(5), 2981–7.

Pelissetto, A. and Vicari, E. (2002). Critical phenomena and renormalization-group theory. *Physics Reports*, **368**(6), 549–727.

Pfeuty, P. (1970). One-dimensional Ising model with a transverse field. *Annals of Physics*, **57**(1), 79.

Politzer, H. D. (1973). Reliable perturbative results for strong interactions. *Physical Review Letters*, **30**(26), 1346–9.

Portengen, T., Ostreich, T. and Sham, L. J. (1996). Theory of electronic ferroelectricity. *Physical Review B*, **54**(24), 17452–63.

Putikka, W. O., Luchini, M. U. and Rice, T. M. (1992). Aspects of the phase-diagram of the 2-dimensional t–J model. *Physical Review Letters*, **68**(4), 538–41.

Putikka, W. O., Luchini, M. U. and Singh, R. R. P. (1998). Violation of Luttinger's theorem in the two-dimensional t–J model. *Physical Review Letters*, **81**(14), 2966–9.

Putikka, W. O. and Luchini, M. U. (2000). Limits on phase separation for two-dimensional strongly correlated electrons. *Physical Review B*, **62**(3), 1684–7.

Rado, G. T. (1970). Magnetoelectric studies of critical behavior in Ising-like antiferromagnet $DyPO_4$. *Solid State Communications*, **8**(17), 1349–52.

Rajan, V. T. and Riseborough, P. S. (1983). High-temperature series expansion for random Ising magnets. *Physical Review B*, **27**(1), 532–43.

Rasetti, M. (1991). *The Hubbard Model: Recent Results*. (Singapore: World Scientific).

Rauchwarger, K., Jafarey, S. and Wang, Y. L. (1979). Induced-moment singlet-triplet model – high-temperature series expansion. *Physical Review B*, **19**(5), 2712–23.

Reznik, D., Bourges, P., Fong, H. F., *et al.* (1996). Direct observation of optical magnons in $YBa_2Cu_3O_{6.2}$. *Physical Review B*, **53**(22), 14741–4.

Robson, D. and Webber, D. M. (1980). Gauge theories on a small lattice. *Zeitschrift Fur Physik C – Particles and Fields*, **7**(1), 53–60.

Robson, D. and Webber, D. M. (1982). Gauge covariance in lattice field-theories. *Zeitschrift Fur Physik C – Particles and Fields*, **15**(3), 199–226.

Röder, H., Singh, R. R. P. and Zang, J. (1997). High-temperature thermodynamics of the ferromagnetic Kondo-lattice model. *Physical Review B*, **56**(9), 5084–7.

Rogiers, J., Betts, D. D. and Lookman, T. (1978). Spin 1/2 XY model. II. Analysis of high-temperature series expansions of some thermodynamic quantities in 3 dimensions. *Canadian Journal of Physics*, **56**(4), 420–37.

Ronnow, H. M., McMorrow, D. F., Coldea, R., *et al.* (2001). Spin dynamics of the 2D spin 1/2 quantum antiferromagnet copper deuteroformate tetradeuterate (CFTD). *Physical Review Letters*, **87**(3), 037202.

Rosner, H., Singh, R. R. P., Zheng, W. H., Oitmaa, J., Drechsler, S. L. and Pickett, W. E. (2002). Realization of a large J_2 quasi-2D spin-half Heisenberg system: Li_2VOSiO_4. *Physical Review Letters*, **88**(18), 186405.

Rosner, H., Singh, R. R. P., Zheng, W. H., Oitmaa, J. and Pickett, W. E. (2003). High-temperature expansions for the J_1–J_2 Heisenberg models: Applications to ab initio calculated models for Li_2VOSiO_4 and Li_2VOGeO_4. *Physical Review B*, **67**(1), 014416.

Rothe, H. J. (1997). *Lattice Gauge Theory: an Introduction*. (Singapore: World Scientific).

Rushbrooke, G. S., Baker, G. A. and Wood, P. J. (1974). *Phase Transitions and Critical Phenomena*. Vol. 3. (New York: Academic Press).

Sachdev, S. (1999). *Quantum Phase Transitions*. (Cambridge: Cambridge University Press).

Sandvik, A. W. (1997). Finite-size scaling of the ground-state parameters of the two-dimensional Heisenberg model. *Physical Review B*, **56**(18), 11678–11690.

Saul, D. M., Wortis, M. and Stauffer, D. (1974). Tricritical behavior of Blume-Capel model. *Physical Review B*, **9**(11), 4964–80.

Schreiber, D. (1993). *t*-expansion of QCD baryons. *Physical Review D*, **48**(11), 5393, and Ph.D. thesis.

Schreiber, D. (1994). *t*-expansion calculation of $\langle \psi \bar{\psi} \rangle$ in the chiral limit. *Physical Review D*, **49**(9), 4751–4.

Schwinger, J. (1962). Gauge invariance and mass. II. *Physical Review*, **128**(5), 2425.

Senthil, T., Balents, L., Sachdev, S., Vishwanath, A. and Fisher, M. P. A. (2004). Quantum criticality beyond the Landau–Ginzburg–Wilson paradigm. *Physical Review B*, **70**(14), 144407.

Seo, K. (1982). Glueball mass estimate by strong coupling expansion in lattice gauge theories. *Nuclear Physics B*, **209**(1), 200–16.

Shapir, Y. and Aharony, A. (1982). High-temperature series and exact relations for the Ising model in a random field. *Journal of Physics C – Solid State Physics*, **15**(7), 1361–80.

Shastry, B. S. and Sutherland, B. (1981a). Exact ground-state of a quantum-mechanical antiferromagnet. *Physica B*, **108**(1–3), 1069–70.

Shastry, B. S. and Sutherland, B. (1981b). Excitation spectrum of a dimerized next-neighbor anti-ferromagnetic chain. *Physical Review Letters*, **47**(13), 964–7.

Shi, Z. P. and Singh, R. R. P. (1995). Ising expansion for the Hubbard model. *Physical Review B*, **52**(13), 9620–8.

Shi, Z. P., Singh, R. R. P., Gelfand, M. P. and Wang, Z. Q. (1995). Phase-transitions in the symmetrical Kondo lattice model in 2 and 3 dimensions. *Physical Review B*, **51**(21), 15630–3.

Shiroishi, M. and Takahashi, M. (2002). Integral equation generates high-temperature expansion of the Heisenberg chain. *Physical Review Letters*, **89**(11), 117201.

Singh, R. R. P. (1989). Thermodynamic parameters of the $T = 0$, spin-$\frac{1}{2}$ square-lattice Heisenberg-antiferromagnet. *Physical Review B*, **39**(13), 9760–3.

Singh, R. R. P. and Chakravarty, S. (1986). Critical behavior of an Ising spin-glass. *Physical Review Letters*, **57**(2), 245–8.

Singh, R. R. P., Gelfand, M. P. and Huse, D. A. (1988). Ground-states of low-dimensional quantum antiferromagnets. *Physical Review Letters*, **61**(21), 2484–7.

Singh, R. R. P. and Gelfand, M. P. (1995). Spin-wave excitation spectra and spectral weights in square lattice antiferromagnets. *Physical Review B*, **52**(22), 15695–8.

Singh, R. R. P. and Glenister, R. L. (1992a). Magnetic properties of the lightly doped $t - J$ model – a study through high-temperature expansions. *Physical Review B*, **46**(18), 11871–83.

Singh, R. R. P. and Glenister, R. L. (1992b). Momentum distribution function for the 2-dimensional $t–J$ model. *Physical Review B*, **46**(21), 14313–16.

Singh, R. R. P. and Huse, D. A. (1989). Microscopic calculation of the spin-stiffness constant for the spin-1/2 square-lattice Heisenberg antiferromagnet. *Physical Review B*, **40**(10), 7247–51.

Singh, R. R. P. and Zheng, W. (1999). Dynamical transition from triplets to spinon excitations: A series expansion study of the $J_1–J_2–\delta$ spin-$\frac{1}{2}$ chain. *Physical Review B*, **59**(15), 9911–15.

Singh, R. R. P., Zheng, W. H., Hamer, C. J. and Oitmaa, J. (1999). Dimer order with striped correlations in the $J_1–J_2$ Heisenberg model. *Physical Review B*, **60**(10), 7278–83.

Smart, J. S. (1966). *Effective Field Theories of Magnetism*. Studies in Physics and Chemistry, (Philadelphia, PA: Saunders).

Smit, J. (1982). Estimate of glueball masses from their strong coupling series in lattice QCD. *Nuclear Physics B*, **206**(2), 309–20.

Smit, J. (2002). *Introduction to Quantum Fields on a Lattice*. (Cambridge; New York: Cambridge University Press).

Soehianie, A. and Oitmaa, J. (1997). Extension of the high temperature susceptibility expansion for general Ising systems. *Modern Physics Letters B*, **11**(14), 609–14.

Stack, J. D. (1984). Heavy-quark potential in SU(3) lattice gauge-theory. *Physical Review D*, **29**(6), 1213–18.

Stauffer, D. and Aharony, A. (1992). *Introduction to Percolation Theory*. (London; Philadelphia, PA: Taylor & Francis).

Stein, J. (1997). Flow equations and the strong-coupling expansion for the Hubbard model. *Journal of Statistical Physics*, **88**(1–2), 487–511.

Stubbins, C. (1988). Methods of extrapolating the t-expansion series. *Physical Review D*, **38**(6), 1942–9.

Styer, D. F. (1985). A 1st-order phase-transition in the face-centered-cubic Ising antiferromagnet. *Physical Review B*, **32**(1), 393–9.

Sushkov, O. P., Oitmaa, J. and Zheng, W. H. (2001). Quantum phase transitions in the two-dimensional $J_1–J_2$ model. *Physical Review B*, **63**(10), 104420.

Sushkov, O. P., Oitmaa, J. and Zheng, W. H. (2002). Critical dynamics of singlet and triplet excitations in strongly frustrated spin systems. *Physical Review B*, **66**(5), 054401.

Suzuki, M. (1971). Equivalence of 2-dimensional Ising model to ground state of linear XY-model. *Physics Letters A*, **34**(2), 94.

Sykes, M. F. and Glen, M. (1976). Percolation processes in 2 dimensions. I. Low-density series expansions. *Journal of Physics A – Mathematical and General*, **9**(1), 87–95.

Sykes, M. F. and Hunter, D. L. (1974). On the determination of weights for the high temperature star cluster expansion of the free energy of the Ising model in zero magnetic field. *Journal of Physics A – Mathematical and General*, **7**(13), 1589–95.

Sykes, M. F. and Wilkinson, M. K. (1986). Derivation of series expansions for a study of percolation processes. *Journal of Physics A – Mathematical and General*, **19**(16), 3415–24.

Takahashi, M. (1971). Exact ground state energy of Lieb and Wu. *Progress of Theoretical Physics*, **45**(3), 756–60.

Takahashi, M. (1977). Half-filled Hubbard model at low-temperature. *Journal of Physics C – Solid State Physics*, **10**(8), 1289–1301.

Takahashi, M. (1999). *Thermodynamics of one-dimensional solvable models*. (Cambridge: Cambridge University Press).

ten Haaf, D. F. B., Brouwer, P. W., Denteneer, P. J. H. and van Leeuwen, J. M. J. (1995). Low-temperature behavior of the large-U Hubbard model from high-temperature expansions. *Physical Review B*, **51**(1), 353–67.

ten Haaf, D. F. B. and van Leeuwen, J. M. J. (1992). High-temperature series expansions for the Hubbard model. *Physical Review B*, **46**(10), 6313–27.

Tennant, D. A., Broholm, C., Reich, D. H., *et al.* (2003). Neutron scattering study of two-magnon states in the quantum magnet copper nitrate. *Physical Review B*, **67**(5), 054414.

Trebst, S., Monien, H., Hamer, C. J., Zheng, W. H. and Singh, R. R. P. (2000). Strong-coupling expansions for multiparticle excitations: Continuum and bound states. *Physical Review Letters*, **85**(20), 4373–6.

Uhrig, G. S. and Schulz, H. J. (1996). Magnetic excitation spectrum of dimerized antiferromagnetic chains. *Physical Review B*, **54**(14), R9624–7.

Umino, Y. (2002). Hamiltonian lattice QCD at finite density: equation of state in the strong coupling limit. *Physical Review D*, **66**(7), 074501.

Van den Doel, C. P. and Horn, D. (1986). The t-expansion and SU(3) lattice gauge theory. *Physical Review D*, **33**(10), 3011–17.

Van den Doel, C. P. and Roskies, R. (1986). Connected moments of the Hamiltonian in SU(3) lattice gauge-theory. *Physical Review D*, **34**(10), 3165–9.

Van Vleck, J. H. (1965). *The theory of Electric and Magnetic Susceptibilities*. (London: Oxford University Press).

Velgakis, M. J. and Ferer, M. (1983). Fluctuation induced, 1st-order transition in a BCC Ising model with competing interactions. *Physical Review B*, **27**(1), 401–12.

Wang, Y. L. and Lee, F. (1977). High-temperature series expansion for complicated level systems. *Physical Review Letters*, **38**(16), 912–15.

Wang, Y. L. and Wentworth, C. (1985). Transverse susceptibility of the Ising-model. *Journal of Applied Physics*, **57**(8), 3329–31.

Wegner, F. J. (1971). Duality in generalized Ising models and phase transitions without local order parameters. *Journal of Mathematical Physics*, **12**(10), 2259.

Wegner, F. J. (1994). Flow equations for hamiltonians. *Annalen Der Physik*, **3**(2), 77–91.

Wentworth, C. D. and Wang, Y. L. (1989). Linked-cluster series expansion for the spin-one Heisenberg-model with easy-axis anisotropy – specific heat. *Journal of Physics – Condensed Matter*, **1**(21), 3349–58.

Wilson, K. G. (1974). Confinement of quarks. *Physical Review D*, **10**(8), 2445–59.

Wilson, K. G. (1975). *New Phenomena in Subnuclear Physics*. (New York: Plenum).

Wortis, M. (1974). *Phase Transitions and Critical Phenomena*, Vol. 3. (New York: Academic Press).

Wu, T. T., McCoy, B. M., Tracy, C. A. and Barouch, E. (1976). Spin-spin correlation-functions for 2-dimensional Ising-model – exact theory in scaling region. *Physical Review B*, **13**(1), 316–74.

Yang, Y. S., Thompson, C. J. and Guttmann, A. J. (1992). High-temperature expansions for the strongly correlated Falicov–Kimball model. *Physica A*, **184**(3–4), 587–97.

Yeomans, J. M. (1992). *Statistical Mechanics of Phase Transitions*. Oxford Science Publications, (Oxford; New York: Clarendon Press; Oxford University Press). 3, 13

Zheng, W. H. (1997). Various series expansions for the bilayer $S = \frac{1}{2}$ Heisenberg antiferromagnet. *Physical Review B*, **55**(18), 12267–75.

Zheng, W. H., Gelfand, M. P., Singh, R. R. P., Oitmaa, J. and Hamer, C. J. (1997). Heisenberg models for CaV_4O_9: Expansions about high-temperature, plaquette, Ising, and dimer limits. *Physical Review B*, **55**(17), 11377–90.

Zheng, W. H. and Hamer, C. J. (1993). Spin-wave theory and finite-size scaling for the Heisenberg antiferromagnet. *Physical Review B*, **47**(13), 7961–70.

Zheng, W. H., Hamer, C. J. and Oitmaa, J. (1999). Series expansions for a Heisenberg antiferromagnetic model for $SrCu_2(BO_3)_2$. *Physical Review B*, **60**(9), 6608–16.

Zheng, W. H., Hamer, C. J., Singh, R. R. P., Trebst, S. and Monien, H. (2001a). Deconfinement transition and bound states in frustrated Heisenberg chains: Regimes of forced and spontaneous dimerization. *Physical Review B*, **63**(14), 144411.

Zheng, W. H., Hamer, C. J., Singh, R. R. P., Trebst, S. and Monien, H. (2001b). Linked cluster series expansions for two-particle bound states. *Physical Review B*, **63**(14), 144410.

Zheng, W. H., Hamer, C. J., Oitmaa, J. and Singh, R. R. P. (2002a). Series expansions for variable electron density. *Physical Review B*, **65**(16), 165117.

Zheng, W. H., Hamer, C. J. and Singh, R. R. P. (2003). Spectral weight contributions of many-particle bound states and continuum. *Physical Review Letters*, **91**(3), 037206.

Zheng, W. H., Oitmaa, J. and Hamer, C. J. (1991). Square lattice Heisenberg antiferromagnet at $T = 0$. *Physical Review B*, **43**(10), 8321–30.

Zheng, W. H., Oitmaa, J. and Hamer, C. J. (1994). Series expansions for the 3D transverse Ising model at $T = 0$. *Journal of Physics A – Mathematical and General*, **27**(16), 5425–44.

Zheng, W. H., Oitmaa, J. and Hamer, C. J. (1998). Metaplaquette expansion for the triplet excitation spectrum in CaV_4O_9. *Physical Review B*, **58**(21), 14147–50.

Zheng, W. H. and Oitmaa, J. (2001). Local spin correlations in Heisenberg antiferromagnets. *Physical Review B*, **63**(6), 064425.

Zheng, W. H., Oitmaa, J., Hamer, C. J. and Bursill, R. J. (2001c). Numerical studies of the two-leg Hubbard ladder. *Journal of Physics – Condensed Matter*, **13**(3), 433–48.

Zheng, W. H., Oitmaa, J. and Hamer, C. J. (2002b). Phase diagram of the Shastry–Sutherland antiferromagnet. *Physical Review B*, **65**(1), 014408.

Zheng, W. H. and Oitmaa, J. (2003). Zero-temperature series expansions for the Kondo lattice model at half filling. *Physical Review B*, **67**(21), 214406.

Zheng, W. H., Oitmaa, J. and Hamer, C. J. (2005a). Series studies of the spin-$\frac{1}{2}$ Heisenberg antiferromagnet at $T = 0$: Magnon dispersion and structure factors. *Physical Review B*, **71**(18), 184440.

Zheng, W. H., Singh, R. R. P., Oitmaa, J., Sushkov, O. P. and Hamer, C. J. (2005b). Magnon and hole excitations in the two-dimensional half-filled Hubbard model. *Physical Review B*, **72**(3), 33107.

Ziman, J. M. (1979). *Models of Disorder*. (Cambridge: Cambridge University Press).

Zinn-Justin, J. (1996). *Quantum Field Theory and Critical Phenomena*, 3rd edn. Oxford Science Publications. (Oxford; New York: Clarendon Press; Oxford University Press).

Index